STUDENT'S SOLUTIONS MANUAL

Gary L. Peterson ■ Peter Kohn
James Madison University

Linear Algebra and Differential Equations

Gary L. Peterson ■ James Sochacki

Addison
Wesley

Boston San Francisco New York
London Toronto Sydney Tokyo Singapore Madrid
Mexico City Munich Paris Cape Town Hong Kong Montreal

ISBN   0-201-66213-2

1 2 3 4 5 6 7 8 9 10 DPC 04 03 02 01

Contents

Chapter 1

Matrices and Determinants

1.1 Systems of Linear Equations

Exercises 1.1, pp. 15-17

1. Express the system of equations as an augmented matrix and use row operations to perform Gauss-Jordan elimination, as shown here:

$$\left[\begin{array}{ccc|c} 1 & 1 & -1 & 0 \\ 2 & 3 & -2 & 6 \\ 1 & 2 & 2 & 10 \end{array}\right] \begin{array}{c} R_2 - 2R_1 \\ R_3 - R_1 \\ \to \end{array} \left[\begin{array}{ccc|c} 1 & 1 & -1 & 0 \\ 0 & 1 & 0 & 6 \\ 0 & 1 & 3 & 10 \end{array}\right] \begin{array}{c} R_3 - R_2 \\ \to \end{array} \left[\begin{array}{ccc|c} 1 & 1 & -1 & 0 \\ 0 & 1 & 0 & 6 \\ 0 & 0 & 3 & 4 \end{array}\right]$$

$$\begin{array}{c} (1/3)R_3 \\ \to \end{array} \left[\begin{array}{ccc|c} 1 & 1 & -1 & 0 \\ 0 & 1 & 0 & 6 \\ 0 & 0 & 1 & 4/3 \end{array}\right] \begin{array}{c} R_1 - R_2 \\ R_1 + R_3 \\ \to \end{array} \left[\begin{array}{ccc|c} 1 & 0 & 0 & -14/3 \\ 0 & 1 & 0 & 6 \\ 0 & 0 & 1 & 4/3 \end{array}\right]$$

We can now read off the solution: $x = -14/3, y = 6$, and $z = 4/3$.

3. Express the system of equations as an augmented matrix, and begin reducing the augmented matrix to row-echelon form:

$$\left[\begin{array}{ccc|c} 2 & 3 & -4 & 3 \\ 2 & 3 & -2 & 3 \\ 4 & 6 & -2 & 7 \end{array}\right] \begin{array}{c} R_2 - R_1 \\ R_3 - 2R_1 \\ \to \end{array} \left[\begin{array}{ccc|c} 2 & 3 & -4 & 3 \\ 0 & 0 & 2 & 0 \\ 0 & 0 & 6 & 1 \end{array}\right] \begin{array}{c} R_3 - 3R_2 \\ \to \end{array} \left[\begin{array}{ccc|c} 2 & 3 & -4 & 3 \\ 0 & 0 & 2 & 0 \\ 0 & 0 & 0 & 1 \end{array}\right]$$

Row 3 now corresponds to the equation $0x + 0y + 0z = 1$. This is impossible, so there is no solution to this system of linear equations.

5. Express the system of linear equations as an augmented matrix and begin reducing it to row-echelon form:

$$\left[\begin{array}{ccc|c} 1 & 0 & 3 & 0 \\ 2 & 1 & -1 & 0 \\ 4 & 1 & 5 & 0 \end{array}\right] \begin{array}{c} R_2 - 2R_1 \\ R_3 - 4R_1 \\ \to \end{array} \left[\begin{array}{ccc|c} 1 & 0 & 3 & 0 \\ 0 & 1 & -7 & 0 \\ 0 & 1 & -7 & 0 \end{array}\right] \begin{array}{c} R_3 - R_2 \\ \to \end{array} \left[\begin{array}{ccc|c} 1 & 0 & 3 & 0 \\ 0 & 1 & -7 & 0 \\ 0 & 0 & 0 & 0 \end{array}\right]$$

Since there are three variables and only two nonzero rows of the matrix, we know there is one free variable, say z. That means that z can take on any value, and we will still be able to find values of x and y that will satisfy the system of equations. Using back substitution, from row 2 we see that $y - 7z = 0$, so $y = 7z$. From row 1, we have $x + 3z = 0$, so $x = 3z$. (Notice that we really didn't need to augment the matrix with a column of zeros, since a column of zeros is never changed by row operations.)

7. Express the system of linear equations as an augmented matrix and reduce it to row-echelon form using row operations:

$$\left[\begin{array}{cccc|c} 3 & 1 & -3 & -1 & 6 \\ 1 & 1 & -2 & 1 & 0 \\ 3 & 2 & -4 & 1 & 5 \\ 1 & 2 & -3 & 3 & 4 \end{array}\right] \begin{array}{c} R_2 \leftrightarrow R_1 \\ \to \end{array} \left[\begin{array}{cccc|c} 1 & 1 & -2 & 1 & 0 \\ 3 & 1 & -3 & -1 & 6 \\ 3 & 2 & -4 & 1 & 5 \\ 1 & 2 & -3 & 3 & 4 \end{array}\right]$$

$$
\begin{matrix}
R_2 - 3R_1 \\
R_3 - 3R_1 \\
R_4 - R_1 \\
\rightarrow
\end{matrix}
\left[\begin{array}{rrrr|r}
1 & 1 & -2 & 1 & 0 \\
0 & -2 & 3 & -4 & 6 \\
0 & -1 & 2 & -2 & 5 \\
0 & 1 & -1 & 2 & 4
\end{array}\right]
\begin{matrix}
R_4 \leftrightarrow R_2 \\
\rightarrow
\end{matrix}
\left[\begin{array}{rrrr|r}
1 & 1 & -2 & 1 & 0 \\
0 & 1 & -1 & 2 & 4 \\
0 & -1 & 2 & -2 & 5 \\
0 & -2 & 3 & -4 & 6
\end{array}\right]
$$

$$
\begin{matrix}
R_3 + R_2 R_4 + 2R_2 \\
\rightarrow
\end{matrix}
\left[\begin{array}{rrrr|r}
1 & 1 & -2 & 1 & 0 \\
0 & 1 & -1 & 2 & 4 \\
0 & 0 & 1 & 0 & 9 \\
0 & 0 & 1 & 0 & 14
\end{array}\right]
\begin{matrix}
R_4 - R_3 \\
\rightarrow
\end{matrix}
\left[\begin{array}{rrrr|r}
1 & 1 & -2 & 1 & 0 \\
0 & 1 & -1 & 2 & 4 \\
0 & -0 & 1 & 0 & 9 \\
0 & 0 & 0 & 0 & 5
\end{array}\right]
$$

Row 4 now corresponds to the equation $0x_1 + 0x_2 + 0x_3 + 0x_4 = 5$. This is impossible, so the system of linear equations has no solution.

9. Express the system of linear equations as an augmented matrix and begin reducing it to row-echelon form.

$$
\left[\begin{array}{rrrr|r}
1 & 2 & -3 & 4 & 2 \\
2 & -4 & 6 & -5 & 10 \\
1 & -6 & 9 & -9 & 8 \\
3 & -2 & 4 & -1 & 12
\end{array}\right]
\begin{matrix}
R_2 - 2R_1 \\
R_3 - R_1 \\
R_4 - 3R_1 \\
\rightarrow
\end{matrix}
\left[\begin{array}{rrrr|r}
1 & 2 & -3 & 4 & 2 \\
0 & -8 & 12 & -13 & 6 \\
0 & -8 & 12 & -13 & 6 \\
0 & -8 & 13 & -13 & 6
\end{array}\right]
$$

$$
\begin{matrix}
R_3 - R_2 \\
R_4 - R_2 \\
\rightarrow
\end{matrix}
\left[\begin{array}{rrrr|r}
1 & 2 & -3 & 4 & 2 \\
0 & -8 & 12 & -13 & 6 \\
0 & 0 & 0 & 0 & 0 \\
0 & 0 & 1 & 0 & 0
\end{array}\right]
\begin{matrix}
R_1 + 3R_4 \\
R_2 - 12R_4 \\
\rightarrow
\end{matrix}
\left[\begin{array}{rrrr|r}
1 & 2 & 0 & 4 & 2 \\
0 & -8 & 0 & -13 & 6 \\
0 & 0 & 0 & 0 & 0 \\
0 & 0 & 1 & 0 & 0
\end{array}\right]
$$

While this is not yet in reduced row-echelon form, we can already find the solution by back substitution. Since there are four variables, but only three nonzero rows, we have one free variable. Let the free variable be x_4. Notice that $x_3 = 0$ from row 4. From row 2, we have $-8x_2 - 13x_4 = 6$, so $x_2 = (-13/8)x_4 - 3/4$. Since row 1 tells us that $x_1 + 2x_2 + 4x_4 = 2$, we solve for x_1 by $x_1 = 2 - 2x_2 - 4x_4 = 2 - 2((-13/8)x_4 - 3/4) - 4x_4 = 7/2 - (3/4)x_4$.

11. Express the system of linear equations as an augmented matrix, and reduce the augmented matrix to row-echelon form:

$$
\left[\begin{array}{rrr|r}
1 & 2 & 1 & -2 \\
2 & 2 & -2 & 3
\end{array}\right]
\begin{matrix}
R_2 - 2R_1 \\
\rightarrow
\end{matrix}
\left[\begin{array}{rrr|r}
1 & 2 & 1 & -2 \\
0 & -2 & -4 & 7
\end{array}\right]
$$

Using back substitution, from row 2 we see that $-2y - 4z = 7$, so $y = -2z - 7/2$. From row 1, we have $x + 2y + z = -2$, so $x = -2y - z - 2 = -2(-2z - 7/2) - z - 2 = 3z + 5$.

13. Express the system of linear equations as an augmented matrix and use row operations to convert it to reduced row-echelon form:

$$
\left[\begin{array}{rr|r}
1 & -2 & 2 \\
1 & 8 & -4 \\
2 & 1 & 1
\end{array}\right]
\begin{matrix}
R_2 - 2R_1 \\
R_3 - 2R_1 \\
\rightarrow
\end{matrix}
\left[\begin{array}{rr|r}
1 & -2 & 2 \\
0 & 10 & -6 \\
0 & 5 & -3
\end{array}\right]
\begin{matrix}
R_2 \leftrightarrow R_3 \\
\rightarrow
\end{matrix}
\left[\begin{array}{rr|r}
1 & -2 & 2 \\
0 & 5 & -3 \\
0 & 10 & -6
\end{array}\right]
$$

$$
\begin{matrix}
R_3 - 2R_2 \\
\rightarrow
\end{matrix}
\left[\begin{array}{rr|r}
1 & -2 & 2 \\
0 & 5 & -3 \\
0 & 0 & 0
\end{array}\right]
\begin{matrix}
(1/5)R_2 \\
\rightarrow
\end{matrix}
\left[\begin{array}{rr|r}
1 & -2 & 2 \\
0 & 1 & -3/5 \\
0 & 0 & 0
\end{array}\right]
\begin{matrix}
R_1 + 2R_2 \\
\rightarrow
\end{matrix}
\left[\begin{array}{rr|r}
1 & 0 & 4/5 \\
0 & 1 & -3/5 \\
0 & 0 & 0
\end{array}\right]
$$

The solution is $x = 4/5, y = -3/5$.

15. Express the system of linear equations as an augmented matrix and reduced the augmented matrix to row-echelon form:

$$
\left[\begin{array}{rrrrr|r}
2 & -1 & -1 & 1 & 1 & 0 \\
1 & -1 & 1 & 2 & -3 & 0 \\
3 & -2 & -1 & -1 & 2 & 0
\end{array}\right]
$$

$$
\begin{matrix}
R_1 \leftrightarrow R_2 \\
\rightarrow
\end{matrix}
\left[\begin{array}{rrrrr|r}
1 & -1 & 1 & 2 & -3 & 0 \\
2 & -1 & -1 & 1 & -1 & 0 \\
3 & -2 & -1 & -1 & 2 & 0
\end{array}\right]
$$

$$
\begin{array}{c}
R_2 - 2R_1 \\
R_3 - 3R_1 \\
\rightarrow
\end{array}
\left[\begin{array}{ccccc|c}
1 & -1 & 1 & 2 & -3 & 0 \\
0 & 1 & -3 & -3 & 7 & 0 \\
0 & 1 & -4 & -7 & 11 & 0
\end{array}\right]
\quad
\begin{array}{c}
R_1 + R_2 \\
R_3 - R_2 \\
\rightarrow
\end{array}
\left[\begin{array}{ccccc|c}
1 & 0 & -2 & -1 & 4 & 0 \\
0 & 1 & -3 & -3 & 7 & 0 \\
0 & 0 & -1 & -4 & 4 & 0
\end{array}\right]
$$

$$
\begin{array}{c}
(-1)R_3 \\
\rightarrow
\end{array}
\left[\begin{array}{ccccc|c}
1 & 0 & -2 & -1 & 4 & 0 \\
0 & 1 & -3 & -3 & 7 & 0 \\
0 & 0 & 1 & 4 & -4 & 0
\end{array}\right]
\quad
\begin{array}{c}
R_1 + 2R_3 \\
R_2 + 3R_3 \rightarrow
\end{array}
\left[\begin{array}{ccccc|c}
1 & 0 & 0 & 7 & -4 & 0 \\
0 & 1 & 0 & 9 & -5 & 0 \\
0 & 0 & 1 & 4 & -4 & 0
\end{array}\right]
$$

We see that x_4 and x_5 are free variables. Using back substitution, we first look at row 3. It tells us that $x_3 + 4x_4 - 4x_5 = 0$, so $x_3 = -4x_4 + 4x_5$. Similarly, from row 2 for x_2 and row 1 for x_1 we find that $x_2 = -9x_4 + 5x_5$ and $x_1 = -7x_4 + 4x_5$. (Notice that it's not really necessary to augment the coefficient matrix when the constants are all zero, since a column of zeros will not be changed by any row operation.)

17. Set up and reduce the augmented matrix for this system, using the variables a, b, and c in the column of constants.

$$
\left[\begin{array}{ccc|c}
2 & -1 & 3 & a \\
1 & -3 & 2 & b \\
1 & 2 & 1 & c
\end{array}\right]
\quad
\begin{array}{c}
R_1 \leftrightarrow R_2 \\
\rightarrow
\end{array}
\left[\begin{array}{ccc|c}
1 & -3 & 2 & b \\
2 & -1 & 3 & a \\
1 & 2 & 1 & c
\end{array}\right]
\quad
\begin{array}{c}
R_2 - 2R_1 \\
R_3 - R_1 \\
\rightarrow
\end{array}
\left[\begin{array}{ccc|c}
1 & -3 & 2 & b \\
0 & 5 & -1 & a - 2b \\
0 & 5 & -1 & c - b
\end{array}\right]
$$

$$
\begin{array}{c}
R_2 - 2R_1 \\
R_3 - R_1 \\
\rightarrow
\end{array}
\left[\begin{array}{ccc|c}
1 & -3 & 2 & b \\
0 & 5 & -1 & a - 2b \\
0 & 0 & 0 & c - b - (a - 2b)
\end{array}\right]
$$

Now the third row is all zeros except for the constant, $c - b - (a - 2b) = -a + b + c$. If $-a + b + c = 0$, we will have z as a free variable and be able to solve for the other two variables by back substitution. On the other hand, if $-a + b + c \neq 0$, the system will have no solution, since $0x + 0y + 0z$ cannot equal a nonzero number. For that reason, the criterion for this system to have a solution is $-a + b + c = 0$.

19. Set up and begin reducing the augmented matrix for this system of linear equations, using the variables a, b, c, and d in the column of constants.

$$
\left[\begin{array}{cccc|c}
1 & 1 & 1 & -1 & a \\
1 & -1 & -1 & 1 & b \\
1 & 1 & 1 & 1 & c \\
1 & -1 & 1 & 1 & d
\end{array}\right]
\quad
\begin{array}{c}
R_2 - R_1 \\
R_3 - R_1 \\
R_4 - R_1 \\
\rightarrow
\end{array}
\left[\begin{array}{cccc|c}
1 & 1 & 1 & -1 & a \\
0 & 2 & 2 & 2 & b - a \\
0 & 0 & 0 & 2 & c - a \\
0 & 2 & 0 & 2 & d - a
\end{array}\right]
$$

At this point we can tell that the coefficient matrix will eventually be reduced to

$$
\left[\begin{array}{cccc}
1 & 0 & 0 & 0 \\
0 & 1 & 0 & 0 \\
0 & 0 & 1 & 0 \\
0 & 0 & 0 & 1
\end{array}\right],
$$

with expressions in the variables a, b, c and d in the constants column. (Try a few more steps of the row reduction if you are not convinced.) Since there will be no rows of zeros in the coefficient matrix, we don't have to worry about a nonzero constant, as we did in Problem 17. This system will have a solution for every a, b, c, and d.

21. A homogeneous linear system of equations with more variables than equations always has nontrivial solutions.

23. Notice that x and z have the same coefficients, so this is actually a system of equations in only two variables, where one of the variables is y and the other is the sum of x and z:

$$
\begin{aligned}
(x + z) - y &= 0 \\
2(x + z) + y &= 0 \\
3(x + z) - 5y &= 0
\end{aligned}
$$

If we let y=0, we can let z be any number; then set $x = -z$ for an infinite number of solutions.

25. There could be no solutions, if the equations are inconsistent. For example,

$$\begin{aligned} x+y &= 5 \\ x+y &= 6 \\ x+y+z+w &= 14 \end{aligned}$$

has no solution, since the first two equations are inconsistent. But if there is at least one solution, there will be many solutions, since there will be at least one free variable. For example, with the system

$$\begin{aligned} x+y &= 5 \\ x+y+z+w &= 14 \end{aligned}$$

we can pick any two numbers whose sum is 9 for the variables z and w.

27. A linear equation in three variables represents a plane in $\Re^3$, the three dimensional coordinate system of real numbers. Two simultaneous linear equations represent two planes. A solution to the system of equations is a point of intersection of the two planes. The two planes might be parallel, so they wouldn't intersect at all–there would be no solution. The two equations might both represent the same plane, so all their points would be points of intersection–there would be infinitely many solutions. Finally, the two planes might intersect in a line, in which case the number of solutions would again be infinite.

1.2   Matrices and Matrix Operations

Exercises 1.2, pp. 26-27

1. To add two matrices of the same size, add corresponding entries.

$$A+B = \begin{bmatrix} 1 & 2 \\ 3 & -1 \\ 2 & -1 \end{bmatrix} + \begin{bmatrix} 2 & -1 \\ -3 & -2 \\ 0 & 4 \end{bmatrix} = \begin{bmatrix} 1+2 & 2+(-1) \\ 3+(-3) & -1+(-2) \\ 2+0 & -1+4 \end{bmatrix} = \begin{bmatrix} 3 & 1 \\ 0 & -3 \\ 2 & 3 \end{bmatrix}.$$

3. To multiply a matrix by a scalar, multiply each entry of the matrix by the scalar.

$$2B = 2 \begin{bmatrix} 2 & -1 \\ -3 & -2 \\ 0 & 4 \end{bmatrix} = \begin{bmatrix} 2\cdot 2 & 2\cdot(-1) \\ 2\cdot(-3) & 2\cdot(-2) \\ 2\cdot 0 & 2\cdot 4 \end{bmatrix} = \begin{bmatrix} 4 & -2 \\ -6 & -4 \\ 0 & 8 \end{bmatrix}.$$

5. To subtract a matrix, multiply the matrix by the scalar -1, then add the resulting matrix.

$$A-4B = \begin{bmatrix} 1 & 2 \\ 3 & -1 \\ 2 & -1 \end{bmatrix} + (-1)(4) \begin{bmatrix} 2 & -1 \\ -3 & -2 \\ 0 & 4 \end{bmatrix} = \begin{bmatrix} 1 & 2 \\ 3 & -1 \\ 2 & -1 \end{bmatrix} + (-1) \begin{bmatrix} 4\cdot 2 & 4\cdot(-1) \\ 4\cdot(-3) & 4\cdot(-2) \\ 4\cdot 0 & 4\cdot 4 \end{bmatrix}$$

$$= \begin{bmatrix} 1 & 2 \\ 3 & -1 \\ 2 & -1 \end{bmatrix} + \begin{bmatrix} -4\cdot 2 & -4\cdot(-1) \\ -4\cdot(-3) & -4\cdot(-2) \\ -4\cdot 0 & -4\cdot 4 \end{bmatrix} = \begin{bmatrix} 1+(-8) & 2+4 \\ 3+12 & -1+8 \\ 2+0 & -1+(-16) \end{bmatrix} = \begin{bmatrix} -7 & 6 \\ 15 & 7 \\ 2 & -17 \end{bmatrix}.$$

7. The product of an $m \times n$ matrix and an $n \times r$ matrix is an $m \times r$ matrix. The (i, j) entry of the product matrix is calculated from the ith row of the left hand matrix and the jth column of the right hand matrix. So the $(2, 1)$ entry of matrix AB would come from the second row of A and the first column of B. It's calculated like the dot product of vectors that you studied in your calculus course.

$$CD = \begin{bmatrix} 2 & -1 \\ 1 & 5 \end{bmatrix} \begin{bmatrix} 0 & 1 \\ 3 & -1 \end{bmatrix} = \begin{bmatrix} 2\cdot 0 + -1\cdot 3 & 2\cdot 1 + (-1)\cdot(-1) \\ 1\cdot 0 + 5\cdot 3 & 1\cdot 1 + 5\cdot(-1) \end{bmatrix} = \begin{bmatrix} -3 & 3 \\ 15 & -4 \end{bmatrix}.$$

9. The product of an $m \times n$ matrix and an $n \times r$ matrix is an $m \times r$ matrix. The (i, j) entry of the product matrix is calculated from the ith row of the left hand matrix and the jth column of the right hand matrix. So the $(2, 1)$ entry of matrix AB would come from the second row of A and the first column of B. It's calculated like the dot product of vectors that you studied in your calculus course.

$$CD = \begin{bmatrix} 1 & -3 & 5 \\ 2 & 1 & -1 \\ 1 & 1 & 0 \end{bmatrix} \begin{bmatrix} 1 & -1 & 4 \\ 2 & -3 & 6 \\ 1 & 0 & 1 \end{bmatrix}$$

$$= \begin{bmatrix} 1\cdot1+-3\cdot2+5\cdot1 & 1\cdot(-1)+(-3)\cdot(-3)+5\cdot0 & 1\cdot4+(-3)\cdot6+5\cdot1 \\ 2\cdot1+1\cdot2+(-1)\cdot1 & 2\cdot(-1)+1\cdot(-3)+(-1)\cdot0 & 2\cdot4+1\cdot6+(-1)\cdot1 \\ 1\cdot1+1\cdot2+0\cdot1 & 1\cdot(-1)+1\cdot(-3)+0\cdot0 & 1\cdot4+1\cdot6+0\cdot1 \end{bmatrix}$$

$$= \begin{bmatrix} 0 & 8 & -9 \\ 3 & -5 & 13 \\ 3 & -4 & 10 \end{bmatrix}.$$

11. A is 3×2; B is 3×3. The product AE is not defined, because the number of columns of A does not match the number of rows of E.

13. Do the operation inside the parentheses first.

$$(E+F)A = \left(\begin{bmatrix} 1 & -3 & 5 \\ 2 & 1 & -1 \\ 1 & 1 & 0 \end{bmatrix} + \begin{bmatrix} 1 & -1 & 4 \\ 2 & -3 & 6 \\ 1 & 0 & 1 \end{bmatrix} \right) \begin{bmatrix} 1 & 2 \\ 3 & -1 \\ 2 & -1 \end{bmatrix}$$

$$= \begin{bmatrix} 2 & -4 & 9 \\ 4 & -2 & 5 \\ 2 & 1 & 1 \end{bmatrix} \begin{bmatrix} 1 & 2 \\ 3 & -1 \\ 2 & -1 \end{bmatrix}$$

$$= \begin{bmatrix} 2\cdot1+(-4)\cdot3+9\cdot2 & 2\cdot2+(-4)\cdot(-1)+9\cdot(-1) \\ 4\cdot1+(-2)\cdot3+5\cdot2 & 4\cdot2+(-2)\cdot(-1)+5\cdot(-1) \\ 2\cdot1+1\cdot3+1\cdot2 & 2\cdot2+1\cdot(-1)+1\cdot(-1) \end{bmatrix} = \begin{bmatrix} 8 & -1 \\ 8 & 5 \\ 7 & 2 \end{bmatrix}.$$

15. You can do the scalar multiplication by 3 either before or after you do the matrix multiplication.

$$3AC = 3 \begin{bmatrix} 1 & 2 \\ 3 & -1 \\ 2 & -1 \end{bmatrix} \begin{bmatrix} 2 & -1 \\ 1 & 5 \end{bmatrix} = \begin{bmatrix} 3\cdot1 & 3\cdot2 \\ 3\cdot3 & 3\cdot(-1) \\ 3\cdot2 & 3\cdot(-1) \end{bmatrix} \begin{bmatrix} 2 & -1 \\ 1 & 5 \end{bmatrix}$$

$$= \begin{bmatrix} 3 & 6 \\ 9 & -3 \\ 6 & -3 \end{bmatrix} \begin{bmatrix} 2 & -1 \\ 1 & 5 \end{bmatrix} = \begin{bmatrix} 3\cdot2+6\cdot1 & 3\cdot(-1)+6\cdot5 \\ 9\cdot2+(-3)\cdot1 & 9\cdot(-1)+(-3)\cdot5 \\ 6\cdot2+(-3)\cdot1 & 6\cdot(-1)+(-3)\cdot5 \end{bmatrix} = \begin{bmatrix} 12 & 27 \\ 15 & -24 \\ 9 & -21 \end{bmatrix}.$$

17. Just as with real numbers, $C^2 = CC$.

$$C^2 = CC = \begin{bmatrix} 2 & -1 \\ 1 & 5 \end{bmatrix} \begin{bmatrix} 2 & -1 \\ 1 & 5 \end{bmatrix} = \begin{bmatrix} 2\cdot2+(-1)\cdot1 & 2\cdot(-1)+(-1)\cdot5 \\ 1\cdot2+5\cdot1 & 1\cdot(-1)+5\cdot5 \end{bmatrix} = \begin{bmatrix} 3 & -7 \\ 7 & 24 \end{bmatrix}.$$

19. The coefficient matrix of the system of equations is A; X is a column matrix consisting of the list of variable names, and B is a column matrix equal to the last column of the augmented matrix that you have already learned to construct. Thus,

$$A = \begin{bmatrix} 2 & -1 & 4 \\ 1 & 1 & -1 \\ 0 & 1 & 3 \\ 1 & 1 & 0 \end{bmatrix}, X = \begin{bmatrix} x \\ y \\ z \end{bmatrix}, \text{ and } B = \begin{bmatrix} 1 \\ 4 \\ 5 \\ 2 \end{bmatrix}. \text{ The matrix equation } AX = B \text{ is}$$

$$\begin{bmatrix} 2 & -1 & 4 \\ 1 & 1 & -1 \\ 0 & 1 & 3 \\ 1 & 1 & 0 \end{bmatrix} \begin{bmatrix} x \\ y \\ z \end{bmatrix} = \begin{bmatrix} 1 \\ 4 \\ 5 \\ 2 \end{bmatrix}.$$

To check your work, multiply out.

21. To write a matrix equation as a system of equations, just multiply out the left hand side, then set corresponding entries equal:

$$\begin{bmatrix} 2 & -2 & 5 & 7 \\ 4 & 5 & -11 & 3 \end{bmatrix} \begin{bmatrix} x_1 \\ x_2 \\ x_3 \\ x_4 \end{bmatrix} = \begin{bmatrix} 2x_1 - 2x_2 + 5x_3 + 7x_4 \\ 4x_1 + 5x_2 - 11x_3 + 3x_4 \end{bmatrix} = \begin{bmatrix} 12 \\ -3 \end{bmatrix}.$$

So the system of equations is $\begin{array}{l} 2x_1 - 2x_2 + 5x_3 + 7x_4 = 12 \\ 4x_1 + 5x_2 - 11x_3 + 3x_4 = 3 \end{array}$.

23. Suppose column k of B is a column of zeros. Since the (i, j) entry of AB is calculated from the ith row of A and the jth column of B, each time we calculate an entry in the kth column of AB, we will use the kth column of B, and the result will be zero. Thus if B has a column of zeros, so will AB. It's a different story if A has a column of zeros. For example, $\begin{bmatrix} 1 & 0 \\ 2 & 0 \end{bmatrix} \begin{bmatrix} 1 & 2 \\ 3 & 4 \end{bmatrix} = \begin{bmatrix} 1 & 2 \\ 2 & 4 \end{bmatrix}$. (But if A has a *row* of zeros, so will AB.)

1.3   Inverses of Matrices

Exercises 1.3, pp. 36-37

1. If matrix A is invertible, we can find its inverse by forming the augmented matrix $[A|I]$ and applying row operations to reduce A to I. If this is possible, the reduced augmented matrix will have the form $[I|A^{-1}]$. If it is not possible to reduce A to I, then A is not invertible. In this case, $A = \begin{bmatrix} 1 & 2 \\ -3 & 1 \end{bmatrix}$, so we have

$$\begin{bmatrix} 1 & 2 & | & 1 & 0 \\ -3 & 1 & | & 0 & 1 \end{bmatrix} \xrightarrow{\substack{R_2 + 3R_1}} \begin{bmatrix} 1 & 2 & | & 1 & 0 \\ 0 & 7 & | & 3 & 1 \end{bmatrix} \xrightarrow{(1/7)R_2} \begin{bmatrix} 1 & 2 & | & 1 & 0 \\ 0 & 1 & | & 3/7 & 1/7 \end{bmatrix}$$

$$\xrightarrow{R_1 - 2R_2} \begin{bmatrix} 1 & 0 & | & 1/7 & -2/7 \\ 0 & 1 & | & 3/7 & 1/7 \end{bmatrix}.$$

So A^{-1} is $\begin{bmatrix} 1/7 & -2/7 \\ 3/7 & 1/7 \end{bmatrix}$. You can check your work by multiplying AA^{-1} to see if you get the identity matrix.

3. If matrix A is invertible, we can find its inverse by forming the augmented matrix $[A|I]$ and applying row operations to reduce A to I. If this is possible, the reduced augmented matrix will have the form $[I|A^{-1}]$. If it is not possible to reduce A to I, then A is not invertible. In this case, $A = \begin{bmatrix} 1 & -2 & 3 \\ 2 & -1 & 4 \\ 1 & 1 & 1 \end{bmatrix}$, so we have

$$\begin{bmatrix} 1 & -2 & 3 & | & 1 & 0 & 0 \\ 2 & -1 & 4 & | & 0 & 1 & 0 \\ 1 & 1 & 1 & | & 0 & 0 & 1 \end{bmatrix} \xrightarrow{\substack{R_2 + -2R_1 \\ R_3 - R_1}} \begin{bmatrix} 1 & -2 & 3 & | & 1 & 0 & 0 \\ 0 & 3 & -2 & | & -2 & 1 & 0 \\ 0 & 3 & -2 & | & -1 & 0 & 1 \end{bmatrix}.$$

At this point, we see that the lefthand side of the matrix has two rows the same, which will give us a row of zeros in the next step. Hence A is not invertible.

5. If matrix A is invertible, we can find its inverse by forming the augmented matrix $[A|I]$ and applying row operations to reduce A to I. If this is possible, the reduced augmented matrix will have the form $[I|A^{-1}]$. If it is not possible to reduce A to I, then A is not invertible. In this case, $A = \begin{bmatrix} 0 & -2 & 1 \\ 2 & 4 & -1 \\ 2 & 1 & 2 \end{bmatrix}$, so we have

$$\begin{bmatrix} 0 & -2 & 1 & | & 1 & 0 & 0 \\ 2 & 4 & -1 & | & 0 & 1 & 0 \\ 2 & 1 & 2 & | & 0 & 0 & 1 \end{bmatrix} \xrightarrow{\text{row reduction}} \begin{bmatrix} 1 & 0 & 0 & | & 3/2 & 5/6 & -1/3 \\ 0 & 1 & 0 & | & -1 & -1/3 & 1/3 \\ 0 & 0 & 1 & | & -1 & -2/3 & 2/3 \end{bmatrix}.$$

So A^{-1} is $\begin{bmatrix} 3/2 & 5/6 & -1/3 \\ -1 & -1/3 & 1/3 \\ -1 & -2/3 & 2/3 \end{bmatrix}$. You can check your work by multiplying AA^{-1} to see if you get the identity matrix.

7. If matrix A is invertible, we can find its inverse by forming the augmented matrix $[A|I]$ and applying row operations to reduce A to I. If this is possible, the reduced augmented matrix will have the form $[I|A^{-1}]$. If it is not possible to reduce A to I, then A is not invertible. In this case, $A = \begin{bmatrix} 1 & -1 & 1 & 2 \\ 1 & 2 & -1 & -1 \\ 1 & -4 & 1 & 5 \\ 3 & 1 & 1 & 6 \end{bmatrix}$, so we have

$$\begin{bmatrix} 1 & -1 & 1 & 2 & | & 1 & 0 & 0 & 0 \\ 1 & 2 & -1 & -1 & | & 0 & 1 & 0 & 0 \\ 1 & -4 & 1 & 5 & | & 0 & 0 & 1 & 0 \\ 3 & 1 & 1 & 6 & | & 0 & 0 & 0 & 1 \end{bmatrix}$$

$$\xrightarrow[\rightarrow]{\text{row reduction}} \begin{bmatrix} 1 & 0 & 0 & 0 & | & 11/12 & 3/4 & 1/12 & -1/4 \\ 0 & 1 & 0 & 0 & | & -1/4 & -1/4 & -1/4 & 1/4 \\ 0 & 0 & 1 & 0 & | & 1 & -1/2 & -1/2 & 0 \\ 0 & 0 & 0 & 1 & | & -7/12 & -1/4 & 1/12 & 1/4 \end{bmatrix}.$$

So A^{-1} is $\begin{bmatrix} 11/12 & 3/4 & 1/12 & -1/4 \\ -1/4 & -1/4 & -1/4 & 1/4 \\ 1 & -1/2 & -1/2 & 0 \\ -7/12 & -1/4 & 1/12 & 1/4 \end{bmatrix}$. You can check your work by multiplying AA^{-1} to see if you get the identity matrix.

9. Write the system of equations as a matrix equation $AX = B$:

$$\begin{bmatrix} 0 & -2 & 1 \\ 2 & 4 & -1 \\ 2 & 1 & 2 \end{bmatrix} \begin{bmatrix} x \\ y \\ z \end{bmatrix} = \begin{bmatrix} 2 \\ -1 \\ 5 \end{bmatrix}.$$

We can multiply both sides of this equation on the left by A^{-1}, obtaining $A^{-1}AX = A^{-1}B$. Since $A^{-1}A = I$, we have $X = A^{-1}B$. Since we already know A^{-1} from Problem 5, the calculation is easy:

$$\begin{bmatrix} x \\ y \\ z \end{bmatrix} = X = A^{-1}B = \begin{bmatrix} 3/2 & 5/6 & -1/3 \\ -1 & -1/3 & 1/3 \\ -1 & -2/3 & 2/3 \end{bmatrix} \begin{bmatrix} 2 \\ -1 \\ 5 \end{bmatrix} = \begin{bmatrix} 1/2 \\ 0 \\ 2 \end{bmatrix}.$$

So $x = 1/2$, $y = 0$, and $z = 2$.

11. We obtain the elementary matrix we want by performing the same operations on the identity matrix that we want to apply to the matrix A.

(a) We can get EA from A by multiplying row 2 of A by 2, so E is the matrix we get by multiplying row 2 of I_2 by 2:
$E = \begin{bmatrix} 1 & 0 \\ 0 & 2 \end{bmatrix}$.

(b) We can get EA from A by multiplying row 2 of A by 2 and adding it to row 1. So E is the matrix we get by multiplying row 2 of I_2 by 2 and adding it to row 1 of I_2: $E = \begin{bmatrix} 1 & 2 \\ 0 & 1 \end{bmatrix}$.

(c) We can get EA from A by switching rows 1 and 2. So E is the matrix we get when we switch rows 1 and 2 of I_2:
$E = \begin{bmatrix} 0 & 1 \\ 1 & 0 \end{bmatrix}$.

1.4   Special Matrices and Additional Properties of Matrices

Exercises 1.4, pp. 41-43

1. $A^2 = \begin{bmatrix} -1 & 0 & 0 \\ 0 & -2 & 0 \\ 0 & 0 & 1 \end{bmatrix}^2$. Since A is a diagonal matrix, its square is the diagonal matrix whose diagonal entries are the squares of the diagonal entries of A, namely $\begin{bmatrix} 1 & 0 & 0 \\ 0 & 4 & 0 \\ 0 & 0 & 1 \end{bmatrix}$.

3. $A^5 = \begin{bmatrix} -1 & 0 & 0 \\ 0 & -2 & 0 \\ 0 & 0 & 1 \end{bmatrix}^5$. Since A is a diagonal matrix, its fifth power is the diagonal matrix whose diagonal entries are the

fifth powers of the diagonal entries of A, namely $\begin{bmatrix} -1 & 0 & 0 \\ 0 & -32 & 0 \\ 0 & 0 & 1 \end{bmatrix}$.

5. $AB = \begin{bmatrix} 1 & 2 & 0 \\ 0 & -1 & 3 \\ 0 & 0 & 1 \end{bmatrix} \begin{bmatrix} 6 & -1 & 3 \\ 0 & 4 & -1 \\ 0 & 0 & 2 \end{bmatrix} = \begin{bmatrix} 6 & 7 & 1 \\ 0 & -4 & 7 \\ 0 & 0 & 2 \end{bmatrix}$. Note that A, B, and AB are all upper triangular matrices.

7. A^T is the matrix whose rows are the columns of A: $\begin{bmatrix} 1 & 1 \\ 2 & -2 \\ -3 & 1 \end{bmatrix}$.

9. See problem 7 for the computation of A^T. Then

$$A^T + 4B = \begin{bmatrix} 1 & 1 \\ 2 & -2 \\ -3 & 1 \end{bmatrix} + 4\begin{bmatrix} -2 & 1 \\ 3 & 5 \\ -4 & 1 \end{bmatrix} = \begin{bmatrix} 1 & 1 \\ 2 & -2 \\ -3 & 1 \end{bmatrix} + \begin{bmatrix} -8 & 4 \\ 12 & 20 \\ -16 & 4 \end{bmatrix} = \begin{bmatrix} -7 & 5 \\ 14 & 18 \\ -19 & 5 \end{bmatrix}.$$

11. $(AB)^T$ is the matrix whose rows are the columns of AB. Since

$$AB = \begin{bmatrix} 1 & 2 & -3 \\ 1 & -2 & 1 \end{bmatrix} \begin{bmatrix} -2 & 1 \\ 3 & 5 \\ -4 & 1 \end{bmatrix} = \begin{bmatrix} 16 & 8 \\ -12 & -8 \end{bmatrix},$$

we see that $(AB)^T = \begin{bmatrix} 16 & -12 \\ 8 & -8 \end{bmatrix}$.

13. $A^T = \begin{bmatrix} 1 & 1 \\ 2 & -2 \\ -3 & 1 \end{bmatrix}$, and $B^T = \begin{bmatrix} -2 & 3 & -4 \\ 1 & 5 & 1 \end{bmatrix}$, so

$$A^T B^T = \begin{bmatrix} 1 & 1 \\ 2 & -2 \\ -3 & 1 \end{bmatrix} \begin{bmatrix} -2 & 3 & -4 \\ 1 & 5 & 1 \end{bmatrix} = \begin{bmatrix} -1 & 8 & -3 \\ -6 & -4 & -10 \\ 7 & -4 & 13 \end{bmatrix}.$$

Notice that $(AB)^T$, which you calculated in problem 11, is *not* the same as $A^T B^T$. They are not even the same size.

15. One way to tell if a matrix is symmetric is to see if its entries are symmetric about the main diagonal. Another way is to see if corresponding rows and columns have the same entries. In this case, row 1 has the same entries as column 1, row 2 has the same entries as column 2, and row 3 has the same entries as column 3. Thus, A *is* symmetric.

17. $A + B = \begin{bmatrix} 1 & -2 & 3 \\ -2 & 0 & 4 \\ 3 & 4 & 5 \end{bmatrix} + \begin{bmatrix} 3 & 0 & -1 \\ 1 & 4 & 2 \\ -1 & 2 & 1 \end{bmatrix} = \begin{bmatrix} 4 & -2 & 2 \\ -1 & 4 & 6 \\ 2 & 6 & 6 \end{bmatrix}$. Notice that the entry in row 1, column 2 is -2, while the entry in row 2, column 1 is -1. Thus, $A + B$ is *not* symmetric, since $a_{ij} \neq a_{ji}$ for some i and j.

19. $BB^T = \begin{bmatrix} 3 & 0 & -1 \\ 1 & 4 & 2 \\ -1 & 2 & 1 \end{bmatrix} \begin{bmatrix} 3 & 1 & -1 \\ 0 & 4 & 2 \\ -1 & 2 & 1 \end{bmatrix} = \begin{bmatrix} 10 & 1 & -4 \\ 1 & 21 & 9 \\ -4 & 9 & 4 \end{bmatrix}$, which *is* symmetric. (In fact, it turns out that BB^T and $B^T B$ are always symmetric, for any matrix B.)

21. Part a: Recall that, in a diagonal matrix $A = \text{diag}(a_1, a_2, \cdots, a_n)$, $\text{ent}_{ij}(A) = a_i$ when $i = j$, and 0 when $i \neq j$. Similarly, for diagonal matrix $B = \text{diag}(b_1, b_2, \cdots, b_n)$, $\text{ent}_{ij}(B) = b_i$ when $i = j$, and 0 when $i \neq j$. So if $i \neq j$, $\text{ent}_{ij}(A + B) = 0 + 0 = 0$, and if $i = j$, $\text{ent}_{ii}(A + B) = a_i + b_i$. Thus, $A + B = \text{diag}(a_1 + b_1, a_2 + b_2, \cdots, a_n + b_n)$.

Part b: Recall that, in a diagonal matrix $A = \text{diag}(a_1, a_2, \cdots, a_n)$, $\text{ent}_{ij}(A) = a_i$ when $i = j$, and 0 when $i \neq j$. Similarly, for diagonal matrix $B=\text{diag}(b_1,b_2,\cdots,b_n)$, $\text{ent}_{ij}(B)=b_i$ when $i=j$, and 0 when $i\neq j$. Then $\text{ent}_{ij}(AB)=\sum_{k=1}^{n}a_{ik}b_{kj}$. Since $a_{ik}=0$ unless $i=k$, and $b_{kj}=0$ unless $k=j$, the only way there can be a nonzero term is when $i=j$. In that case, $\text{ent}_{ii}(AB)=\sum_{k=1}^{n}a_{ik}b_{ki}$ of which all terms are zero except $a_{ii}b_{ii}=a_ib_i$. Thus, $A+B=\text{diag}(a_1b_1,a_2b_2,\cdots,a_nb_n)$.

Part c: We prove the "further" statement first. Suppose all of the diagonal entries of A are nonzero, and

$$B=\text{diag}(1/a_1,1/a_2,\cdots,1/a_n).$$

By part b of this problem,

$$AB=BA=\text{diag}(a_1(1/a_1),a_2(1/a_2),\cdots,a_n(1/a_n))$$
$$=\text{diag}(1,1,\cdots,1)=I_n$$

. So $B=A^{-1}$ by the definition of the inverse of a matrix. This also proves that if all diagonal entries of a diagonal matrix are nonzero, that matrix is invertible. To show that is A is invertible, then each a_i is nonzero, we prove the contrapositive: suppose one of the a_i *is* zero. Then one of the rows of A is a row of zeros, so A is not invertible.

23. Part a: To show that $(A^T)^T=A$, we show that they have the same entries. Let $A=[a_{ij}]$. Then the (i,j) entry of A^T is a_{ji}, and the (i,j) entry of $(A^T)^T$ is a_{ij}, the same as the (i,j) entry of A.

Part b: To show that $(A+B)^T=A^T+B^T$, we show that the two matrices have the same entries. Suppose $A=[a_{ij}]$, and $B=[b_{ij}]$. Then $A+B=[a_{ij}+b_{ij}]$, so the (i,j) entry of $(A+B)^T$ is $a_{ji}+b_{ji}$. The (i,j) entry of A^T is a_{ji}, and the (i,j) entry of B^T is b_{ji}, so the (i,j) entry of A^T+B^T is $a_{ji}+b_{ji}$.

Part c: Let $A=[a_{ij}]$. Then the (i,j) entry of cA is ca_{ij}, and the (i,j) entry of $(cA)^T$ is ca_{ji}. The (i,j) entry of cA^T is c times the (i,j) entry of A^T—ca_{ji}. Since they have the same entries, $(cA)^T=cA^T$.

Part d: To show that $(A^T)^{-1}=(A^{-1})^T$, consider

$$I=I^T=(AA^{-1})^T=(A^{-1})^TA^T,$$

and

$$I=I^T=(A^{-1}A)^T=(A^TA^{-1})^T.$$

Thus $(A^{-1})^T$ is the inverse of A^T.

25. Part a: Apply the elementary row operations cR_1 and $(1/c)R_1$ to A for any nonzero scalar c. We obtain A, so A is row-equivalent to A.

Part b: Suppose A is row-equivalent to B. The, by a finite number of elementary row operations, we transform A to B. For each elementary row operation r there is an elementary row operation that reverses or "undoes" the effect of r. Applying these inverse elementary row operations in reverse order to B, we obtain A. Thus B is row-equivalent to A.

Part c: Suppose A is row-equivalent to B abd B is row-equivalent to C. Then there is a finite sequence of elementary row operations, say $r_1,r_2,\cdots,r_k$, that transforms A to B. Similarly, there is a finite sequence of elementary row operations, say $s_1,s2,\cdots,s_m$, that transforms B to C. Then the finite sequence of elementary row operations $r_1,r_2,\cdots,r_k,s_1,s_2,\cdots,s_m$ transforms A to C, so A is row-equivalent to C.

27. ($\Rightarrow$) Suppose A and B are row-equivalent. Then we can do elementary row operations to transform A to B and then continue doing elementary row operations until it is transformed to the reduced row-echelon form of B. Thus A and B have the same reduced row echelon form.

($\Leftarrow$) Suppose A and B have the same reduced row-echelon form, R. Then A is row-equivalent to R, and R is row-equivalent to B, so by Theorem 1.15, Part 3, A us row-equivalent to B.

29. Suppose A and B are invertible matrices of the same size. Since we can calculate the inverse of A by reducing A to the identity matrix via elementary row operations, A is row-equivalent to I. By Theorem 1.15, Part 2, since A is row-equivalent to I, I is row-equivalent to A. Similarly, since B is invertible, B is row-equivalent to I. Since B is row-equivalent to I and I is row-equivalent to A, by Theorem 1.15 Part 3, B is row-equivalent to A (and by the same theorem, Part 2, A is row-equivalent to B).

31. To find the reduced column echelon form of a matrix A, calculate A^T, find *its* reduced *row* echelon form, and transpose that.

For $A=\begin{bmatrix} -3 & -1 & 4 \\ 2 & 3 & -1 \\ 1 & -2 & -3 \end{bmatrix}$, $A^T=\begin{bmatrix} -3 & 2 & 1 \\ -1 & 3 & -2 \\ 4 & -1 & 3 \end{bmatrix}$. The reduced row echelon form of A^T is the identity matrix, which when transposed is still the identity matrix. (The answer in the back of the textbook (first edition) is incorrect.)

33. Let A be an $n \times n$ upper triangular matrix whose diagonal entries are all zero.

Claim: In A^m, all rows whose row number is greater than $n - m$ are zero rows. (For example, if A is a 5×5 upper triangular matrix with zeros on the diagonal, in A^2, all rows with numbers greater than $5 - 2 = 3$ will be zero rows. So rows 4 and 5 will be zero rows.) If we can prove the claim, we will have proved that $A^n = 0$, because in A^n, *all* rows of the matrix will be zero rows.

To prove the claim, we use mathematical induction on m.

If $A^1 = A$, the claim is true, because an upper triangular matrix with zeros on the diagonal has all zeros in its last row, row n. Now for the induction hypothesis, we assume that for the matrix A^{m-1}, all rows with row numbers greater than

$$n - (m - 1) = n - m + 1$$

are zero rows. Consider A^m. First, let's view A^m as $A^{m-1}A$. Since the (i,j) entry of A^m is calculated from row i of A^{m-1} and column j of A), and since all rows of A^{m-1} with row numbers higher than $n - m + 1$ are zero, we see that any element of A^m in row greater than $n - m + 1$ must be zero also. So we just need to show that row $n - m + 1$ is itself a zero row. To see this, let's view A^m as AA^{m-1}. Since A is upper triangular, with zeros on the diagonal, the entries of any row are all zero up to and including the diagonal entry. For row $n - m + 1$ of A, all entries are zero up to and including $a_{n-m+1,n-m+1}$, the entry on the main diagonal. But since A^{m-1} has all zeros in all rows with row numbers higher than $n - m + 1$, when we multiply row $n - m + 1$ of A by *any* column of A^{m-1}, we'll get 0. Thus, A^m has zero rows in all rows with numbers higher than $n - m$.

We have shown that the claim holds for A^1, and that whenever the claim holds for A^{m-1}, it also holds for A^m. By the principle of mathematical induction, the claim holds for all natural numbers n.

By the claim, A^n has zero rows in all rows with numbers greater than $n - n = 0$. So $A^n = 0$.

For a lower triangular matrix A, note that A^T is upper triangular, and $(A^n)^T = (A^T)^n = 0$, so $A^n = 0$.

1.5   Determinants

Exercises 1.5, pp. 50-51

1. Expanding about row 1, $\det(A) = 2C_{11} + (-1)C_{12} + 3C_{13}$

$$= 2(-1)^{1+1}\begin{vmatrix} 1 & -2 \\ 2 & 1 \end{vmatrix} + (-1)(-1)^{1+2}\begin{vmatrix} 4 & -2 \\ -3 & 1 \end{vmatrix} + 3(-1)^{1+3}\begin{vmatrix} 4 & 1 \\ -3 & 2 \end{vmatrix}$$

$$= 2(1+4) + 1(4-6) + 3(8+3) = 41$$

3. Expanding about row 3, $\det(A) = -3C_{31} + 2C_{32} + 1C_{33}$

$$= -3(-1)^{3+1}\begin{vmatrix} -1 & 3 \\ 1 & -2 \end{vmatrix} + 2(-1)^{3+2}\begin{vmatrix} 2 & 3 \\ 4 & -2 \end{vmatrix} + 1(-1)^{3+3}\begin{vmatrix} 2 & -1 \\ 4 & 1 \end{vmatrix}$$

$$= -3(2-3) - 2(-4-12) + 1(2+4) = 41$$

5. Expanding about column 2, $\det(A) = (-1)C_{12} + 1C_{22} + 2C_{33}$

$$= (-1)(-1)^{1+2}\begin{vmatrix} 4 & -2 \\ -3 & 1 \end{vmatrix} + 1(-1)^{2+2}\begin{vmatrix} 2 & 3 \\ -3 & 1 \end{vmatrix} + 2(-1)^{3+2}\begin{vmatrix} 2 & 3 \\ 4 & -2 \end{vmatrix}$$

$$= (4-6) + (2+9) - 2(-4-12) = 41$$

7. Expand about column 2, since that column contains mostly zeros. The determinant is

$$0C_{12} - 1C_{22} + 0C_{32} = -1(-1)^{2+2}\begin{vmatrix} -3 & 4 \\ 4 & 5 \end{vmatrix} = -(-15-16) = -(-31) = 31$$

9. Expand about row 3, since that row contains mostly zeros. The determinant is

$$0C_{31} + 1C_{32} + 0C_{33} = -1(-1)^{3+2} \begin{vmatrix} 4 & 2 & 1 \\ -2 & -1 & -2 \\ 0 & 0 & -2 \end{vmatrix}.$$

To get the determinant of the 3 by 3 matrix, expand about *its* third row, which is mostly zeros, and obtain

$$\begin{vmatrix} 4 & 2 & 1 \\ -2 & -1 & -2 \\ 0 & 0 & -2 \end{vmatrix} = -2C_{33} = -2(-1)^{3+3} \begin{vmatrix} 4 & 2 \\ -2 & -1 \end{vmatrix} = -2(-4+4) = 0.$$

Substituting 0 for the determinant of the 3 by 3 matrix in the original computation, we see that the determinant of the original 4 by 4 matrix is 0.

11. We convert the matrix into an upper triangular matrix, whose determinant is the product of entries on the main diagonal. We must modify the determinant depending on the row operations we use, but the determinant remains unchanged if we add a multiple of another row to a row, which is the only type of row operation used in this problem.

$$\begin{vmatrix} 1 & -2 & 1 \\ 2 & 1 & 3 \\ -1 & 4 & 5 \end{vmatrix} \begin{matrix} R_2 - 2R_1 \\ R_3 + R_1 \\ \rightarrow \end{matrix} \begin{vmatrix} 1 & -2 & 1 \\ 0 & 5 & 5 \\ 0 & 2 & 6 \end{vmatrix} \begin{matrix} R_3 - 2/5R_1 \\ R_3 + R_1 \\ \rightarrow \end{matrix} \begin{vmatrix} 1 & -2 & 1 \\ 0 & 5 & 5 \\ 0 & 0 & 28/5 \end{vmatrix} = 28.$$

13. We convert the matrix into an upper triangular matrix, whose determinant is the product of entries on the main diagonal. We must modify the determinant depending on the row operations we use. The determinant remains unchanged if we add a multiple of another row to a row, but if we exchange two rows the determinant is multiplied by (-1).

$$\begin{vmatrix} 1 & -2 & 1 & -1 \\ 2 & -4 & 3 & 2 \\ 5 & -11 & 2 & -6 \\ 1 & -1 & 1 & 3 \end{vmatrix} \begin{matrix} R_2 - 2R_1 \\ R_3 - 5R_1 \\ R_4 - R_1 \\ = \end{matrix} \begin{vmatrix} 1 & -2 & 1 & -1 \\ 0 & 0 & 1 & 4 \\ 0 & -1 & -3 & -1 \\ 0 & 1 & 0 & 4 \end{vmatrix}$$

$$\begin{matrix} R_2 \leftrightarrow R_4 \\ = \end{matrix} (-1) \begin{vmatrix} 1 & -2 & 1 & -1 \\ 0 & 1 & 0 & 4 \\ 0 & -1 & -3 & -1 \\ 0 & 0 & 1 & 4 \end{vmatrix} \begin{matrix} R_3 + R_2 \\ = \end{matrix} (-1) \begin{vmatrix} 1 & -2 & 1 & -1 \\ 0 & 1 & 0 & 4 \\ 0 & 0 & -3 & 3 \\ 0 & 0 & 1 & 4 \end{vmatrix}$$

$$\begin{matrix} R_4 + (1/3)R_2 \\ = \end{matrix} (-1) \begin{vmatrix} 1 & -2 & 1 & -1 \\ 0 & 1 & 0 & 4 \\ 0 & 0 & -3 & 3 \\ 0 & 0 & 0 & 5 \end{vmatrix} = (-1)(-15) = 15.$$

15. A lower triangular $n \times n$ matrix $A = [a_{ij}]$ has all entries above the main diagonal equal to zero. That means $a_{ij} = 0$ if $i < j$. Look at this lower triangular matrix to see that $a_{ij} = 0$ when $i < j$. (There are subscripts on the zeros in the matrix to show their entry numbers.):

$$A = \begin{bmatrix} a_{11} & 0_{12} & 0_{13} & 0_{14} & \cdots & 0_{1n} \\ a_{21} & a_{22} & 0_{23} & 0_{24} & \cdots & 0_{2n} \\ a_{31} & a_{32} & a_{33} & 0_{34} & \cdots & 0_{3n} \\ \vdots & \vdots & \vdots & \vdots & \ddots & \vdots \\ a_{n1} & a_{n2} & a_{n3} & a_{n4} & \cdots & a_{nn} \end{bmatrix}$$

If we calculate the determinant of this matrix by expanding about row 1 (which is all zeros except for a_{11}), we have

$$\det(A) = a_{11}C_{11} + 0C_{12} + 0C_{13} + \cdots + 0C_{1n}$$

$$= a_{11} \begin{vmatrix} a_{22} & 0_{23} & 0_{24} & \cdots & 0_{2n} \\ a_{32} & a_{33} & 0_{34} & \cdots & 0_{3n} \\ \vdots & \vdots & \vdots & \ddots & \vdots \\ a_{n2} & a_{n3} & a_{n4} & \cdots & a_{nn} \end{vmatrix}.$$

Continuing to expand each remaining determinant about the first row (which will be all zeros except for the first entry, a_{ii}, we obtain $\det(A) = a_{11}a_{22}\cdots a_{nn}$ as we wanted.

1.6   Further Properties of Determinants

Exercises 1.6, pp. 57-58

1. The determinant of this matrix is $6(2) - (-4)(3) = 12 - 12 = 0$. Since the determinant is 0, this matrix is not invertible.

3. Expanding about the first row, we calculate the determinant of this matrix to be

$$(-1)((1)(1) - (0)(1)) - 0((1)(1) - (3)(-1)) + 2((1)(0) - (3)(1)) = (-1)(1) - 0 + 2(-3) = -6.$$

Since the determinant is nonzero, this matrix is invertible.

5. By Corollary 1.27, $A^{-1} = \frac{1}{\det(A)}\text{adj}(A)$. For $A = \begin{bmatrix} 1 & 3 \\ -2 & 1 \end{bmatrix}$, $\det(A) = (1)(1) - (-2)(3) = 7$. The adjoint of A is the

transpose of the matrix of cofactors, so $\text{adj}(A) = \begin{bmatrix} C_{11} & C_{12} \\ C_{21} & C_{22} \end{bmatrix}^T = \begin{bmatrix} C_{11} & C_{21} \\ C_{12} & C_{22} \end{bmatrix} = \begin{bmatrix} 1 & -3 \\ 2 & 1 \end{bmatrix}$, and

$$A^{-1} = \frac{1}{7}\begin{bmatrix} 1 & -3 \\ 2 & 1 \end{bmatrix} = \begin{bmatrix} \frac{1}{7} & -\frac{3}{7} \\ \frac{2}{7} & \frac{1}{7} \end{bmatrix}.$$

7. For Cramer's Rule, A is the coefficient matrix, the matrix A_1 is obtained by replacing the first column of A by the column of constants in the system of equations (the "answer column"), and the matrix A_2 is obtained by replacing the second column of A by the answer column. In this case, $A = \begin{bmatrix} 3 & -4 \\ 2 & 3 \end{bmatrix}$, $A_1 = \begin{bmatrix} 1 & -4 \\ 2 & 3 \end{bmatrix}$, and $A_2 = \begin{bmatrix} 3 & 1 \\ 2 & 2 \end{bmatrix}$. Then by Cramer's Rule, $x = \frac{\det(A_1)}{\det(A)} = \frac{11}{17}$, and $y = \frac{\det(A_2)}{\det(A)} = \frac{4}{17}$.

9. For Cramer's Rule, A is the coefficient matrix, the matrix A_1 is obtained by replacing the first column of A by the column of constants in the system of equations (the "answer column"), the matrix A_2 is obtained by replacing the second column of A by the answer column, and the matrix A_3 is obtained by replacing the third column of A by the answer column. In this case, $A = \begin{bmatrix} 3 & -1 & 1 \\ 2 & 1 & -3 \\ 1 & -2 & 1 \end{bmatrix}$, $A_1 = \begin{bmatrix} 1 & -1 & 1 \\ 3 & 1 & -3 \\ 7 & -2 & 1 \end{bmatrix}$, $A_2 = \begin{bmatrix} 3 & 1 & 1 \\ 2 & 3 & -3 \\ 1 & 7 & 1 \end{bmatrix}$, and $A_3 = \begin{bmatrix} 3 & -1 & 1 \\ 2 & 1 & -3 \\ 1 & -2 & 1 \end{bmatrix}$ Then by Cramer's Rule, $x = \frac{\det(A_1)}{\det(A)} = \frac{6}{-15} = \frac{-2}{5}$, $y = \frac{\det(A_2)}{\det(A)} = \frac{78}{-15} = \frac{-26}{5}$, and $z = \frac{\det(A_3)}{\det(A)} = \frac{45}{-15} = -3$.

11. For Cramer's Rule, A is the coefficient matrix, the matrix A_1 is obtained by replacing the first column of A by the column of constants in the system of equations (the "answer column"), and the matrix A_2 is obtained by replacing the second column of A by the answer column. In this case,

$$A = \begin{bmatrix} e^t \sin 2t & e^t \cos 2t \\ 2e^t \cos 2t & -2e^t \sin 2t \end{bmatrix}, A_1 = \begin{bmatrix} t & e^t \cos 2t \\ t^2 & -2e^t \sin 2t \end{bmatrix}, \text{ and } A_2 = \begin{bmatrix} e^t \sin 2t & t \\ 2e^t \cos 2t & t^2 \end{bmatrix}.$$

The determinant of A is $-2e^{2t} \sin^2 2t - 2e^{2t} \cos^2 2t$, which becomes $-2e^{2t}(1)$ because $\sin^2 \theta + \cos^2 \theta = 1$ for any θ. The other determinants are straightforward. Then by Cramer's Rule,

$$x = \frac{\det(A_1)}{\det(A)} = \frac{-2te^t \sin 2t - t^2 e^t \cos 2t}{-2e^{2t}} = \frac{2e^t \sin 2t + t^2 \cos 2t}{2e^t}, \text{ and}$$

$$y = \frac{\det(A_2)}{\det(A)} = \frac{t^2 e^t \sin 2t - 2te^t \cos 2t}{-2e^{2t}} = \frac{2t \cos 2t - t^2 \sin 2t}{2e^t}.$$

13. Part a: Since E was obtained from I by multiplying a row of I by a nonzero scalar, by Theorem 1.20, Part 2,

$$\det(E) = c\det(I) = c(1) = c.$$

Part b: Since E was obtained from I by replacing a row of I by itself plus a multiple of another row of I, by Theorem 1.20, Part 3, $\det(E) = \det(I) = 1$.

15. Part a: We calculate $\det(A) = (3)(4) - (1)(-2) = 12 + 2 = 14$, and $\det(B) = (1)(3) - (-2)(2) = 3 + 4 = 7$.

Part b: Since $\det(AB) = \det(A)\det(B)$, we have $\det(AB) = (14)(7) = 98$. Since

$$\det(A^{-1}) = 1/\det(A), \text{ we have } \det(A^{-1}) = 1/14.$$

Since $\det(B^T) = \det(B)$ and $\det(A^{-1}) = 1/\det(A)$, we have $\det(B^T A^{-1}) = (7)(1/14) = 1/2$.

Part c: Note that $\det(A + B) = \begin{vmatrix} 3+1 & -2+2 \\ 1-2 & 4+3 \end{vmatrix} = \begin{vmatrix} 4 & 0 \\ -1 & 7 \end{vmatrix} = 28$, which is not the same as $\det(A) + \det(B) = 14 + 7 = 21$.

1.7 Proofs of Theorems on Determinants

Exercises 1.7, p. 64

1. By definition of determinant, expansion about row one gives the determinant, which equals

$$a_{11}C_{11} + a_{12}C_{12} = a_{11}a_{22} - a_{12}a_{21}.$$

If we expand about row two, we have $a_{21}C_{21} + a_{22}C_{22} = a_{21}(-a_{12}) + a_{22}a_{11} = a_{11}a_{22} - a_{12}a_{21} = \det(A)$.

3. $\sum_{i=1}^{2} a_{i1}C_{i1} = a_{11}C_{11} + a_{21}C_{21} = a_{11}a_{22} - a_{21}a_{12}$, which is the determinant of A.

5. We imitate the proof that $A\text{adj}(A) = \det(A)I$. Notice that $\text{ent}_{ij}(\text{adj}(A)A) = \sum_{k=1}^{n} C_{ki}a_{kj}$. If $i = j$, then

$$\text{ent}_{ii}(\text{adj}(A)A) = \sum_{k=1}^{n} a_{ki}C_{ki} = \det(A).$$

If $i \neq j$, then $\text{ent}_{ij}(\text{adj}(A)A) = \sum_{k=1}^{n} a_{kj}C_{ki}$ is the determinant obtained from replacing the ith column of A by the jth column of A. Since this determinant contains two columns with the same entries, we have $\text{ent}_{ij}(\text{adj}(A)A) = \sum_{k=1}^{n} a_{kj}C_{ki} = 0$ when $i \neq j$. This gives us $\text{adj}(A)A = \det(A)I$.

7. Part a: The determinant of $\begin{bmatrix} 1 & 1 \\ x_1 & x_2 \end{bmatrix}$ is $(1)x_2 - x_1(1) = x_2 - x_1$.

Part b: Using row operations, we convert the Vandermonde matrix to an upper triangular matrix whose determinant is the product of entries on the main diagonal:

$$\begin{bmatrix} 1 & 1 & 1 \\ x_1 & x_2 & x_3 \\ x_1^2 & x_2^2 & x_3^2 \end{bmatrix} \begin{array}{c} R_3 - x_1R_2 \\ R_2 - x_1R_1 \\ \rightarrow \end{array} \begin{bmatrix} 1 & 1 & 1 \\ 0 & x_2 - x_1 & x_3 - x_1 \\ 0 & x_2(x_2 - x_1) & x_3(x_3 - x_1) \end{bmatrix}$$

$$\begin{array}{c} R_3 - x_2R_2 \\ \rightarrow \end{array} \begin{bmatrix} 1 & 1 & 1 \\ 0 & x_2 - x_1 & x_3 - x_1 \\ 0 & 0 & (x_3 - x_1)(x_3 - x_2) \end{bmatrix}$$

Since none of our row operations changed the determinant, the determinant of the original Vandermonde matrix is

$$(x_2 - x_1)(x_3 - x_1)(x_2 - x_3).$$

Part c: Using row operations, we convert the Vandermonde matrix to an upper triangular matrix whose determinant is the product of entries on the main diagonal:

$$\begin{bmatrix} 1 & 1 & 1 & 1 \\ x_1 & x_2 & x_3 & x_4 \\ x_1^2 & x_2^2 & x_3^2 & x_4^2 \\ x_1^3 & x_2^3 & x_3^3 & x_4^3 \end{bmatrix} \begin{array}{c} R_4 - x_1R_3 \\ R_3 - x_1R_2 \\ R_2 - x_1R_1 \\ \rightarrow \end{array} \begin{bmatrix} 1 & 1 & 1 & 1 \\ 0 & x_2 - x_1 & x_3 - x_1 & x_4 - x_1 \\ 0 & x_2(x_2 - x_1) & x_3(x_3 - x_1) & x_4(x_4 - x_1) \\ 0 & x_2^2(x_2 - x_1) & x_3^2(x_3 - x_1) & x_4^2(x_4 - x_1) \end{bmatrix}$$

$$\begin{array}{c} R_4 - x_2R_3 \\ R_3 - x_2R_2 \\ \rightarrow \end{array} \begin{bmatrix} 1 & 1 & 1 & 1 \\ 0 & x_2 - x_1 & x_3 - x_1 & x_4 - x_1 \\ 0 & 0 & (x_3 - x_2)(x_3 - x_1) & (x_4 - x_2)(x_4 - x_1) \\ 0 & 0 & x_3(x_3 - x_2)(x_3 - x_1) & x_4(x_4 - x_2)(x_4 - x_1) \end{bmatrix}$$

$$R_4 - x_3R_3 \rightarrow \begin{bmatrix} 1 & 1 & 1 & 1 \\ 0 & x_2 - x_1 & x_3 - x_1 & x_4 - x_1 \\ 0 & 0 & (x_3 - x_2)(x_3 - x_1) & (x_4 - x_2)(x_4 - x_1) \\ 0 & 0 & 0 & (x_4 - x_3)(x_4 - x_2)(x_4 - x_1) \end{bmatrix}.$$

Since none of these row operations changed the determinant, the determinant of the Vandermonde matrix is

$$(x_2 - x_1)(x_3 - x_1)(x_4 - x_1)(x_4 - x_2)(x_3 - x_2)(x_4 - x_3).$$

Chapter 2

Vector Spaces

2.1 Vector Spaces

Exercises 2.1, pp. 73-74

1. Properties 7 and 8 both follow from the definition of scalar multiplication of functions.

3. For part a, we will check each of the vector space properties and determine which ones hold and which ones fail to hold. The numbers in parentheses indicate the property number in the definition of vector space in your textbook. In all the parts of this problem, we will let $u = (x_1, y_1), v = (x_2, y_2), w = (x_3, y_3)$, and c and d are scalars.
Part a: (1) $u + v = (x_1, y_1) + (x_2, y_2) = (x_1, y_2)$, while $v + u = (x_2, y_2) + (x_1, y_1) = (x_2, y_1)$. Notice that $u + v \neq v + u$, so property 1 fails to hold.
(2)

$$u + (v + w) = (x_1, y_1) + ((x_2, y_2) + (x_3, y_3)) = (x_1, y_1) + (x_2, y_3) = (x_1, y_3),$$

and

$$(u + v) + w = ((x_1, y_1) + (x_2, y_2)) + (x_3, y_3) = (x_1, y_2) + (x_3, y_3) = (x_1, y_3).$$

Notice that $u + (v + w) = (u + v) + w$, so property 2 holds.
(3)   For a zero element, we need an element (z_1, z_2) so that $(x_1, y_1) + (z_1, z_2) = (x_1, y_1)$. But $(x_1, y_1) + (z_1, z_2) = (x_1, z_2)$, which means z_2 would have to be equal to the second coordinate of every ordered pair in $\Re^2$. This is impossible, so no zero element exists. Property 3 fails.
(4)   Since there is no zero element, property 4 cannot hold.
(5)

$$c(u + v) = c((x_1, y_1) + (x_2, y_2)) = c(x_1, y_2) = (cx_1, cy_2),$$

and

$$cu + cv = c(x_1, y_1) + c(x_2, y_2) = (cx_1, cy_1) + (cx_2, cy_2) = (cx_1, cy_2).$$

Notice that $c(u + v) = cu + cv$, so property 5 holds.
(6)

$$(c + d)v = (c + d)(x_2, y_2) = ((c + d)x_2, (c + d)y_2),$$

while

$$cv + dv = c(x_2, y_2) + d(x_2, y_2) = (cx_2, cy_2) + (dx_2, dy_2) = (cx_2, dy_2).$$

Notice that $(c + d)v \neq cv + dv$, so property 6 fails to hold.
(7)

$$c(dv) = c(d(x_2, y_2)) = c(dx_2, dy_2) = (cdx_2, cdy_2) = (cd)(x_2, y_2) = (cd)v.$$

So property 7 holds. (8) $1v = 1(x_2, y_2) = (1x_2, 1y_2) = (x_2, y_2) = v$. Property 8 holds.
Since at least one of the vector space properties does not hold, this is not a vector space.
Part b: (1)

$$u + v = (x_1, y_1) + (x_2, y_2) = (x_1 + x_2, y_1 + y_2) = (x_2 + x_1, y_2 + y_1) = v + u.$$

So property 1 holds.

(2)
$$u + (v + w) = (x_1, y_1) + ((x_2, y_2) + (x_3, y_3)) = (x_1, y_1) + (x_2 + x_3, y_2 + y_3) = (x_1 + x_2 + x_3, y_1 + y_2 + y_3).$$

Since addition of real numbers is associative, this expression is the same as what we get when we add $(u + v) + w$. Since $u + (v + w) = (u + v) + w$, so property 2 holds.

(3)   For a zero element, we need an element (z_1, z_2) so that $(x_1, y_1) + (z_1, z_2) = (x_1, y_1)$. Since

$$(x_1, y_1) + (0, 0) = (x_1 + 0, y_1 + 0) = (x_1, y_1),$$

$(0, 0)$ is the zero element.

(4)   Let $-v = (-x_2, -y_2)$. Then

$$v + (-v) = (x_2, y_2) + (-x_2, -y_2) = (x_2 + (-x_2, y_2 + -y_2) = (0, 0).$$

(5)
$$c(u + v) = c((x_1, y_1) + (x_2, y_2) = c(x_1 + x_2, y_1 + y_2) = (c + x_1 + x_2, c + y_1 + y_2),$$

and
$$cu + cv = c(x_1, y_1) + c(x_2, y_2) = (c + x_1, c + y_1) + (c + x_2, c + y_2) = (2c + x_1 + x_2, 2c + y_1 + y_2).$$

Notice that $c(u + v) \neq cu + cv$, so property 5 does not hold.

(6)
$$(c + d)v = (c + d)(x_2, y_2) = ((c + d) + x_2, (c + d) + y_2),$$

while
$$cv + dv = c(x_2, y_2) + d(x_2, y_2) = (c + x_2, c + y_2) + (d + x_2, d + y_2) = (c + d + 2x_2, c + d + 2y_2).$$

Notice that $(c + d)v \neq cv + dv$, so property 6 fails to hold.

(7)
$$c(dv) = c(d(x_2, y_2)) = c(d + x_2, d + y_2) = (c + d + x_2, c + d + y_2)$$
$$\neq (cd)(x_2, y_2) = (cd + x_2, cd + y_2) = (cd)v.$$

So property 7 does not hold. (8) $1v = 1(x_2, y_2) = (1 + x_2, 1 + y_2) \neq (x_2, y_2) = v$. Property 8 does not hold. Since at least one vector space property fails to hold, this is not a vector space.

Part c: (1)
$$u + v = (x_1, y_1) + (x_2, y_2) = (x_1 + y_2, x_2 + y_1)$$

while
$$v + u = (x_2, y_2) + (x_1, y_1) = (x_2 + y_1, x_1 + y_2) \neq u + v.$$

So property 1 does not hold.

(2)
$$u + (v + w) = (x_1, y_1) + ((x_2, y_2) + (x_3, y_3))$$

and
$$((x_1, y_1) + (x_2, y_2)) + (x_3, y_3) = (x_1 + y_2 + x_2, x_3 + y_1 + x_2).$$

They are not equal. Since $u + (v + w) \neq (u + v) + w$, so property 2 does not hold.

(3)   For a zero element, we need an element (z_1, z_2) so that $(x_1, y_1) + (z_1, z_2) = (x_1, y_1)$. Since

$$(x_1, y_1) + (0, 0) = (x_1 + 0, y_1 + 0) = (x_1, y_1),$$

$(0, 0)$ is the zero element.

(4)   Let $-v = (-y_2, -x_2)$. Then

$$v + (-v) = (x_2, y_2) + (-y_2, -x_2) = (x_2 + (-x_2, y_2 + -y_2) = (0, 0).$$

(5)
$$c(u + v) = c((x_1, y_1) + (x_2, y_2) = c(x_1 + x_2, y_1 + y_2) = (cx_1 + cx_2, cy_1 + cy_2),$$

and
$$cu + cv = c(x_1, y_1) + c(x_2, y_2) = (cx_1, cy_1) + (cx_2, cy_2).$$

Notice that $c(u+v) = cu+cv$, so property 5 holds.

(6)
$$(c+d)v = (c+d)(x_2,y_2) = ((c+d)x_2,(c+d)y_2),$$

while

$$cv+dv = c(x_2,y_2)+d(x_2,y_2) = (cx_2,cy_2)+(dx_2,dy_2) = (cdx_2,cdy_2).$$

Notice that $(c+d)v = cv+dv$, so property 6 holds.

(7)
$$c(dv) = c(d(x_2,y_2)) = c(dx_2,dy_2) = (cdx_2,cdy_2)$$
$$= (cd)(x_2,y_2) = (cdx_2,cdy_2) = (cd)v.$$

So property 7 holds. (8) $1v = 1(x_2,y_2) = (1x_2,1y_2) = (x_2,y_2) = v$. Property 8 holds. Since at least one vector space property fails to hold, this is not a vector space.

5. Verify all the properties in the same ways as you did in Problem 3.

7. C is a vector space. Properties 1 and 2 hold because addition takes place inside the individual terms of the sequence, and addition of real numbers is commutative and associative. Property 3 holds because the zero sequence (every term is 0) converges to 0. Property 4 holds because if $\{a_n\}$ converges to r, then $\{-a_n\}$ converges to $-r$. This gives us the negative of a sequence. Properties 5-8 hold because, once we take multiplication inside the terms of a sequence, the properties required ar just properties of the real numbers.

2.2 Subspaces and Spanning Sets

Exercises 2.2, pp. 81-83

1. By Theorem 2.3, a subset of $\Re^2$ is a subspace if and only if it is closed under addition and scalar multiplication.

Part a: Let $\begin{bmatrix} 0 \\ y_1 \end{bmatrix}$ and $\begin{bmatrix} 0 \\ y_2 \end{bmatrix}$ be arbitrary elements of this subset, and let c be a scalar. Since $\begin{bmatrix} 0 \\ y_1 \end{bmatrix} + \begin{bmatrix} 0 \\ y_2 \end{bmatrix} = \begin{bmatrix} 0 \\ y_1+y_2 \end{bmatrix}$ has the form $\begin{bmatrix} 0 \\ y \end{bmatrix}$ for $y = y_1+y_2$, the set is closed under addition. Since $c\begin{bmatrix} 0 \\ y_1 \end{bmatrix} = \begin{bmatrix} 0 \\ cy_1 \end{bmatrix}$ has the form $\begin{bmatrix} 0 \\ y \end{bmatrix}$ for $y = cy_1$, the set is closed under scalar multiplication. Because the set of vectors of the form $\begin{bmatrix} 0 \\ y \end{bmatrix}$ is closed under addition and scalar multiplication, it is a subspace of $\Re^2$.

Part b: Let $\begin{bmatrix} x_1 \\ 3x_1 \end{bmatrix}$ and $\begin{bmatrix} x_2 \\ 3x_2 \end{bmatrix}$ be arbitrary elements of this subset, and let c be a scalar. Since

$$\begin{bmatrix} x_1 \\ 3x_1 \end{bmatrix} + \begin{bmatrix} x_2 \\ 3x_2 \end{bmatrix} = \begin{bmatrix} x_1+x_2 \\ 3(x_1+x_2) \end{bmatrix}$$

has the form $\begin{bmatrix} x \\ 3x \end{bmatrix}$ for $x = x_1+x_2$, the set is closed under addition. Since $c\begin{bmatrix} x \\ 3x_1 \end{bmatrix} = \begin{bmatrix} cx_1 \\ 3cx_1 \end{bmatrix}$ has the form $\begin{bmatrix} x \\ 3x \end{bmatrix}$ for $x = cx_1$, the set is closed under scalar multiplication. Because the set of vectors of the form $\begin{bmatrix} x \\ 3x \end{bmatrix}$ is closed under addition and scalar multiplication, it is a subspace of $\Re^2$.

Part c: Let $\begin{bmatrix} x_1 \\ 2-5x_1 \end{bmatrix}$ and $\begin{bmatrix} x_2 \\ 2-5x_2 \end{bmatrix}$ be arbitrary elements of this subset. Since

$$\begin{bmatrix} x_1 \\ 2-5x_1 \end{bmatrix} + \begin{bmatrix} x_2 \\ 2-5x_2 \end{bmatrix} = \begin{bmatrix} x_1+x_2 \\ 4-5(x_1+x_2) \end{bmatrix}$$

does not have the form $\begin{bmatrix} x \\ 2-5x \end{bmatrix}$, this set is *not* closed under addition and hence is not a subspace.

Part d: Let $\begin{bmatrix} x_1 \\ y_1 \end{bmatrix}$ and $\begin{bmatrix} x_2 \\ y_2 \end{bmatrix}$ be arbitrary elements of this subset, and let c be a scalar. Notice that because $\begin{bmatrix} x_1 \\ y_1 \end{bmatrix}$ and

$\begin{bmatrix} x_2 \\ y_2 \end{bmatrix}$ are elements of this set, $x_1 + y_1 = 0$ and $x_2 + y_2 = 0$. Consider

$$\begin{bmatrix} x_1 \\ y_1 \end{bmatrix} + \begin{bmatrix} x_2 \\ y_2 \end{bmatrix} = \begin{bmatrix} x_1 + x_2 \\ y_1 + y_2 \end{bmatrix},$$

and $(x_1 + x_2) + (y_1 + y_2) = (x_1 + y_1) + (x_2 + y_2) = 0 + 0 = 0$. So the sum of two arbitrary vectors in this set is also an element of the set; thus this set is closed under addition. Since $c \begin{bmatrix} x_1 \\ y_1 \end{bmatrix} = \begin{bmatrix} cx_1 \\ cy_1 \end{bmatrix}$, and since $cx_1 + cx_2 = c(x_1 + x_2) = c \cdot 0 = 0$, the set is also closed under scalar multiplication. Because the set of vectors of this form is closed under addition and scalar multiplication, it is a subspace of $\Re^2$.

3. By Theorem 2.3, a subset of $\Re^2$ is a subspace if and only if it is closed under addition and scalar multiplication.
Part a: Let f_1 and f_2 be arbitrary functions in $F[a,b]$, and suppose $f_1(a) = f_2(a) = 0$. Then

$$f_1 + f_2(a) = f_1(a) + f_2(a) = 0 + 0 = 0,$$

so the sum of two arbitrary functions in this subset is also in the subset. Also, if c is a scalar, then $cf_1(a) = c(f_1(a)) = c \cdot 0 = 0$, so the scalar multiple of a function in this set is also in the set.
Because the set of functions of this form is closed under addition and scalar multiplication, by Theorem 2.3 it is a subspace of $F[a,b]$.
Part b: Let f_1 and f_2 be arbitrary functions in $F[a,b]$, and suppose $f_1(a) = f_2(a) = 1$. Then

$$f_1 + f_2(a) = f_1(a) + f_2(a) = 1 + 1 = 2,$$

so the sum of two arbitrary functions in this subset is *not* in the subset. Since the set is *not* closed under addition, by Theorem 2.3, it is *not* a subspace of $F[a,b]$.
Part c: Let f_1 and f_2 be arbitrary functions in $C[a,b]$, and suppose $\int_a^b f_1(x)dx = \int_a^b f_2(x)dx = 0$. Then

$$\int_a^b (f_1 + f_2)(x)dx = \int_a^b f_1(x)dx + \int_a^b f_2(x)dx = 0 + 0 = 0,$$

so the sum of two arbitrary functions in this subset is also in the subset. Also, if c is a scalar, then

$$\int_a^b (cf_1(x))dx = c \int_a^b (f_1(x))dx = c \cdot 0 = 0,$$

so the scalar multiple of a function in this set is also in the set.
Because the set of functions of this form is closed under addition and scalar multiplication, by Theorem 2.3 it is a subspace of $C[a,b]$.
Part d: Let f_1 and f_2 be arbitrary functions in $D[a,b]$, and suppose $f_1'(x) = f_1(x)$ and $f_2'(x) = f_2x$. Then

$$f_1' + f_2'(x) = f_1'(x) + f_2'(x) = f_1x + f_2x = f_1 + f_2x.$$

Also, if c is a scalar, then $cf_1'(x) = c(f_1'(x)) = cf_1(x)$, so the scalar multiple of a function in this set is also in the set.
Because the set of functions of this form is closed under addition and scalar multiplication, by Theorem 2.3 it is a subspace of $D[a,b]$.
Part e: Let f_1 and f_2 be arbitrary functions in $D[a,b]$, and suppose $f_1'(x) = f_2'(x) = e^x$. Then

$$f_1' + f_2'(x) = f_1'(x) + f_2'(x) = e^x + e^x = 2e^x,$$

so the sum of two arbitrary functions in this subset is *not* in the subset. Since the set is *not* closed under addition, by Theorem 2.3, it is *not* a subspace of $D[a,b]$.

5. Let X_1 and X_2 be arbitrary solutions to $AX = B$. That means $AX_1 = B$ and $AX_2 = B$. Then

$$A(X_1 + X_2) = AX_1 + AX_2 = B + B = 2B \neq B.$$

Since the sum of two solutions to $AX = B$ is not a solution to $AX = B$, the set of solutions is not closed under addition, so (by Theorem 2.3) the set of solutions is not a subspace of $\Re^n$. (But notice that if B is the zero vector, then the sum of two solutions is $2B = 20 = 0$. It turns out that if $B = 0$, the set of solutions *is* a vector space. See Theorem 2.4 for the details.)

7. If sequence $\{a_n\}$ converges to r, and sequence $\{b_n\}$ converges to s, then the sum of the two sequences

$$\{a_n\} + \{b_n\} = \{a_n + b_n\}$$

converges to $r+s$. (See your calculus book if you need to remind yourself of the details of why this is true.) So the sum of two sequences that each converge to zero is a sequence that converges to $0+0=0$. Hence the set of all sequences that converge to zero is closed under addition. Also, if we multiply the sequence $\{a_n\}$ by the scalar c, we get $c\{a_n\} = \{ca_n\}$, which converges to cr. If $r = 0$, then $cr = 0$, so the set of sequences that converge to zero is also closed under scalar multiplication. Since the set of sequences that converge to zero is closed under addition and scalar multiplication, it is a subspace of the vector space of convergent sequences, by Theorem 2.3.

9. To see if $\begin{bmatrix} 2 \\ 1 \end{bmatrix}$ is in $\text{Span}\left\{ \begin{bmatrix} -1 \\ 1 \end{bmatrix}, \begin{bmatrix} 2 \\ 3 \end{bmatrix} \right\}$, we see if we can write $\begin{bmatrix} 2 \\ 1 \end{bmatrix}$ as a linear combination of $\begin{bmatrix} -1 \\ 1 \end{bmatrix}$ and $\begin{bmatrix} 2 \\ 3 \end{bmatrix}$.

To solve $c_1 \begin{bmatrix} -1 \\ 1 \end{bmatrix} + c_2 \begin{bmatrix} 2 \\ 3 \end{bmatrix}$ for the coefficients of the linear combination, we can use the augmented matrix

$$\begin{bmatrix} 1 & -1 & | & 2 \\ 1 & 3 & | & 1 \end{bmatrix} \xrightarrow{\text{row reduction}} \begin{bmatrix} 1 & 0 & | & -8/5 \\ 0 & 1 & | & 1/5 \end{bmatrix}.$$

Since there is a solution, the vector $\begin{bmatrix} 2 \\ 1 \end{bmatrix}$ is in $\text{Span}\left\{ \begin{bmatrix} -1 \\ 1 \end{bmatrix}, \begin{bmatrix} 2 \\ 3 \end{bmatrix} \right\}$.

11. To see if $\begin{bmatrix} 1 \\ -5 \\ -3 \end{bmatrix}$ is in $\text{Span}\left\{ \begin{bmatrix} -1 \\ 1 \\ 0 \end{bmatrix}, \begin{bmatrix} 1 \\ 1 \\ 1 \end{bmatrix}, \begin{bmatrix} 2 \\ 0 \\ 1 \end{bmatrix} \right\}$, we see if we can write $\begin{bmatrix} 1 \\ -5 \\ -3 \end{bmatrix}$ as a linear combination of $\begin{bmatrix} -1 \\ 1 \\ 0 \end{bmatrix}, \begin{bmatrix} 1 \\ 1 \\ 1 \end{bmatrix}$, and $\begin{bmatrix} 2 \\ 0 \\ 1 \end{bmatrix}$. To solve

$$c_1 \begin{bmatrix} -1 \\ 1 \\ 0 \end{bmatrix} + c_2 \begin{bmatrix} 1 \\ 1 \\ 1 \end{bmatrix} + c_3 \begin{bmatrix} 2 \\ 0 \\ 1 \end{bmatrix} = \begin{bmatrix} 1 \\ -5 \\ -3 \end{bmatrix}$$

for the coefficients of the linear combination, we can use the augmented matrix

$$\begin{bmatrix} 1 & 1 & 2 & | & 1 \\ -1 & 1 & 0 & | & -5 \\ 0 & 1 & 1 & | & -3 \end{bmatrix} \xrightarrow{\text{row reduction}} \begin{bmatrix} 1 & 1 & 2 & | & 1 \\ 0 & 1 & 1 & | & -2 \\ 0 & 0 & 0 & | & -1 \end{bmatrix}.$$

Since there is no solution, the vector $\begin{bmatrix} 1 \\ -5 \\ -3 \end{bmatrix}$ is *not* in the span of the three vectors given.

13. To see if $3x^2$ is in $\text{Span}\{x^2 - x, x^2 + x + 1, x^2 - 1\}$, we see if we can write $3x^2$ as a linear combination of these three polynomials. In other words, we need to solve

$$c_1(x^2 - x) + c_2(x^2 + x + 1) + c_3(x^2 - 1) = 3x^2.$$

Multiplying out and collecting terms, we have:

$$(c_1 + c_2 + c_3)x^2 + (-c_1 + c_2)x + (c_2 - c_3) = 3x^2 + 0x + 0.$$

This leads to a system of linear equations:

$$c_1 + c_2 + c_3 = 3,$$
$$-c_1 + c_2 = 0, \text{and}$$
$$c_2 - c_3 = 0.$$

To solve for c_1, c_2, and c_2, we can use the augmented matrix

$$\begin{bmatrix} 1 & 1 & 1 & | & 3 \\ -1 & 1 & 0 & | & 0 \\ 0 & 1 & -1 & | & 0 \end{bmatrix} \xrightarrow{\text{row reduction}} \begin{bmatrix} 1 & 0 & 0 & | & 1 \\ 0 & 1 & 0 & | & 1 \\ 0 & 0 & 1 & | & 1 \end{bmatrix}.$$

Since we can find a linear combination of the three vectors that gives $3x^2$, $3x^2$ is in $\text{Span}\{x^2 - x, x^2 + x + 1, x^2 - 1\}$.

15. To see if these three vectors span $\Re^2$, we see if an arbitrary element $\begin{bmatrix} a \\ b \end{bmatrix}$ is in $\text{Span}\left\{ \begin{bmatrix} 1 \\ 1 \end{bmatrix}, \begin{bmatrix} 2 \\ 1 \end{bmatrix}, \begin{bmatrix} -1 \\ 1 \end{bmatrix} \right\}$. That is, we see if we can write $\begin{bmatrix} a \\ b \end{bmatrix}$ as a linear combination of the three vectors given. To solve

$$c_1 \begin{bmatrix} 1 \\ 1 \end{bmatrix} + c_2 \begin{bmatrix} 2 \\ 1 \end{bmatrix} + c_3 \begin{bmatrix} -1 \\ 1 \end{bmatrix} = \begin{bmatrix} a \\ b \end{bmatrix}$$

for the coefficients of the linear combination, we can use the augmented matrix

$$\begin{bmatrix} 1 & 2 & -1 & | & a \\ 1 & 1 & 1 & | & b \end{bmatrix} \xrightarrow[\rightarrow]{\text{row reduction}} \begin{bmatrix} 1 & 2 & -1 & | & a \\ 0 & 1 & -2 & | & -b \end{bmatrix}.$$

Since there are more variables than equations, we can solve this system for all a and b, so the three vectors given do span $\Re^2$.

17. To see if these three vectors span $\Re^4$, we see if an arbitrary element $\begin{bmatrix} a \\ b \\ c \\ d \end{bmatrix}$ is in $\text{Span}\left\{ \begin{bmatrix} 1 \\ 0 \\ 1 \\ 1 \end{bmatrix}, \begin{bmatrix} -1 \\ -1 \\ 1 \\ 1 \end{bmatrix}, \begin{bmatrix} 0 \\ 1 \\ 1 \\ 0 \end{bmatrix} \right\}$. That is, we see if we can write $\begin{bmatrix} a \\ b \\ c \\ d \end{bmatrix}$ as a linear combination of the three vectors given. To solve

$$c_1 \begin{bmatrix} 1 \\ 0 \\ 1 \\ 1 \end{bmatrix} + c_2 \begin{bmatrix} -1 \\ -1 \\ 1 \\ 1 \end{bmatrix} + c_3 \begin{bmatrix} 0 \\ 1 \\ 1 \\ 0 \end{bmatrix} = \begin{bmatrix} a \\ b \\ c \\ d \end{bmatrix}$$

for the coefficients of the linear combination, we can use the augmented matrix

$$\begin{bmatrix} 1 & -1 & 0 & | & a \\ 0 & -1 & 1 & | & b \\ 1 & 1 & 1 & | & c \\ 1 & 1 & 0 & | & d \end{bmatrix} \xrightarrow[\rightarrow]{\text{row reduction}} \begin{bmatrix} 1 & 0 & 0 & | & a + d/2 \\ 0 & 1 & 0 & | & d/2 \\ 0 & 0 & 1 & | & b + d/2 \\ 0 & 0 & 0 & | & a - b + c - 3d/2 \end{bmatrix}.$$

Since there is a solution only if a-b+c-3d/2=0, we cannot solve this system for all a, b, c, and d, so the three vectors given do *not* span $\Re^4$. (We will see in the next section that we need at least four vectors to span $\Re^4$).

19. To see if $x^2 - 1, x^2 + 1, x^2 + x$ span P_2, see if we can write any element $ax^2 + bx + c$ as a linear combination of these three polynomials. In other words, we need to solve

$$c_1(x^2 - 1) + c_2(x^2 + 1) + c_3(x^2 + x) = ax^2 + bx + c.$$

Multiplying out and collecting terms, we have:

$$(c_1 + c_2 + c_3)x^2 + (c_3)x + (-c_1 + c_2) = ax^2 + bx + c.$$

This leads to a system of linear equations:

$$c_1 + c_2 + c_3 = a,$$
$$c_3 = b, \text{and}$$
$$-c_1 + c_2 = c.$$

To solve for c_1, c_2, and c_2, we can use the augmented matrix

$$\begin{bmatrix} 1 & 1 & 1 & | & a \\ 0 & 1 & 0 & | & b \\ -1 & 1 & 0 & | & c \end{bmatrix} \xrightarrow[\rightarrow]{\text{row reduction}} \begin{bmatrix} 1 & 0 & 0 & | & b - c \\ 0 & 1 & 0 & | & b \\ 0 & 0 & 1 & | & a - 2b + c \end{bmatrix}.$$

There is a solution no matter what a, b, and c are. Since we can find a linear combination of the three vectors that gives any element of P_2, these three vectors do span P_2.

2.3 Linear Independence and Bases

Exercises 2.3, pp. 93-95

1. To determine if a set of vectors is linearly independent, set a linear combination of them equal to the zero vector, and solve for the coefficients of the linear combination. If the only solution is the trivial solution, the vectors are linearly independent; otherwise they are linearly dependent. Suppose $c_1 \begin{bmatrix} 1 \\ -1 \end{bmatrix} + c_2 \begin{bmatrix} 2 \\ 1 \end{bmatrix} = \begin{bmatrix} 0 \\ 0 \end{bmatrix}$. Then

$$1 \cdot c_1 + 2 \cdot c_2 = 0, \text{ and}$$
$$-1 \cdot c_1 + 1 \cdot c_2 = 0.$$

To solve for c_1 and c_2, we can use the augmented matrix

$$\begin{bmatrix} 1 & 2 & | & 0 \\ -1 & 1 & | & 0 \end{bmatrix} \xrightarrow[\rightarrow]{\text{row reduction}} \begin{bmatrix} 1 & 0 & | & 0 \\ 0 & 1 & | & 0 \end{bmatrix}.$$

We see from the reduced row-echelon form of the matrix that $c_1 = c_2 = 0$, so the two vectors are linearly independent.

3. To determine if a set of vectors is linearly independent, set a linear combination of them equal to the zero vector, and solve for the coefficients of the linear combination. If the only solution is the trivial solution, the vectors are linearly independent; otherwise they are linearly dependent. Suppose $c_1 \begin{bmatrix} 6 \\ -4 \\ 2 \end{bmatrix} + c_2 \begin{bmatrix} -9 \\ 6 \\ -3 \end{bmatrix} = \begin{bmatrix} 0 \\ 0 \\ 0 \end{bmatrix}$. Then

$$6c_1 - 9c_2 = 0,$$
$$-4c_1 + 6c_2 = 0, \text{ and}$$
$$2c_1 - 3 \cdot c_2 = 0.$$

To solve for c_1 and c_2, we can use the augmented matrix

$$\begin{bmatrix} 6 & -9 & | & 0 \\ -4 & 6 & | & 0 \\ 2 & -3 & | & 0 \end{bmatrix} \xrightarrow[\rightarrow]{\text{row reduction}} \begin{bmatrix} 1 & -3/2 & | & 0 \\ 0 & 0 & | & 0 \\ 0 & 0 & | & 0 \end{bmatrix}.$$

We see from the reduced row-echelon form of the matrix that there are nontrivial solutions, so the three vectors are linearly dependent.

5. To determine if a set of vectors is linearly independent, set a linear combination of them equal to the zero vector, and solve for the coefficients of the linear combination. If the only solution is the trivial solution, the vectors are linearly independent; otherwise they are linearly dependent. Suppose $c_1 \begin{bmatrix} 1 \\ -1 \\ 0 \end{bmatrix} + c_2 \begin{bmatrix} 1 \\ 2 \\ -1 \end{bmatrix} + c_3 \begin{bmatrix} 1 \\ 1 \\ 0 \end{bmatrix} = \begin{bmatrix} 0 \\ 0 \\ 0 \end{bmatrix}$. Then

$$c_1 + c_2 + c_3 = 0,$$
$$-c_1 + 2c_2 + c_3 = 0, \text{ and}$$
$$0c_1 - c_2 + 0c_3 = 0.$$

To solve for c_1 and c_2, we can use the augmented matrix

$$\begin{bmatrix} 1 & 1 & 1 & | & 0 \\ -1 & 2 & 1 & | & 0 \\ 0 & -1 & 0 & | & 0 \end{bmatrix} \xrightarrow[\rightarrow]{\text{row reduction}} \begin{bmatrix} 1 & 0 & 0 & | & 0 \\ 0 & 1 & 0 & | & 0 \\ 0 & 0 & 1 & | & 0 \end{bmatrix}.$$

We see from the reduced row-echelon form of the matrix that the only solution is the trivial one, so the three vectors are linearly independent.

7. To determine if a set of vectors is linearly independent, set a linear combination of them equal to the zero vector, and solve for the coefficients of the linear combination. If the only solution is the trivial solution, the vectors are linearly independent; otherwise they are linearly dependent. Suppose $c_1 \begin{bmatrix} 1 & 0 \\ -1 & 1 \end{bmatrix} + c_2 \begin{bmatrix} 0 & 1 \\ 1 & 0 \end{bmatrix} + c_3 \begin{bmatrix} 1 & 1 \\ 1 & 1 \end{bmatrix} = \begin{bmatrix} 0 & 0 \\ 0 & 0 \end{bmatrix}$. Then

$$c_1 + 0c_2 + c_3 = 0,$$
$$0c_1 + 1c_2 + 1c_3 = 0,$$
$$-1c_1 + c_2 + c_3 = 0, \text{ and}$$
$$0c_1 - c_2 + 0c_3 = 0.$$

To solve for c_1 and c_2, we can use the augmented matrix

$$\begin{bmatrix} 1 & 0 & 1 & | & 0 \\ 0 & 1 & 1 & | & 0 \\ -1 & 1 & 1 & | & 0 \\ 1 & 0 & 1 & | & 0 \end{bmatrix} \xrightarrow[\rightarrow]{\text{row reduction}} \begin{bmatrix} 1 & 0 & 0 & | & 0 \\ 0 & 1 & 0 & | & 0 \\ 0 & 0 & 1 & | & 0 \\ 0 & 0 & 0 & | & 0 \end{bmatrix}.$$

Although there is a row of zeros at the end of this 4×3 matrix, there are only three variables, not four, and we see from the reduced row-echelon form of the matrix that the only solution is the trivial one, so the three vectors are linearly independent.

9. To determine if a set of vectors is linearly independent, set a linear combination of them equal to the zero vector, and solve for the coefficients of the linear combination. If the only solution is the trivial solution, the vectors are linearly independent; otherwise they are linearly dependent. Suppose $c_1(x^2 + x + 2) + c_2(x^2 + 2x + 1) + c_3(2x^2 + 5x + 1) = 0 = 0x^2 + 0x + 0$. Then, multiplying out and collecting terms, we have

$$c_1 + c_2 + 2c_3 = 0,$$
$$c_1 + 2c_2 + 5c_3 = 0, \text{ and}$$
$$2c_1 + c_2 + c_3 = 0.$$

To solve for c_1 and c_2, we can use the augmented matrix

$$\begin{bmatrix} 1 & 1 & 2 & | & 0 \\ 1 & 2 & 5 & | & 0 \\ 2 & 1 & 1 & | & 0 \end{bmatrix} \xrightarrow[\rightarrow]{\text{row reduction}} \begin{bmatrix} 1 & 0 & -1 & | & 0 \\ 0 & 1 & 3 & | & 0 \\ 0 & 0 & 0 & | & 0 \end{bmatrix}.$$

We see from the reduced row-echelon form of the matrix that there are nontrivial solutions, so the three vectors are linearly dependent.

11. Hint: To show that a set of vectors forms a basis for a vector space, show (1) the vectors are linearly independent and (2) the vectors span the vector space.

13. Hint: To show that a set of vectors forms a basis for a vector space, show (1) the vectors are linearly independent and (2) the vectors span the vector space.

15. Hint: To show that a set of vectors forms a basis for a vector space, show (1) the vectors are linearly independent and (2) the vectors span the vector space.

17. Hint: To show that a set of vectors forms a basis for a vector space, show (1) the vectors are linearly independent and (2) the vectors span the vector space.

19. Hint: To show that a set of vectors does not form a basis for a vector space, show that the set of vectors fails to be linearly independent *or* fails to span the vector space.

21. Hint: To show that a set of vectors does not form a basis for a vector space, show that the set of vectors fails to be linearly independent *or* fails to span the vector space.

23. Part a: We must find $c_1, c_2,$ and c_3 so that $c_1 \begin{bmatrix} 1 \\ 3 \\ -1 \end{bmatrix} + c_2 \begin{bmatrix} 0 \\ -1 \\ 2 \end{bmatrix} + c_3 \begin{bmatrix} 2 \\ 1 \\ 3 \end{bmatrix} = \begin{bmatrix} 1 \\ 2 \\ -1 \end{bmatrix}$. Then

$$c_1 + 0c_2 + 2c_3 = 1,$$
$$3c_1 - c_2 + c_3 = 2, \text{ and}$$
$$-c_1 + 2c_2 + 3c_3 = -1.$$

To solve for c_1 and c_2, we can use the augmented matrix

$$\begin{bmatrix} 1 & 0 & 2 & | & 1 \\ 3 & -1 & 1 & | & 2 \\ -1 & 2 & 3 & | & -1 \end{bmatrix} \xrightarrow[\rightarrow]{\text{row reduction}} \begin{bmatrix} 1 & 0 & 0 & | & 1/5 \\ 0 & 1 & 0 & | & -1 \\ 0 & 0 & 1 & | & 2/5 \end{bmatrix}.$$

We see from the reduced row-echelon form of the matrix that $c_1 = 1/5$, $c_2 = -1$, and $c_3 = 2/5$, so $[v]_\alpha = \begin{bmatrix} 1/5 \\ -1 \\ 2/5 \end{bmatrix}$.

Part b: $1\begin{bmatrix} 1 \\ 3 \\ -1 \end{bmatrix} + 2\begin{bmatrix} 0 \\ -1 \\ 2 \end{bmatrix} - 1\begin{bmatrix} 2 \\ 1 \\ 3 \end{bmatrix} = \begin{bmatrix} -1 \\ 0 \\ 0 \end{bmatrix} = v.$

25. Part a: We must find c_1, c_2, and c_3 so that $c_1(x^2 + x + 1) + c_2(x^2 - x + 1) + c_3(x^2 - 1) = 2x^2 + 3x$. Multiplying out and collecting terms, we see that

$$c_1 + c_2 + c_3 = 2,$$
$$c_1 - c_2 = 3, \text{and}$$
$$c_1 + c_2 - c_3 = 0.$$

To solve for c_1, c_2, and c_3, we can use the augmented matrix

$$\begin{bmatrix} 1 & 1 & 1 & | & 2 \\ 1 & -1 & 0 & | & 3 \\ 1 & 1 & -1 & | & 0 \end{bmatrix} \xrightarrow[\rightarrow]{\text{row reduction}} \begin{bmatrix} 1 & 0 & 0 & | & 2 \\ 0 & 1 & 0 & | & -1 \\ 0 & 0 & 1 & | & 1 \end{bmatrix}.$$

We see from the reduced row-echelon form of the matrix that $c_1 = 2$, $c_2 = -1$, and $c_3 = 1$, so $[v]_\alpha = \begin{bmatrix} 2 \\ -1 \\ 1 \end{bmatrix}$.

Part b: $1(x^2 + x + 1) + 2(x^2 - x + 1) + 3(x^2 - 1) = (x^2 + 2x^2 + 3x^2) + (x - 2x) + (1 + 2 - 3) = 6x^2 - x = v.$

2.4 Dimension: Nullspace, Row Space, and Column Space

Exercises 2.4, pp. 104-106

1. Part a: Since the dimension of R^2 is 2, one vector cannot form a basis for R^2.

   Part b: Since $\begin{bmatrix} 2 \\ -1 \end{bmatrix}$ is not a scalar multiple of $\begin{bmatrix} 1 \\ 1 \end{bmatrix}$, these two vectors are linearly independent. Since the dimension of R^2 is 2, two linearly independent vectors will span R^2. Hence these two vectors *do* form a basis for R^2.

   Part c: The second vector is a scalar multiple of the first vector, so these two vectors are *not* linearly independent. Since the dimension of R^2 is 2, we need two linearly independent vectors to form a basis, so these vectors do *not* form a basis for R^2.

   Part d: Since there are three vectors and the dimension of R^2 is two, these *cannot* form a basis for R^2.

3. Part a: Since the dimension of P_2 is 3, these three vectors will form a basis if and only if they are linearly independent. To determine that, suppose

   $$c_1(x^2 + x - 1) + c_2(2x^2 - 3) + c_3(x^2 + x + 2) = 0.$$

   Then (distributing the coefficients and collecting terms), we see that

   $$(c_1 + 2c_2 + c_3)x^2 + (c_1 + c_3)x + (-c_1 - 3c_2 + 2c_3) = 0,$$

   so

   $$c_1 + 2c_2 + c_3 = 0,$$
   $$c_1 + c_3 = 0, \text{ and}$$
   $$-c_1 - 3c_2 + 2c_3 = 0.$$

   Solving this system of equations (by whatever method you like best), we see that $c_1 = c_2 = c_3 = 0$. Thus, these polynomials are linearly independent and do form a basis for P_2.

   Part b: Since the dimension of P_2 is 3, two polynomials cannot form a basis for it.

Part c: Three polynomials will form a basis for P_2 if they are linearly independent, since the dimension of P_2 is 3. To see if these polynomials are linearly independent, set a linear combination of the polynomials equal to the zero polynomial and solve for the coefficients, as follows: suppose

$$c_1(x^2 + x - 1) + c_2(x^2 + x + 2) + c_3(x^2 + x + 14) = 0.$$

Then (distributing the coefficients and collecting terms), we see that

$$(c_1 + c_2 + c_3)x^2 + (c_1 + c_2 + c_3)x + (-c_1 + 2c_2 + 14c_3) = 0,$$

so

$$c_1 + c_2 + c_3 = 0,$$
$$c_1 + c_2 + c_3 = 0, \text{ and}$$
$$6c_1 + 2c_2 + 14c_3 = 0.$$

The determinant of the associated matrix is zero (since the matrix has two equal rows), so this system has nontrivial solutions. Hence the polynomials are *not* linearly independent and do *not* form a basis for P_2.
Part d: Since the dimension of P_2 is 3, four polynomials cannot form a basis for it.

5. Part a: The nullspace of a matrix A is the set of vectors $\mathbf{v}$ such that $A\mathbf{v} = \mathbf{0}$. In this case

$$\begin{bmatrix} 1 & 1 \\ 1 & -2 \end{bmatrix} \begin{bmatrix} x \\ y \end{bmatrix} = \begin{bmatrix} 0 \\ 0 \end{bmatrix}.$$

Since the determinant of A is $-3 \neq 0$, the only solution is $x = y = 0$, so the nullspace is just the vector $\begin{bmatrix} 0 \\ 0 \end{bmatrix}$. There is no basis for this nullspace.
Part b: A basis for the rowspace of a matrix A is found by converting A to reduced row-echelon form and taking the nonzero rows of that form. In this case, the reduced row-echelon form is the 2×2 identity matrix, so a basis for the nullspace consists of the two vectors $[1, 0]$ and $[0, 1]$.
Part c: A basis for the column space of a matrix A is found by taking the transpose of A and finding *its* rowspace as described in part b of this problem. The row vectors of the reduced row-echelon form of A^T are then transposed individually into column vectors to form a basis for the column space of A. Since this particular matrix is symmetric, $A^T = A$, so we can just transpose the vectors we found for the basis of A's row space: $\begin{bmatrix} 1 \\ 0 \end{bmatrix}$ and $\begin{bmatrix} 0 \\ 1 \end{bmatrix}$.
Part d: Since the matrix is 2×2 and the dimension of its nullspace is 0, the rank of the matrix is $2 - 0 = 2$.

7. Part a: The nullspace of a matrix A is the set of vectors $\mathbf{v}$ such that $A\mathbf{v} = \mathbf{0}$. In this case

$$\begin{bmatrix} 1 & -1 & 1 \\ -1 & 1 & 0 \\ 1 & -1 & 2 \end{bmatrix} \begin{bmatrix} x \\ y \\ z \end{bmatrix} = \begin{bmatrix} 0 \\ 0 \\ 0 \end{bmatrix}.$$

Solving this matrix equation, we obtain a parametric solution: all vectors of the form $\begin{bmatrix} t \\ t \\ 0 \end{bmatrix} = t \begin{bmatrix} 1 \\ 1 \\ 0 \end{bmatrix}$. So the vector $\begin{bmatrix} 1 \\ 1 \\ 0 \end{bmatrix}$ forms a basis for the nullspace of this matrix.

Part b: To find a basis for the rowspace, row reduce the matrix until it is in reduced row-echelon form. The matrix $\begin{bmatrix} 1 & -1 & 1 \\ -1 & 1 & 0 \\ 1 & -1 & 2 \end{bmatrix}$ reduces to $\begin{bmatrix} 1 & -1 & 0 \\ 0 & 0 & 1 \\ 0 & 0 & 0 \end{bmatrix}$, so, taking the nonzero rows of the reduced row-echelon matrix, we find that a basis for the rowspace consists of the vectors $[1, -1, 0]$ and $[0, 0, 1]$.
Part c: A basis for the column space of a matrix A is found by taking the transpose of A and finding *its* rowspace as described in part b of this problem. The row vectors of the reduced row-echelon form of A^T are then transposed individually into column vectors to form a basis for the column space of A. Taking the transpose of the original matrix: $\begin{bmatrix} 1 & -1 & 1 \\ -1 & 1 & -1 \\ 1 & 0 & 2 \end{bmatrix},$

we row-reduce it to $\begin{bmatrix} 1 & 0 & 2 \\ 0 & 1 & 1 \\ 0 & 0 & 0 \end{bmatrix}$. Thus, a basis for the column space consists of the transposes of the two nonzero row

vectors in the reduced row-echelon form: $\begin{bmatrix} 1 \\ 0 \\ 2 \end{bmatrix}$ and $\begin{bmatrix} 0 \\ 1 \\ 1 \end{bmatrix}$.

Part d: The rank of a matrix is the same as the dimension of its rowspace–in this case, 2.

9. Part a: The nullspace of a matrix A is the set of vectors $\mathbf{v}$ such that $A\mathbf{v} = \mathbf{0}$. In this case

$$\begin{bmatrix} 1 & 1 & 0 & 3 \\ 1 & 1 & 1 & -2 \\ 3 & 3 & 2 & -1 \end{bmatrix} \begin{bmatrix} w \\ x \\ y \\ z \end{bmatrix} = \begin{bmatrix} 0 \\ 0 \\ 0 \\ 0 \end{bmatrix}.$$

When we solve this matrix equation, we obtain a parametric solution:

$$\begin{bmatrix} 1 & 1 & 0 & 3 \\ 1 & 1 & 1 & -2 \\ 3 & 3 & 2 & -1 \end{bmatrix} \text{ row reduces to } \begin{bmatrix} 1 & 1 & 0 & 3 \\ 0 & 0 & 1 & -5 \\ 0 & 0 & 0 & 0 \end{bmatrix}.$$

There are two free variables: x and z. Letting $z = t$ and $x = s$, we see that the nullspace consists of all vectors of the form $\begin{bmatrix} -3t - s \\ s \\ 5t \\ t \end{bmatrix}$. Since there are two free variables, the basis for the nullspace will consist of exactly two vectors. Write

each entry of the column vector expressing the solution set as a linear combination of t and s, then separate it into a linear combination of two vectors, one with a coefficient of t, the other with a coefficient of s:

$$\begin{bmatrix} -3t - s \\ s \\ 5t \\ t \end{bmatrix} = \begin{bmatrix} -3t - s \\ 0t + s \\ 5t + 0s \\ t + 0s \end{bmatrix} = t \begin{bmatrix} -3 \\ 0 \\ 5 \\ 1 \end{bmatrix} + s \begin{bmatrix} -1 \\ 1 \\ 0 \\ 0 \end{bmatrix}.$$

So the vectors $\begin{bmatrix} -3 \\ 0 \\ 5 \\ 1 \end{bmatrix}$ and $\begin{bmatrix} -1 \\ 1 \\ 0 \\ 0 \end{bmatrix}$ form a basis for the nullspace of this matrix.

Part b: To find a basis for the rowspace, row reduce the matrix until it is in reduced row-echelon form. The matrix $\begin{bmatrix} 1 & 1 & 0 & 3 \\ 1 & 1 & 1 & -2 \\ 3 & 3 & 2 & -1 \end{bmatrix}$ row reduces to $\begin{bmatrix} 1 & 1 & 0 & 3 \\ 0 & 0 & 1 & -5 \\ 0 & 0 & 0 & 0 \end{bmatrix}$, so, taking the nonzero rows of the reduced row-echelon matrix,

we find that a basis for the rowspace consists of the vectors $[1, 1, 0, 3]$ and $[0, 0, 1, -5]$.

Part c: A basis for the column space of a matrix A is found by taking the transpose of A and finding *its* rowspace as described in part b of this problem. The row vectors of the reduced row-echelon form of A^T are then transposed individually into

column vectors to form a basis for the column space of A. Taking the transpose of the original matrix: $\begin{bmatrix} 1 & 1 & 3 \\ 1 & 1 & 3 \\ 0 & 1 & 2 \\ 3 & -2 & -1 \end{bmatrix}$,

we row-reduce it to $\begin{bmatrix} 1 & 1 & 3 \\ 0 & 1 & 2 \\ 0 & 0 & 0 \\ 0 & 0 & 0 \end{bmatrix}$. Thus, a basis for the column space consists of the transposes of the two nonzero row

vectors in the reduced row-echelon form: $\begin{bmatrix} 1 \\ 0 \\ 1 \end{bmatrix}$ and $\begin{bmatrix} 0 \\ 1 \\ 2 \end{bmatrix}$.

Part d: The rank of a matrix is the same as the dimension of its rowspace–in this case, 2.

11. Part a: The nullspace of a matrix A is the set of vectors $\mathbf{v}$ such that $A\mathbf{v} = \mathbf{0}$. In this case

$$\begin{bmatrix} 1 & -1 & 2 & 1 \\ 2 & 1 & -1 & 0 \\ 1 & 2 & -3 & -1 \\ 0 & 3 & -5 & -2 \\ 4 & -1 & 3 & 2 \end{bmatrix} \begin{bmatrix} w \\ x \\ y \\ z \end{bmatrix} = \begin{bmatrix} 0 \\ 0 \\ 0 \\ 0 \end{bmatrix}.$$

When we solve this matrix equation, we obtain a parametric solution:

$$\begin{bmatrix} 1 & -1 & 2 & 1 \\ 2 & 1 & -1 & 0 \\ 1 & 2 & -3 & -1 \\ 0 & 3 & -5 & -2 \\ 4 & -1 & 3 & 2 \end{bmatrix} \text{ row reduces to } \begin{bmatrix} 1 & -1 & 2 & 1 \\ 0 & 1 & -5/3 & -2/3 \\ 0 & 0 & 0 & 0 \\ 0 & 0 & 0 & 0 \\ 0 & 0 & 0 & 0 \end{bmatrix}.$$ There are two free variables: y and z. Letting $z = t$ and

$y = s$, we see that the nullspace consists of all vectors of the form $\begin{bmatrix} (-1/3)s - (1/3)t \\ (5/3)s + (2/3)t \\ s \\ t \end{bmatrix}$. Since there are two free variables,

the basis for the nullspace will consist of exactly two vectors. Write each entry of the column vector expressing the solution set as a linear combination of r and s, then separate it into a linear combination of two vectors, one with a coefficient of t and one with a coefficient of s:

$$\begin{bmatrix} (-1/3)s - (1/3)t \\ (5/3)s + (2/3)t \\ s \\ t \end{bmatrix} = \begin{bmatrix} (-1/3)s + (-1/3)t \\ (5/3)s + (2/3)t \\ s + 0t \\ 0s + t \end{bmatrix} = s\begin{bmatrix} -1/3 \\ 5/3 \\ 1 \\ 0 \end{bmatrix} + t\begin{bmatrix} -1/3 \\ 2/3 \\ 0 \\ 1 \end{bmatrix}.$$

So the vectors $\begin{bmatrix} -1/3 \\ 5/3 \\ 1 \\ 0 \end{bmatrix}$ and $\begin{bmatrix} -1/3 \\ 2/3 \\ 0 \\ 1 \end{bmatrix}$ form a basis for the nullspace of this matrix.

Part b: To find a basis for the rowspace, row reduce the matrix until it is in reduced row-echelon form. The matrix

$$\begin{bmatrix} 1 & -1 & 2 & 1 \\ 2 & 1 & -1 & 0 \\ 1 & 2 & -3 & -1 \\ 0 & 3 & -5 & -2 \\ 4 & -1 & 3 & 2 \end{bmatrix} \text{ row reduces to } \begin{bmatrix} 1 & -1 & 2 & 1 \\ 0 & 1 & -5/3 & -2/3 \\ 0 & 0 & 0 & 0 \\ 0 & 0 & 0 & 0 \\ 0 & 0 & 0 & 0 \end{bmatrix},$$ so, taking the nonzero rows of the reduced row-

echelon matrix, we find that a basis for the rowspace consists of the vectors $[1, 0, 1/3, 1/3]$ and $[0, 1, -5/3, -2/3]$.

Part c: A basis for the column space of a matrix A is found by taking the transpose of A and finding *its* rowspace as described in part b of this problem. The row vectors of the reduced row-echelon form of A^T are then transposed individually into column vectors to form a basis for the column space of A. Taking the transpose of the original matrix:

$$\begin{bmatrix} 1 & 2 & 1 & 0 & 4 \\ -1 & 1 & 2 & 3 & -1 \\ 2 & -1 & -3 & -5 & 3 \\ 1 & 0 & -1 & -2 & 2 \end{bmatrix}, \text{ we row-reduce it to } \begin{bmatrix} 1 & 0 & -1 & -2 & 2 \\ 0 & 1 & 1 & 1 & 1 \\ 0 & 0 & 0 & 0 & 0 \\ 0 & 0 & 0 & 0 & 0 \end{bmatrix}.$$ Thus, a basis for the column space consists

of the transposes of the two nonzero row vectors in the reduced row-echelon form: $\begin{bmatrix} 1 \\ 0 \\ -1 \\ -2 \\ 2 \end{bmatrix}$ and $\begin{bmatrix} 0 \\ 1 \\ 1 \\ 1 \\ 1 \end{bmatrix}$.

Part d: The rank of a matrix is the same as the dimension of its rowspace–in this case, 2.

13. To find a basis for the subspace spanned by a set of vectors, construct a matrix whose rows are the given vectors, then find a basis for its rowspace. For this problem, we transpose each of the given column vectors (to obtain a row vector) and make

them into the matrix $\begin{bmatrix} 1 & -2 & 1 \\ 0 & 1 & 3 \\ 5 & -6 & -7 \end{bmatrix}$. When we row reduce this matrix, we get the 3×3 identity matrix. Its rows form a basis for the rowspace of the original matrix. To make these row vectors into column vectors, we transpose them, obtaining

$$\begin{bmatrix} 1 \\ 0 \\ 0 \end{bmatrix}, \begin{bmatrix} 0 \\ 1 \\ 0 \end{bmatrix}, \text{ and } \begin{bmatrix} 0 \\ 0 \\ 1 \end{bmatrix}.$$

15. To find a basis for the subspace spanned by a set of vectors, construct a matrix whose rows are the given vectors, then find a basis for its rowspace. For this problem, we transpose each of the given column vectors (to obtain a row vector) and make them into the matrix $\begin{bmatrix} -2 & 1 & 3 & -4 \\ 1 & -2 & 1 & -1 \\ -1 & -4 & 9 & -11 \end{bmatrix}$. When we row reduce this matrix, we get $\begin{bmatrix} 1 & 0 & -7/3 & 3 \\ 0 & 1 & -5/3 & 2 \\ 0 & 0 & 0 & 0 \end{bmatrix}$. The nonzero rows of the reduced row-echelon form of the matrix form a basis for its row space. To make these row vectors into column vectors, we transpose them, obtaining $\begin{bmatrix} 1 \\ 0 \\ -7/3 \\ 3 \end{bmatrix}$ and $\begin{bmatrix} 0 \\ 1 \\ -5/3 \\ 2 \end{bmatrix}$. These column vectors form a basis for the subspace spanned by the original vectors.

17. If $\text{rank}(A) = n$, then the reduced row-echelon form of A is I_n, so $\det(A) \neq 0$. Conversely, if $\det(A) \neq 0$, then A is invertible, so A is row-equivalent to I_n, and by Theorem 2.13, $\text{RS}(A) = \text{RS}(I_n)$, so $\text{rank}(A) = n$.

19. The column space of A is the row space of A^T, which is an $n \times m$ matrix. By the result of Problem 18, if $n > m$, the rows of A^T, (hence the columns of A) are linearly dependent.

21. Part a: Let $p(x) = ax^2 + bx + d$ be an arbitrary element of P_2. To show that the given polynomials span P_2, we must show that we can write this arbitrary element of P_2 as a linear combination of the four polynomials $p_1(x), p_2(x), p_3(x)$, and $p_4(x)$. That is, for some real coefficients c_1, c_2, c_3, and c_4,

$$c_1 p_1(x) + c_2 p_2(x) + c_3 p_3(x) + c_4 p_4(x) = p(x) = ax^2 + bx + d.$$

Substituting the actual polynomials for the $p_i(x)$, we have

$$c_1(x^2 - 1) + c_2(x^2 + 1) + c_3(x - 1) + c_4(x + 1) = ax^2 + bx + d.$$

Multiplying out and collecting coefficients of x^2, x, and constants, we have

$$(c_1 + c_2)x^2 + (c_3 + c_4)x + (-c_1 + c_2 + c_4) = ax^2 + bx + d,$$

which implies

$$c_1 + c_2 = a$$
$$c_3 + c_4 = b$$
$$-c_1 + c_2 + c_4 = d$$

Since we have three equations in four variables, we know that there are solutions to this system of linear equations. So we can write an arbitrary element of P_2 as a linear combination of the four polynomials, and therefore the four polynomials span P_2.

Part b: Following the proof of Lemma 2.11, we see if $p_4(x)$ is a linear combination of the polynomials $p_1(x), p_2(x)$, and $p_3(x)$. If

$$c_1(x^2 - 1) + c_2(x^2 + 1) + c_3(x - 1) = x + 1,$$

we have (multiplying out and collecting coefficients of x^2 and x),

$$(c_1 + c_2)x^2 + c_3 x + (-c_1 + c_2 - c_3) = x + 1.$$

Therefore,

$$c_1 + c_2 = 0$$
$$c_3 = 1$$
$$-c_1 + c_2 - c_3 = 1$$

Solve this system of linear equations to obtain $c_1 = -1$, $c_2 = 1$, and $c_3 = 1$. So $x + 1 = -1(x^2 - 1) + 1(x^2 + 1) + 1(x - 1)$. By Part a of this problem, we know that p_1, p_2, p_3, and p_4 span P_2. By the work we just did, p_4 is a linear combination of p_1, $p_{2]}$, and p_3, so the three polynomials p_1, p_2, and p_3 also span P_2. Since we know that the dimension of P_2 is 3, these three polynomials must form a basis for P_2.

23. Suppose there are two distinct reduced row-echelon forms of the same matrix, M_1 and M_2. Since the nonzero rows of the reduced row-echelon form of a matrix form a basis for the row space of that matrix, both M_1 and M_2 have the same number of nonzero rows, and any vector in the row space can be written as a linear combination of the nonzero rows of either reduced row-echelon form. Suppose the dimension of the row space is 1. Then M_1 and M_2 each have one nonzero row, and the nonzero row of M_2 is a linear combination of the nonzero row of M_1. But the first nonzero entry of each of these rows must begin with 1; therefore they must be equal. If the dimension of the row space is greater than 1, we can make a similar argument for the nonzero row whose leading 1 is farthest to the right in both matrices. Working backwards, we see that both M_1 and M_2 must have exactly the same nonzero rows.

The reduced column-echelon form of a matrix is the reduced row-echelon form of the transpose of that matrix. Since the reduced row-echelon form of a matrix is unique, so is the reduced column-echelon form.

2.5   Wronskians

Exercises 2.5, p. 110

1. To show that e^{3x} and e^{-2x} are linearly independent on $(-\infty, \infty)$, apply Theorem 2.15. Since e^{3x} and e^{-2x} are both differentiable, we can use the Wronskian of these two functions. The Wronskian of a list of n functions is the determinant of the $n \times n$ matrix whose (i,j) entry is the $(i-1)^{\text{st}}$ derivative of the jth function in the list.

$$\begin{vmatrix} e^{3x} & e^{-2x} \\ 3e^{3x} & -2e^{-2x} \end{vmatrix} = -2e^x - 3e^x = -5e^x.$$

Since $-5e^x$ is nonzero for some x in $(-\infty, \infty)$ (in fact, it is nonzero for every such x), the functions e^{3x} and e^{-2x} are linearly independent on that interval.

3. To show that $e^{2x}\cos x$ and $e^{2x}\sin x$ are linearly independent on $(-\infty, \infty)$, apply Theorem 2.15. Since $e^{2x}\cos x$ and $e^{2x}\sin x$ are both differentiable, we can use the Wronskian of these two functions. The Wronskian of a list of n functions is the determinant of the $n \times n$ matrix whose (i,j) entry is the $(i-1)^{\text{st}}$ derivative of the jth function in the list. We must use the product rule to calculate these derivatives.

$$\begin{vmatrix} e^{2x}\cos x & e^{2x}\sin x \\ e^{2x}(2\cos x - \sin x) & e^{2x}(\cos x + 2\sin x) \end{vmatrix} = e^{4x}(\cos^2 x + \sin^2 x) = e^{4x}.$$

Since e^{4x} is nonzero for some x in $(-\infty, \infty)$ (in fact, it is nonzero for every such x), the functions $e^{2x}\cos x$ and $e^{2x}\sin x$ are linearly independent on that interval.

5. To show that $f(x) = x^2 - 1$, $g(x) = x^2 + 1$, and $h(x) = x + 1$ are linearly independent on $(-\infty, \infty)$, apply Theorem 2.15. Since $f(x) = x^2 - 1$, $g(x) = x^2 + 1$, and $h(x) = x + 1$ are all twice differentiable, we can use the Wronskian of these three functions. The Wronskian of a list of n functions is the determinant of the $n \times n$ matrix whose (i,j) entry is the $(i-1)^{\text{st}}$ derivative of the jth function in the list. The Wronskian of the three functions is

$$\begin{vmatrix} x^2 - 1 & x^2 + 1 & x + 1 \\ 2x & 2x & 1 \\ 2 & 2 & 0 \end{vmatrix}.$$

Expanding about row 3, we calculate this determinant as

$$2\begin{vmatrix} x^2 + 1 & x + 1 \\ 2x & 1 \end{vmatrix} - 2\begin{vmatrix} x^2 - 1 & x + 1 \\ 2x & 1 \end{vmatrix} + 0\begin{vmatrix} x^2 - 1 & x^2 + 1 \\ 2x & 2x \end{vmatrix}$$

$$= 2((x^2 + 1) - (2x)(x + 1)) - 2(x^2 - 1) - (2x)(x + 1) = 4.$$

Since the constant function 4 is nonzero for some x (in fact, for every x) in $(-\infty, \infty)$, the three functions are linearly independent on that interval.

7. To show that $f(x) = e^{4x}$, $g(x) = xe^{4x}$, and $h(x) = x^2 x^{4x}$ are linearly independent on $(-\infty, \infty)$, apply Theorem 2.15. Since $f(x), g(x)$, and $h(x)$ are all twice differentiable, we can use the Wronskian of these three functions. The Wronskian of a list of n functions is the determinant of the $n \times n$ matrix whose (i, j) entry is the $(i-1)^{\text{st}}$ derivative of the jth function in the list. The Wronskian of the three functions is

$$\begin{vmatrix} e^{4x} & xe^{4x} & x^2 e^{4x} \\ 4e^{4x} & 4xe^{4x} + e^{4x} & 4x^2 e^{4x} + 2xe^{4x} \\ 16e^{4x} & e^{4x}(16x+8) & e^{4x}(16x^2 + 16x + 2) \end{vmatrix}.$$

Expanding about column 1, we calculate this determinant as

$$2\begin{vmatrix} x^2+1 & x+1 \\ 2x & 1 \end{vmatrix} - 2\begin{vmatrix} x^2-1 & x+1 \\ 2x & 1 \end{vmatrix} + 0\begin{vmatrix} x^2-1 & x^2+1 \\ 2x & 2x \end{vmatrix}$$

$$= 2((x^2+1) - (2x)(x+1)) - 2(x^2-1) - (2x)(x+1) = 4.$$

Since the constant function 4 is nonzero for some x (in fact, for every x) in $(-\infty, \infty)$, the three functions are linearly independent on that interval.

9. To show that x^3 and $|x^3|$ are linearly independent on $(-\infty, \infty)$, we attempt to apply Theorem 2.15. We need the derivative of each of the functions. The derivative of x^3 is $3x^2$, of course. To find the derivative of $g(x) = |x^3|$, note that we can rewrite $g(x)$ as

$$g(x) = \left\{ \begin{array}{l} x^3 \text{ if } x \geq 0 \\ -x \text{ if } x < 0 \end{array} \right\}.$$

So

$$g'(x) = \left\{ \begin{array}{l} 3x^2 \text{ if } x > 0 \\ -3x^2 \text{ if } x < 0 \end{array} \right\},$$

and at $x = 0$,

$$g'(0) = \lim_{h \to 0} \frac{g(0+h) - g(0)}{h}.$$

Finding the limit from the right, we have

$$\lim_{h \to 0^+} \frac{g(0+h) - g(0)}{h} = \lim_{h \to 0^+} \frac{h^3}{h} = \lim_{h \to 0^+} h^2 = 0,$$

and from the left, we have

$$\lim_{h \to 0^-} \frac{g(0+h) - g(0)}{h} = \lim_{h \to 0^-} \frac{-h^3}{h} = \lim_{h \to 0^-} -h^2 = 0.$$

Since the limit from the right and the limit from the left are equal,

$$g'(0) = \lim_{h \to 0} \frac{g(0+h) - g(0)}{h} = 0.$$

Thus we can say

$$g'(x) = \left\{ \begin{array}{l} 3x^2 \text{ if } x \geq 0 \\ -3x^2 \text{ if } x < 0 \end{array} \right\}.$$

Now we can calculate the Wronskian of x^3 and $|x^3|$. If $x \geq 0$,

$$w(x^3, |x^3|) = \begin{vmatrix} x^3 & x^3 \\ 3x^2 & 3x^2 \end{vmatrix} = 0.$$

If $x < 0$,

$$w(x^3, |x^3|) = \begin{vmatrix} x^3 & -x^3 \\ 3x^2 & -3x^2 \end{vmatrix} = 0.$$

Hence we have that the Wronskian is zero for every x. However, x^3 and $|x^3|$ are linearly independent. Suppose

$$c_1 x^3 + c_2 |x^3| = 0.$$

Substituting $x = 1$ and $x = -1$ into this equation, we arrive at the system

$$c_1 + c_2 = 0$$
$$-c_1 + c_2 = 0 \text{ ,}$$

which has only the trivial solution. Thus x^3 and $|x^3|$ are linearly independent.

11. To show that $x + 1$, $x - 1$, and x are linearly dependent, we simply need to find real numbers c_1, c_2, and c_3, not all zero, so that

$$c_1(x + 1) + c_2(x - 1) + c_3 x = 0$$

for all x. Collecting terms, we see that the coefficient of x, $c_1 + c_2 + c_3$, must equal 0, and so must the constant term $c_1 - c_2$. Solving this system of two equations in three variables, we choose one of the infinite number of solutions, say $c_1 = -1$, $c_2 = -1$, and $c_3 = 2$.

13. To show that $f(x) = x^2$ and $g(x) = x|x|$ are linearly dependent of $[0, \infty)$, it is enough to note that $f(x) = g(x)$ on this interval. Thus, $f(x) - g(x) = 0$ for all x in this interval, and we have found a linear combination, whose coefficients are not all zero, of the two functions whose sum is the zero function.

15. Part a: To find the Wronskian of $1, x, x^2, \cdots, x^{n-1}$: First let's write the Wronskian for a few ns to see if we notice a pattern.

For $n = 1$, $|1| = 1$. For $n = 2$, we calculate this determinant: $\begin{vmatrix} 1 & x \\ 0 & 1 \end{vmatrix} = 1$. For $n = 3$, we calculate $\begin{vmatrix} 1 & x & x^2 \\ 0 & 1 & 2x \\ 0 & 0 & 2 \end{vmatrix} = 2$. For

$n = 4$, we have

$$\begin{vmatrix} 1 & x & x^2 & x^3 \\ 0 & 1 & 2x & 3x^2 \\ 0 & 0 & 2 & 6x \\ 0 & 0 & 0 & 6 \end{vmatrix} = 12$$

Notice that we always get an upper triangular matrix whose diagonal entries are what we might call the last nonzero derivatives of the functions in the first row. That final nonzero derivative for x^k is $k! = (k)(k-1)\cdots(1)$. When $k = 0$, we use $0!$ which is defined to be 1. Since the matrix we are getting is upper triangular, its determinant is the product of its diagonal entries, so the Wronskian of $1, x, x^2, \cdots, x^{n-1}$ is $(n-1)!(n-2)!\cdots 1! 0!$, (which is nonzero).

Part b: You can show this by showing that $w(g(x)f_1(x), g(x)f_2(x), \cdots, g(x)f_n(x)) = |GF| = |G||F|$ shown here:

$$|G||F| = \begin{vmatrix} g & 0 & 0 & 0 & \cdots & 0 \\ g' & g & 0 & 0 & \cdots & 0 \\ g'' & c_{2,2}g' & g & 0 & \cdots & 0 \\ \vdots & \vdots & \vdots & \vdots & \ddots & \vdots \\ g^{(n-1)} & c_{n,1}g^{(n-2)} & c_{n,3}g^{(n-3)} & c_{n,4}g^{(n-4)} & \cdots & g \end{vmatrix} \begin{vmatrix} f_1 & f_2 & \cdots & f_n \\ f_1' & f_2' & \cdots & f_n' \\ \vdots & \vdots & \ddots & \vdots \\ f_1^{(n-1)} & f_2^{(n-1)} & \cdots & f_n^{(n-1)} \end{vmatrix} = g^n w(f_1, f_2, \cdots, f_n).$$

(In the lower triangular matrix G with the derivatives of g, the coefficients $c_{i,j}$ are the correct coefficients so that GF is the matrix whose (i, j) entry is the $i - 1$ derivative of gf_j.)

Part c: Applying Part b with $g(x) = e^{rx}$, we know that the Wronskian of $e^{rx}, xe^{rx}, x^2e^{rx}, \cdots, x^{n-1}e^{rx}$ is

$$(e^{rx})^n w(1, x, x^2, \cdots, x^{n-1}).$$

From Part a, we know that $w(1, x, x^2, \cdots, x^{n-1})$ is a product of factorials, and $k!$ is nonzero for any k. The important thing is that $w(1, x, x^2, \cdots, x^{n-1})$ is nonzero. Since e^{rx} is nonzero for any x, we know that the Wronskian of $e^{rx}, xe^{rx}, x^2e^{rx}, \cdots, x^{n-1}e^{rx}$ is $(e^{rx})^n w(1, x, x^2, \cdots, x^{n-1})$, which is nonzero. Thus, applying Theorem 2.15, the functions $e^{rx}, xe^{rx}, x^2e^{rx}, \cdots, x^{n-1}e^{rx}$ are linearly independent.

Chapter 3

First Order Ordinary Differential Equations

3.1   Introduction to Differential Equations

Exercises 3.1, pp. 119-120

1. The highest order derivative is 2.

3. The highest order derivative is 2.

5. If $y = e^{2x} - 2$, then $y' = 2e^{2x}$. Substituting these into $y' = 2y + 4$ we see that

$$2e^{2x} = 2(e^{2x} - 2) + 4.$$

So the given function is a solution of the differential equation.

However, $y(0) = e^0 - 2 = -1 \neq -2$, so the given function does *not* satisfy the indicated initial condition.

7. If $y = e^x$, then $y' = e^x$ and $y'' = e^x$. Substituting these into $y'' - 2y' + y = 0$ we see that

$$e^x - 2e^x + e^x = 0.$$

So the given function is a solution of the differential equation.

However, $y(0) = e^0 = 1 \neq -1$, so the given function does *not* satisfy the indicated initial condition.

9. For the differential equation $y' = 2xy$ the slope of the tangent line at $(1,0)$ is $y' = 2 \cdot 1 \cdot 0 = 0$.
   At $(2,4)$, $y' = 2 \cdot 2 \cdot 4 = 16$.
   At $(-2,-2)$, $y' = 2(-2)(-2) = 8$.
   To generate the direction field using Maple, type:

   ```
   dfieldplot(diff(y(x),x)=2*x*y(x),y(x),x=-3..3,y=-5..5);
   ```

   at the prompt.

11. For the differential equation $y' = xy^2 + 2x$ the slope of the tangent line at $(1,-1)$ is
    $y' = 1(-1)^2 + 2(1) = 3$.
    At $(4,2)$, $y' = 4(2)^2 + 2(4) = 24$.
    At $(-4,2)$, $y' = -4(2)^2 + 2(-4) = -24$.
    To use Maple to generate the direction field, type:

    ```
    dfieldplot(diff(y(x),x)=x*(y(x))^2+2*x,y(x),x=-2..2,y=-1..1);
    ```

    at the prompt.

13.   1. To find the equilibrium solutions set $y' = 2y + 4 = 0$. This differential equation has the single equilibrium solution
      $y = -2$.

2. For $y > -2$, $y' > 0$, so the graphs of the solutions will be increasing. For $y < -2$, $y' < 0$, so the graphs of the solutions will be decreasing.

To understand the concavity we take the derivative of each side of the original differential equation to get: $y'' = 2y'$. The graphs of the solutions will be concave up when $y'' > 0$; here that is when $y > -2$. Similarly, the graphs of the solutions will be concave down when $y'' < 0$, i.e., when $y < -2$.

3. To use Maple to sketch the phase portrait, type:

```
phaseportrait(diff(y(x),x)=2*y(x)+4,y(x),x=-1..1,[seq([y(0)=k/2],k=-7..-1)],
arrows=none);
```

at the prompt.

15. 1. To find the equilibrium solutions set $y' = y^2 - 4y = y(y-4) = 0$. This differential equation has two equilibria, one at $y = 0$, the other at $y = 4$.

2. When $y < 0$ or $y > 4$, we see that $y' = y^2 - 4y > 0$, so the graphs of the solutions will be increasing. For $0 < y < 4$, we'll have $y' = y^2 - 4y < 0$, so the graphs of the solutions will be decreasing.

To understand the concavity we take the derivative of each side of the original differential equation to get:

$$y'' = 2yy' - 4y' = 2(y-2)y' = 2(y-2)y(y-4).$$

The graphs of the solutions will be concave up when $y'' > 0$. This happens when $0 < y < 2$ and $y > 4$. The graphs of the solutions will be concave down when $y'' < 0$. This happens when $y < 0$ and $2 < y < 4$.

3. To use Maple to sketch the phase portrait, type:

```
phaseportrait(diff(y(x),x)=(y(x))^2-4*y(x),y(x),x=-1..1,[seq([y(0)=.5*k],
k=-2..10)],y=-1..5,arrows=none);
```

at the prompt.

17. 1. To find the equilibrium solutions set $y' = y^3 - 9y = (y+3)y(y-3) = 0$. This differential equation has three equilibria, one is at $y = 0$. The others are at $y = \pm 3$.

2. When $-3 < y < 0$ or $y > 3$, the derivative $y' = y^3 - 9y > 0$, so the graphs of the solutions will be increasing. For $y < -3$ or $0 < y < 3$, $y' = y^3 - 9y < 0$, so the graphs of the solutions will be decreasing.

To understand the concavity we take the derivative of each side of the original differential equation to get:

$$y'' = 3y^2y' - 9y' = 3(y^2-3)y' = 3(y^2-3)(y+3)y(y-3).$$

The graphs of the solutions will be concave up when $y'' > 0$. This happens when $-3 < y < -\sqrt{3}$, $0 < y < \sqrt{3}$ and $y > 3$. The graphs of the solutions will be concave down when $y'' < 0$. This happens when $y < -3$, $-\sqrt{3} < y < 0$ and $\sqrt{3} < y < 3$.

3. To use Maple to sketch the phase portrait, type:

```
phaseportrait(diff(y(x),x)=(y(x))^3-9*y(x),y(x),x=-0.4..0.4,[seq([y(0)=k],
k=-4..4)],y=-4..4,arrows=none);
```

at the prompt.

19. Let $y = \sqrt{1 - e^{-2x}}$. Then $y(0) = \sqrt{1 - e^0} = 0$. We also have,

$$-y + y^{-1} = -\sqrt{1 - e^{-2x}} + \frac{1}{\sqrt{1 - e^{-2x}}} = \frac{-(1 - e^{-2x}) + 1}{\sqrt{1 - e^{-2x}}} = \frac{e^{-2x}}{\sqrt{1 - e^{-2x}}}.$$

But, the derivative

$$y' = \frac{1}{2}(1 - e^{-2x})^{-1/2} \cdot 2e^{-2x} = \frac{e^{-2x}}{\sqrt{1 - e^{-2x}}}.$$

Thus, $y' = -y + y^{-1}$.

Similarly, when $y = -\sqrt{1 - e^{-2x}}$ we may also check that this function satisfies the given initial value problem. Thus we have two different solutions to the initial value problem.

The existence of these two functions, however, does not contradict Theorem 3.1 because for this differential equation $f(x,y) = -y + y^{-1}$ is not continuous on any rectangle containing the point $(0,0)$, i.e., $y(0) = 0$. As this hypothesis of the theorem is violated, there is no reason that the differential equation should have a unique solution.

21. Use the maple command

```
dsolve(diff(y(x),x)=2*y(x));
```

to find the solution to the differential equation.

23. Use the maple command

```
dsolve(2*x*y(x)+(x^2-y(x)^2)*diff(y(x),x)=0);
```

to find the solution to the differential equation.

25. Use the maple command

```
dsolve({diff(y(x),x)=x*y(x)^2+2*x,y(1)=-1});
```

to find the solution to the initial value problem.

3.2   Separable Differential Equations

Exercises 3.2, pp. 124

1. Factoring $y' = x\cos y + 2xy^2$ to $y' = x(\cos y + 2y^2)$, we see that this differential equation is separable.

3. The differential equation $\frac{dy}{dx} = x^2 + 2xy$ cannot be factored to the form $\frac{dy}{dx} = M(x)N(y)$, therefore this differential equation is *not* separable.

5. The separable differential equation $\frac{dy}{dx} = 2x^2y - 4y$ can be factored to $\frac{dy}{dx} = y(2x^2 - 4)$. Thus we have

$$\frac{dy}{y} = (2x^2 - 4)\,dx.$$

We integrate each side

$$\int \frac{dy}{y} = \int (2x^2 - 4)\,dx.$$

To obtain

$$\ln|y| = \frac{2}{3}x^3 - 4x + C.$$

7. The separable differential equation $(x^2 + 1)\frac{dy}{dx} = xy^2 + x$ can be rewritten as

$$(x^2 + 1)\frac{dy}{dx} = x(y^2 + 1)$$

and then as

$$\frac{1}{y^2 + 1}\,dy = \frac{x}{x^2 + 1}\,dx.$$

Integrating

$$\int \frac{dy}{y^2 + 1} = \int \frac{x}{x^2 + 1}\,dx$$

we obtain solutions given by the equation $\arctan y = \frac{1}{2}\ln(x^2 + 1) + C$.
Equivalently, $2\arctan y = \ln(x^2 + 1) + C$.

9. The separable differential equation $e^x(y^2-4y)dx+4dy=0$ can be rewritten as

$$\frac{dy}{y^2-4y}=-\frac{1}{4}e^x dx.$$

Integrating

$$\int\frac{dy}{y^2-4y}=-\frac{1}{4}\int e^x\,dx.$$

Using the method of partial fractions, we obtain

$$\frac{1}{4}\int\left(\frac{1}{y-4}-\frac{1}{y}\right)dy=-\frac{1}{4}e^x dx.$$

Therefore, our solutions are given by the equation

$$\frac{1}{4}(ln|y-4|-\ln|y|)=-\frac{1}{4}e^x+c,$$

or equivalently,

$$ln|y-4|-\ln|y|=-e^x+C.$$

11. The differential equation $t\frac{dy}{dt}+\frac{dy}{dt}=te^y$ is separable. After factoring and separating the variables we obtain: $e^{-y}\,dy=\frac{t}{t+1}\,dt$. Also note that $\frac{t}{t+1}=1-\frac{1}{t+1}$. Integrating

$$\int e^{-y}\,dy=\int\left(1-\frac{1}{t+1}\right)dt$$

we obtain, $-e^{-y}=t-\ln|t+1|+c$. We may solve for y to obtain: $y=-\ln(\ln|t+1|-t+C)$.

13. The differential equation $y'+\frac{x^2}{y}=0$ is separable. After separating the variables we obtain $y\,dy=-x^2\,dx$. Integrating $\int y\,dy=-\int x^2\,dx$, we obtain:

$$\frac{y^2}{2}=-\frac{x^3}{3}+c. \tag{3.1}$$

We are given the initial condition $y(0)=1$. Letting $x=0$ and $y=1$ in equation (3.1), we obtain: $\frac{1}{2}=-\frac{1}{3}\cdot 0+c$. Therefore, $c=\frac{1}{2}$. Thus, $\frac{y^2}{2}=-\frac{x^3}{3}+\frac{1}{2}$, or equivalently: $2x^3+3y^2=3$.

15. The differential equation $\frac{dy}{dx}=\frac{4xy}{y^2+4}$ is separable. After separating the variables we obtain $\frac{y^2+4}{y}\,dy=4x\,dx$. Integrating $\int\frac{y^2+4}{y}\,dy=\int 4x\,dx$, we obtain:

$$\frac{y^2}{2}+4\ln|y|=2x^2+c. \tag{3.2}$$

We are given the initial condition $y(1)=1$. Letting $x=1$ and $y=1$ in equation (3.2), we obtain: $\frac{1}{2}+0=2+c$. Therefore, $c=-\frac{3}{2}$. Thus, $\frac{y^2}{2}+4\ln|y|=2x^2-\frac{3}{2}$, or equivalently: $y^2+8\ln|y|=4x^2-3$.

17. (a) The solution in problem 13 was $2x^3+3y^2=3$. Therefore, $3y^2=3-2x^3$ or $y^2=\frac{9-6x^3}{9}$. Taking the square root, $y=\pm\frac{\sqrt{9-6x^3}}{3}$. Given that the initial condition is $y(0)=1>0$ we choose that positive root, so $y=\frac{\sqrt{9-6x^3}}{3}$.

    (b) From the form of the solution in (a) we need $9-6x^3\geq 0$, however from the form of the original differential equation , $y\neq 0$. Thus we must insist that $9-6x^3>0$. This will occur if $x<(\frac{3}{2})^{\frac{1}{3}}$.

21. (a) The differential equation $y'=(y+x+1)^2$ cannot be factored to the form $\frac{dy}{dx}=M(x)N(y)$, therefore this differential equation is *not* separable.

    (b) Let $v=y+x+1$. Differentiating we obtain, $\frac{dv}{dx}=\frac{dy}{dx}+1$, or $\frac{dy}{dx}=\frac{dv}{dx}-1$. Substituting in the original differential equation we obtain $\frac{dv}{dx}-1=v^2$, which is a separable differential equation. Separating the variables we see $\frac{dv}{v^2+1}=dx$.

    (c) Integrating we have, $\int\frac{dv}{v^2+1}=\int dx$, or $\arctan v=x+c$. Therefore $v=\tan(x+c)$. Since $v=y+x+1$, we now have $y+x+1=\tan(x+c)$, so $y=\tan(x+c)-x-1$.

3.3 Exact Differential Equations

Exercises 3.3, pp. 129-130

1. In the differential equation

$$(3x^2 - 4y^2)\,dx - (8xy - 12y^3)\,dy = 0,$$

$M(x,y) = 3x^2 - 4y^2$ and $N(x,y) = -(8xy - 12y^3)$. Thus, $M_y = -8y$ and $N_x = -8y$. Therefore the differential equation is exact.

To construct $F(x,y)$, we first integrate $M(x,y)$ with respect to x.

$$F(x,y) = \int (3x^2 - 4y^2)\,dx = x^3 - 4xy^2 + h(y).$$

We need to determine $h(y)$ so that $F_y(x,y) = N(x,y)$.
Here, $F_y = -8xy + h'(y) = -(8xy - 12y^3)$. Thus, $h'(y) = 12y^3$, so we can let $h(y) = 3y^4$. Therefore the solutions to the differential equation are given by $x^3 - 4xy^2 + 3y^4 = C$.

3. In the differential equation

$$(2xy + ye^x)\,dx + (x^2 + e^x)\,dy = 0,$$

$M(x,y) = 2xy + ye^x$ and $N(x,y) = x^2 + e^x$. Thus, $M_y = 2x + e^x$ and $N_x = 2x + e^x$. Therefore the differential equation is exact.
To construct $F(x,y)$, we first integrate $M(x,y)$ with respect to x.

$$F(x,y) = \int (2xy + ye^x)\,dx = x^2y + ye^x + h(y).$$

We need to determine $h(y)$ so that $F_y(x,y) = N(x,y)$. Here, $F_y = x^2 + e^x + h'(y) = x^2 + e^x$. Thus, $h'(y) = 0$, so we can let $h(y) = 0$. Therefore the solutions to the differential equation are given by $x^2y + ye^x = C$.

5. In the differential equation

$$(2x\cos y - x^2)\,dx + x^2\sin y\,dy = 0,$$

$M(x,y) = 2x\cos y - x^2$ and $N(x,y) = x^2\sin y$. Thus, $M_y = -2x\sin y$ and $N_x = 2x\sin y$. Since $M_y \neq N_x$ the differential equation is not exact.

7. The differential equation $\frac{dy}{dx} = \frac{1-2xy}{x^2}$ can be rewritten as

$$(2xy - 1)\,dx + x^2\,dy = 0.$$

Here, $M(x,y) = 2xy - 1$ and $N(x,y) = x^2$. Thus, $M_y = 2x$ and $N_x = 2x$. Therefore the differential equation is exact.
To construct $F(x,y)$, we first integrate $M(x,y)$ with respect to x.

$$F(x,y) = \int (2xy - 1)\,dx = x^2y - x + h(y).$$

We need to determine $h(y)$ so that $F_y(x,y) = N(x,y)$. Here, $F_y = x^2 + h'(y) = x^2$. Thus, $h'(y) = 0$, so we can let $h(y) = 0$. Therefore the solutions to the differential equation are given by $x^2y - x = C$.

9. The differential equation $\frac{ds}{dt} = \frac{e^t - 2t\cos s}{e^s - t^2\sin s}$ can be rewritten as

$$(e^s - t^2\sin s)\,ds + (2t\cos s - e^t)\,dt = 0.$$

Here, $M(s,t) = e^s - t^2\sin s$ and $N(s,t) = 2t\cos s - e^t$. Thus, $M_t = -2t\sin s$ and $N_s = -2t\sin s$. Therefore the differential equation is exact.
To construct $F(s,t)$, we first integrate $M(s,t)$ with respect to s.

$$F(s,t) = \int (e^s - t^2\sin s)\,ds = e^s + t^2\cos s + h(t).$$

We need to determine $h(t)$ so that $F_t(s,t) = N(s,t)$. Here, $F_t = 2t\cos s + h'(t) = 2t\cos s - e^t$. Thus, $h'(t) = -e^t$, so we can let $h(t) = -e^t$. Therefore the solutions to the differential equation are given by $e^s + t^2\cos s - e^t = C$.

11. In the differential equation

$$2xy\,dx + (x^2 + 3y^2)\,dy = 0,$$

$M(x,y) = 2xy$ and $N(x,y) = x^2 + 3y^2$. Thus, $M_y = 2x$ and $N_x = 2x$. Therefore the differential equation is exact. To construct $F(x,y)$, we first integrate $M(x,y)$ with respect to x.

$$F(x,y) = \int 2xy\,dx = x^2y + h(y).$$

We need to determine $h(y)$ so that $F_y(x,y) = N(x,y)$. Here, $F_y = x^2 + h'(y) = x^2 + 3y^2$. Thus, $h'(y) = 3y^2$, so we can let $h(y) = y^3$. Therefore the general solution to the differential equation is given by $x^2y + y^3 = C$.

We are also given the initial condition, $y(1) = 1$. Substituting $x = 1$ and $y = 1$ in the general solution we see that $1 + 1 = C$, i.e., $C = 2$. Therefore the solution to the initial value problem is $x^2y + y^3 = 2$.

13. In the differential equation

$$2y\sin(xy)\,dx + (2x\sin(xy) + 3y^2)\,dy = 0,$$

$M(x,y) = 2y\sin(xy)$ and $N(x,y) = 2x\sin(xy) + 3y^2$. Thus, $M_y = 2\sin(xy) + 2xy\cos(xy)$ and $N_x = 2\sin(xy) + 2xy\cos(xy)$. Therefore the differential equation is exact. To construct $F(x,y)$, we first integrate $M(x,y)$ with respect to x.

$$F(x,y) = \int 2y\sin(xy)\,dx = -2\cos(xy) + h(y).$$

We need to determine $h(y)$ so that $F_y(x,y) = N(x,y)$. Here, $F_y = 2x\sin(xy) + h'(y) = 2x\sin(xy) + 3y^2$. Thus, $h'(y) = 3y^2$, so we can let $h(y) = y^3$. Therefore the general solution to the differential equation is given by $-2\cos(xy) + y^3 = C$.

We are also given the initial condition, $y(0) = 1$. Substituting $x = 0$ and $y = 1$ in the general solution we see that $-2 + 1 = C$, i.e., $C = -1$. Therefore the solution to the initial value problem is $-2\cos(xy) + y^3 = -1$, or equivalently, $2\cos(xy) - y^3 = 1$.

17. A separable differential equation has the form $\frac{dy}{dx} = f(x)g(y)$ for some functions $f(x)$ and $g(y)$. This equation may be rewritten as $f(x)\,dx - \frac{1}{g(y)}\,dy = 0$. Letting $M(x,y) = f(x)$ and $N(x,y) = -\frac{1}{g(y)}$, we see that $M_y = N_x = 0$, so the original separable differential equation is exact.

19. Let $y' = \frac{ax+by}{cx+dy}$. This differential equation may be rewritten in the form

$$(ax+by)\,dx - (cx+dy)\,dy = 0.$$

Letting $M(x,y) = ax + by$ and $N(x,y) = -(cx+dy)$ we see that $M_y = b$ and $N_x = -c$. Therefore, the differential equation is exact if and only if $c = -b$. In this case we have $(ax+by)\,dx + (bx - dy)\,dy = 0$.

To construct $F(x,y)$, we first integrate $M(x,y)$ with respect to x.

$$F(x,y) = \int (ax+by)\,dx = \frac{a}{2}x^2 + bxy + h(y).$$

We need to determine $h(y)$ so that $F_y(x,y) = N(x,y)$. Here, $F_y = bx + h'(y) = bx - dy$. Thus, $h'(y) = -dy$, so we can let $h(y) = -\frac{d}{2}y^2$. Therefore the general solution to the differential equation is given by $\frac{a}{2}x^2 + bxy - \frac{d}{2}y^2 = k$.

3.4   Linear Differential Equations

Exercises 3.4, pp. 135-136

1. The differential equation $y' + \frac{1}{x^2}y = 0$ is linear. We find that an integrating factor is

$$u = e^{\int \frac{1}{x^2}\,dx} = e^{-\frac{1}{x}}.$$

Multiplying by this integrating factor gives $e^{-\frac{1}{x}}y' + \frac{1}{x^2}e^{-\frac{1}{x}}y = 0$. Since the left-hand side of this equation is

$$\frac{d}{dx}(uy) = \frac{d}{dx}\left(e^{-\frac{1}{x}}y\right),$$

we have $\frac{d}{dx}\left(e^{-\frac{1}{x}}y\right) = 0$. Integrating each side we obtain $e^{-\frac{1}{x}}y = C$. Multiplying through by $u^{-1} = e^{\frac{1}{x}}$ we see that

$$y = Ce^{\frac{1}{x}}.$$

3. The differential equation $y' - 2xy = x$ is linear. An integrating factor is

$$u = e^{\int -2x\,dx} = e^{-x^2}.$$

Multiplying through by this integrating factor gives $e^{-x^2}y' - 2xe^{-x^2}y = xe^{-x^2}$. Therefore $\frac{d}{dx}\left(e^{-x^2}y\right) = xe^{-x^2}$. Thus,

$$e^{-x^2}y = \int xe^{-x^2}\,dx.$$

The antiderivative on the right may be found using substitution, so

$$e^{-x^2}y = -\frac{1}{2}e^{-x^2} + C.$$

Multiplying through by e^{x^2} gives

$$y = -\frac{1}{2} + Ce^{x^2}.$$

5. The linear differential equation $y' = 1 + \frac{y}{1+2x}$, $x > 0$ may also be expressed in the form $y' - \frac{1}{1+2x}y = 1$. An integrating factor is

$$u = e^{\int -\frac{1}{1+2x}\,dx} = e^{-\frac{1}{2}\ln(1+2x)} = (1+2x)^{-\frac{1}{2}}.$$

Multiplying through by the integrating factor u gives $\frac{d}{dx}\left((1+2x)^{-\frac{1}{2}}y\right) = (1+2x)^{-\frac{1}{2}}$. Thus,

$$(1+2x)^{-\frac{1}{2}}y = \int (1+2x)^{-\frac{1}{2}}\,dx.$$

Finding the antiderivative we have, $(1+2x)^{-\frac{1}{2}}y = (1+2x)^{\frac{1}{2}} + C$. Multiplying through by $(1+2x)^{\frac{1}{2}}$ gives

$$y = 1 + 2x + C(1+2x)^{\frac{1}{2}}.$$

7. We multiply each side of the linear differential equation $xy' + y = x^2$, $x > 0$ by $\frac{1}{x}$ to obtain $y' - \frac{1}{x}y = x$. An integrating factor is

$$u = e^{\int \frac{1}{x}\,dx} = e^{\ln x} = x.$$

Multiplying through by the integrating factor u gives us our original equation! So, please note that the left-hand side of differential equation as originally stated equals $\frac{d}{dx}(xy)$. In other words, observing that the left-hand side of the original differential equation is already the derivative of a product would have saved us some work.

Thus, $\frac{d}{dx}(xy) = x^2$ and $xy = \int x^2\,dx$. Finding the antiderivative we have, $xy = \frac{1}{3}x^3 + C$. Multiplying through by $\frac{1}{x}$ gives

$$y = \frac{1}{3}x^2 + \frac{C}{x}.$$

9. The linear differential equation $(1+xy)\,dx - x^2\,dy = 0$, $x < 0$ may also be expressed in the form $\frac{dy}{dx} - \frac{1}{x}y = \frac{1}{x^2}$. An integrating factor is

$$u = e^{\int -\frac{1}{x}\,dx} = e^{-\ln|x|} = \frac{1}{|x|}.$$

Since $x < 0$, $|x| = -x$, so $u = -\frac{1}{x}$. However, whenever a function $f(x)$ is an integrating factor, any constant multiple of $f(x)$ will also be an integrating factor. Therefore we shall use the integrating factor $\frac{1}{x}$. Multiplying through by this integrating factor gives $\frac{d}{dx}\left(\frac{1}{x}y\right) = \frac{1}{x^3}$. Thus, $\frac{1}{x}y = \int \frac{1}{x^3}\,dx$. Finding the antiderivative we have, $\frac{1}{x}y = -\frac{1}{2x^2} + C$. Multiplying through by x gives

$$y = -\frac{1}{2x} + Cx.$$

11. The differential equation $\frac{dy}{dt} + e^t y = e^t$ is linear. An integrating factor is

$$u = e^{\int e^t \, dt} = e^{e^t}.$$

Multiplying through by this integrating factor gives $e^{e^t} \frac{dy}{dt} + e^t e^{e^t} y = e^t e^{e^t}$.

Therefore $\frac{d}{dt}\left(e^{e^t} y\right) = e^t e^{e^t}$. Thus, $e^{e^t} y = \int e^t e^{e^t} \, dt$. This antiderivative may be found using substitution, so $e^{e^t} y = e^{e^t} + C$. Multiplying through by e^{-e^t} gives

$$y = 1 + Ce^{-e^t}.$$

13. The differential equation $y' + 4y = 2$ is linear. An integrating factor is

$$u = e^{\int 4 \, dx} = e^{4x}.$$

Multiplying through by this integrating factor gives $e^{4x} y' + 4e^{4x} y = 2e^{4x}$. Therefore $\frac{d}{dx}\left(e^{4x} y\right) = 2e^{4x}$. Thus, $e^{4x} y = \int 2e^{4x} \, dx$. So $e^{4x} y = \frac{1}{2}e^{4x} + C$. Multiplying through by e^{-4x} gives

$$y = \frac{1}{2} + Ce^{-4x}. \tag{3.3}$$

Using the initial condition $y(1) = 2$, i.e., $x = 1$ and $y = 2$ in equation (3.3), we obtain $2 = \frac{1}{2} + Ce^{-4}$, so $C = \frac{3}{2}e^4$. Therefore,

$$y = \frac{1}{2} + \frac{3}{2}e^{4-4x}.$$

15. The differential equation $y' + \frac{1}{x+1}y = 2$ is linear. An integrating factor is

$$u = e^{\int \frac{1}{x+1} \, dx} = e^{\ln(x+1)} = x + 1.$$

Multiplying through by this integrating factor gives $(x+1)y' + y = 2(x+1)$. Therefore $\frac{d}{dx}((x+1)y) = 2x + 2$. Thus,

$$(x+1)y = \int (2x+2) \, dx.$$

So $(x+1)y = x^2 + 2x + C$. Multiplying through by $\frac{1}{x+1}$ gives

$$y = \frac{x^2 + 2x + C}{x+1}. \tag{3.4}$$

Using the initial condition $y(0) = 2$, i.e., $x = 0$ and $y = 2$ in equation (3.4), we obtain $C = 2$. Therefore,

$$y = \frac{x^2 + 2x + 2}{x+1}.$$

17. The differential equation $y' = \cos 2x - \frac{y}{x}$ is linear. It can be expressed in the form

$$y' + \frac{1}{x}y = \cos 2x.$$

As we had in problem 7, an integrating factor is $u = x$. Multiplying through by this integrating factor gives $xy' + y = x\cos 2x$. Therefore $\frac{d}{dx}(xy) = x\cos 2x$. Thus, $xy = \int x\cos 2x \, dx$. Using integration by parts this antiderivative can be found to be $xy = \frac{1}{2}x\sin 2x + \frac{1}{4}\cos 2x + C$. Multiplying through by $\frac{1}{x}$ gives

$$y = \frac{1}{2}\sin 2x + \frac{1}{4x}\cos 2x + \frac{C}{x}. \tag{3.5}$$

Using the initial condition $y(\frac{\pi}{2}) = 0$, i.e., $x = \frac{\pi}{2}$ and $y = 0$ in equation (3.5), we obtain $C = \frac{1}{4}$. Therefore,

$$y = \frac{\cos 2x + 1}{4x} + \frac{1}{2}\sin 2x.$$

21. If $y(x) = e^{-\int_{x_0}^{x} p(t)\,dt} \int_{x_0}^{x} u(t) q(t)\,dt + y_0 e^{-\int_{x_0}^{x} p(t)\,dt}$ then by the product rule and the Fundamental Theorem of Calculus

$$y' = -p(x) e^{-\int_{x_0}^{x} p(t)\,dt} \int_{x_0}^{x} u(t) q(t)\,dt + u(x) q(x) e^{-\int_{x_0}^{x} p(t)\,dt} - y_0 p(x) e^{-\int_{x_0}^{x} p(t)\,dt}.$$

Since, $u(t) = e^{\int_{x_0}^{t} p(s)\,ds}$, we have $u(x) = e^{\int_{x_0}^{x} p(s)\,ds}$ and

$$
\begin{aligned}
y' &= -p(x) e^{-\int_{x_0}^{x} p(t)\,dt} \int_{x_0}^{x} u(t) q(t)\,dt + e^{\int_{x_0}^{x} p(s)\,ds} q(x) e^{-\int_{x_0}^{x} p(t)\,dt} - y_0 p(x) e^{-\int_{x_0}^{x} p(t)\,dt} \\
&= -p(x) e^{-\int_{x_0}^{x} p(t)\,dt} \int_{x_0}^{x} u(t) q(t)\,dt + q(x) - y_0 p(x) e^{-\int_{x_0}^{x} p(t)\,dt}.
\end{aligned}
\tag{3.6}
$$

We also have

$$p(x) y = p(x) e^{-\int_{x_0}^{x} p(t)\,dt} \int_{x_0}^{x} u(t) q(t)\,dt + y_0 p(x) e^{-\int_{x_0}^{x} p(t)\,dt}. \tag{3.7}$$

Therefore from equations (3.6) and (3.7),

$$y' + p(x) y = q(x),$$

so the given function is a solution to the differential equation.

Since $y(x_0) = e^{-\int_{x_0}^{x_0} p(t)\,dt} \int_{x_0}^{x_0} u(t) q(t)\,dt + y_0 e^{-\int_{x_0}^{x_0} p(t)\,dt} = e^0 \cdot 0 + y_0 e^0 = y_0$, the given function also satisfies the specified initial condition.

3.5   More Techniques for Solving First Order Differential Equations

Exercises 3.5, pp. 143

1. Consider the differential equation $(x^2 + y^2)\,dx - xy\,dy = 0$. This equation is *not* exact. We will try to find an integrating factor of the form $I(x,y) = x^m y^n$. Multiplying the differential equation by this, we have

$$(x^{m+2} y^n + x^m y^{n+2})\,dx - x^{m+1} y^{n+1}\,dy = 0. \tag{3.8}$$

With $M(x,y) = x^{m+2} y^n + x^m y^{n+2}$ and $N(x,y) = -x^{m+1} y^{n+1}$. In order for this new form of the differential equation to be exact we must have $M_y = n x^{m+2} y^{n-1} + (n+2) x^m y^{n+1} = N_x = -(m+1) x^m y^{n+1}$. For this to occur the coefficients of the $x^{m+2} y^{n-1}$ terms need to be the same and the coefficients of the $x^m y^{n+1}$ terms need be the same in these two expressions. This gives us the system of linear equations

$$n = 0 \text{ and } n + 2 = -(m+1).$$

The solution to this system is $m = -3$ and $n = 0$. Substituting these values into equation (3.8), we have the exact equation

$$(x^{-1} + x^{-3} y^2)\,dx - x^{-2} y\,dy = 0.$$

To solve this exact equation, first we integrate

$$F(x,y) = \int (x^{-1} + x^{-3} y^2)\,dx = \ln|x| - \frac{1}{2} x^{-2} y^2 + h(y).$$

We need $F_y(x,y) = N(x,y)$. Therefore, $-x^{-2} y + h'(y) = -x^{-2} y$. So that $h'(y) = 0$ and $h(y) = 0$. Our solutions are given by

$$\ln|x| - \frac{1}{2} x^{-2} y^2 = k$$

or equivalently

$$2x^2 \ln|x| - y^2 = C x^2.$$

3. Consider the differential equation $(x - 3x^2y)\frac{dy}{dx} = xy^2 + y$. This equation may be rewritten in the form

$$(xy^2 + y)\,dx + (3x^2y - x)\,dy = 0.$$

This equation is *not* exact. We will try to find an integrating factor of the form $I(x,y) = x^m y^n$. Multiplying the differential equation by this, we have

$$(x^{m+1}y^{n+2} + x^m y^{n+1})\,dx + (3x^{m+2}y^{n+1} - x^{m+1}y^n)\,dy = 0. \tag{3.9}$$

With $M(x,y) = x^{m+1}y^{n+2} + x^m y^{n+1}$ and $N(x,y) = 3x^{m+2}y^{n+1} - x^{m+1}y^n$. In order for this new form of the differential equation to be exact we must have

$$M_y = (n+2)x^{m+1}y^{n+1} + (n+1)x^m y^n$$

equal to

$$N_x = 3(m+2)x^{m+1}y^{n+1} - (m+1)x^m y^n.$$

For this to occur the coefficients of the $x^{m+1}y^{n+1}$ terms need to be the same and the coefficients of the $x^m y^n$ terms need to be the same in these two expressions. This gives us the system of linear equations

$$n+2 = 3(m+2) \text{ and } n+1 = -(m+1).$$

The solution to this system is $m = -\frac{3}{2}$ and $n = -\frac{1}{2}$. Substituting these values into equation (3.9), we have the exact equation

$$(x^{-\frac{1}{2}}y^{\frac{3}{2}} + x^{-\frac{3}{2}}y^{\frac{1}{2}})\,dx + (3x^{\frac{1}{2}}y^{\frac{1}{2}} - x^{-\frac{1}{2}}y^{-\frac{1}{2}})\,dy = 0.$$

To solve this exact equation, first we integrate

$$F(x,y) = \int (x^{-\frac{1}{2}}y^{\frac{3}{2}} + x^{-\frac{3}{2}}y^{\frac{1}{2}})\,dx = 2x^{\frac{1}{2}}y^{\frac{3}{2}} - 2x^{-\frac{1}{2}}y^{\frac{1}{2}} + h(y).$$

We need $F_y(x,y) = N(x,y)$. Therefore, $3x^{\frac{1}{2}}y^{\frac{1}{2}} - x^{-\frac{1}{2}}y^{-\frac{1}{2}} + h'(y) = 3x^{\frac{1}{2}}y^{\frac{1}{2}} - x^{-\frac{1}{2}}y^{-\frac{1}{2}}$. So that $h'(y) = 0$ and $h(y) = 0$. Our solutions are given by

$$2x^{\frac{1}{2}}y^{\frac{3}{2}} - 2x^{-\frac{1}{2}}y^{\frac{1}{2}} = k$$

or equivalently

$$2x^{-\frac{1}{2}}y^{\frac{1}{2}} - 2x^{\frac{1}{2}}y^{\frac{3}{2}} = C.$$

5. Consider the differential equation $(2x^2 + 3y^2 - 2)\,dx - 2xy\,dy = 0$. This equation is *not* exact. We will try to find an integrating factor of the form $I(x,y) = x^m y^n$. Multiplying the differential equation by this, we have

$$(2x^{m+2}y^n + 3x^m y^{n+2} - 2x^m y^n)\,dx - 2x^{m+1}y^{n+1}\,dy = 0. \tag{3.10}$$

With $M(x,y) = 2x^{m+2}y^n + 3x^m y^{n+2} - 2x^m y^n$ and $N(x,y) = -2x^{m+1}y^{n+1}$. In order for this new form of the differential equation to be exact we must have

$$M_y = 2nx^{m+2}y^{n-1} + 3(n+2)x^m y^{n+1} - 2nx^m y^{n-1}$$

equal to

$$N_x = -2(m+1)x^m y^{n+1}.$$

For this to occur the coefficients of the $x^{m+2}y^{n-1}$ terms need to be the same, the coefficients of the $x^m y^{n+1}$ terms need to be the same and the coefficients of the $x^m y^{n-1}$ terms need to be the same in these two expressions. This gives us the system of linear equations

$$2n = 0, \text{ and } 3(n+2) = -2(m+1).$$

The solution to this system is $m = -4$ and $n = 0$. Substituting these values into equation (3.10), we have the exact equation

$$(2x^{-2} + 3x^{-4}y^2 - 2x^{-4})\,dx - 2x^{-3}y\,dy = 0.$$

To solve this exact equation, first we integrate

$$F(x,y) = \int (2x^{-2} + 3x^{-4}y^2 - 2x^{-4})\,dx = -2x^{-1} - x^{-3}y^2 + \frac{2}{3}x^{-3} + h(y).$$

We need $F_y(x,y) = N(x,y)$. Therefore, $-2x^{-3}y + h'(y) = -2x^{-3}y$. So that $h'(y) = 0$ and $h(y) = 0$. Our solutions are given by

$$-2x^{-1} - x^{-3}y^2 + \frac{2}{3}x^{-3} = C.$$

7. Consider the differential equation $y' + p(x)y = q(x)$. If we multiply both sides of this equation by $e^{\int p(x)\,dx}$, we obtain $\frac{dy}{dx}e^{\int p(x)\,dx} + p(x)ye^{\int p(x)\,dx} = q(x)e^{\int p(x)\,dx}$ which may be rewritten in the form

$$\left(p(x)ye^{\int p(x)\,dx} - q(x)e^{\int p(x)\,dx}\right)dx + e^{\int p(x)\,dx}\,dy = 0. \tag{3.11}$$

With $M(x,y) = p(x)ye^{\int p(x)\,dx} - q(x)e^{\int p(x)\,dx}$ and $N(x,y) = e^{\int p(x)\,dx}$. Here,

$$M_y(x,y) = p(x)e^{\int p(x)\,dx} = N_x(x,y)$$

by the chain rule and the Fundamental Theorem of Calculus. Therefore equation (3.11) is exact.

9. Consider the differential equation
$$(xy+x)\,dx + (x^2+y^2-1)\,dy = 0. \tag{3.12}$$

With $r(x,y) = xy+x$ and $s(x,y) = x^2+y^2-1$, we have

$$\frac{s_x(x,y) - r_y(x,y)}{r(x,y)} = \frac{2x-x}{x(y+1)} = \frac{1}{y+1}$$

which is a function of y only. Now we solve the separable equation $\frac{du}{dy} = \frac{1}{y+1}u$ to find the integrating factor u. This gives

$$\frac{du}{u} = \frac{1}{y+1}\,dy$$

$$\ln u = \ln(x+1)$$

or $u = y+1$. Multiplying the differential equation (3.12) by $u = y+1$ we have the exact equation

$$x(y+1)^2\,dx + (x^2(y+1) + y^3 + y^2 - y - 1)\,dy = 0.$$

To solve this exact equation, first we integrate

$$F(x,y) = \int x(y+1)^2\,dx = \frac{1}{2}x^2(y+1)^2 + h(y).$$

We need $F_y(x,y) = N(x,y)$. Therefore, $x^2(y+1) + h'(y) = x^2(y+1) + y^3 + y^2 - y - 1$. So that $h'(y) = y^3 + y^2 - y - 1$ and $h(y) = \frac{1}{4}y^4 + \frac{1}{3}y^3 - \frac{1}{2}y^2 - y$. Our solutions are given by

$$\frac{1}{2}x^2(y+1)^2 + \frac{1}{4}y^4 + \frac{1}{3}y^3 - \frac{1}{2}y^2 - y = C.$$

11. Consider the differential equation $(x^2 - y^2)\,dx + xy\,dy$.
With $M(x,y) = x^2 - y^2$ and $N(x,y) = xy$, we have

$$\frac{M(ax,ay)}{N(ax,ay)} = \frac{(ax)^2 - (ay)^2}{(ax)(ay)} = \frac{a^2(x^2-y^2)}{a^2xy} = \frac{x^2-y^2}{xy} = \frac{M(x,y)}{N(x,y)}$$

therefore the differential equation has homogeneous coefficients.

To solve the equation, we first write the equation in the form $\frac{dy}{dx} = -\frac{M(x,y)}{N(x,y)}$, here

$$\frac{dy}{dx} = \frac{y^2-x^2}{xy} = \frac{\left(\frac{y}{x}\right)^2 - 1}{\frac{y}{x}}. \tag{3.13}$$

We also set
$$\frac{y}{x} = u, \tag{3.14}$$

so $y = ux$ and
$$\frac{dy}{dx} = u + x\frac{du}{dx}. \tag{3.15}$$

Making the substitutions from equations (3.14) and (3.15) in equation (3.13), we arrive at the differential equation

$$u + x\frac{du}{dx} = \frac{u^2 - 1}{u}.$$

This is a separable differential equation and can be rewritten in the form $u\,du = -\frac{1}{x}\,dx$. Integrating each side of this equation we obtain $\frac{1}{2}u^2 = -\ln|x| + k$. Since $u = \frac{y}{x}$, the solutions are given by

$$\frac{1}{2}\left(\frac{y}{x}\right)^2 = -\ln|x| + k,$$

or equivalently

$$y^2 = -2x^2 \ln|x| + Cx^2.$$

13. Consider the differential equation $\left(y + xe^{\frac{y}{x}}\right) dx - x\,dy$.

    With $M(x,y) = y + xe^{\frac{y}{x}}$ and $N(x,y) = -x$, we have

    $$\frac{M(ax, ay)}{N(ax, ay)} = \frac{ay + axe^{\frac{ay}{ax}}}{-ax} = \frac{a\left(y + xe^{\frac{y}{x}}\right)}{-ax} = \frac{y + xe^{\frac{y}{x}}}{-x} = \frac{M(x,y)}{N(x,y)}$$

    therefore the differential equation has homogeneous coefficients.

    To solve the equation, we first write the equation in the form $\frac{dy}{dx} = -\frac{M(x,y)}{N(x,y)}$, here

    $$\frac{dy}{dx} = \frac{y + xe^{\frac{y}{x}}}{-x} = \frac{y}{x} + e^{\frac{y}{x}}. \tag{3.16}$$

    We also set

    $$\frac{y}{x} = u, \tag{3.17}$$

    so $y = ux$ and

    $$\frac{dy}{dx} = u + x\frac{du}{dx}. \tag{3.18}$$

    Making the substitutions from equations (3.17) and (3.18) in equation (3.16), we arrive at the differential equation

    $$u + x\frac{du}{dx} = u + e^u.$$

    This is a separable differential equation and can be rewritten in the form $e^{-u}\,du = \frac{1}{x}\,dx$. Integrating each side of this equation we obtain $-e^{-u} + C = \ln|x|$. Since $u = \frac{y}{x}$, the solutions are given by

    $$\ln|x| + e^{-\frac{y}{x}} = C.$$

17. The differential equation $xy' + y = y^2$ can be rewritten in the form

    $$y' + \frac{1}{x}y = \frac{1}{x}y^2. \tag{3.19}$$

    This is a Bernoulli equation with $n = 2$. We make the substitution

    $$v = y^{1-2} = y^{-1}, \tag{3.20}$$

    so $y = \frac{1}{v}$. We have $\frac{dy}{dx} = -\frac{1}{v^2}\frac{dv}{dx}$. We substitute this and equation (3.20) in equation (3.19) to obtain

    $$-\frac{1}{v^2}\frac{dv}{dx} + \frac{1}{x}\frac{1}{v} = \frac{1}{x}\frac{1}{v^2}.$$

This can be rewritten in the form $\frac{dv}{dx} - \frac{1}{x}v = -\frac{1}{x}$, which is linear. We use the integrating factor

$$u = e^{-\int \frac{1}{x}\,dx} = e^{-\ln x} = \frac{1}{x}.$$

Multiplying each side of the linear differential equation by u we obtain $\frac{d}{dx}\left(\frac{1}{x}v\right) = -\frac{1}{x^2}$. Integrating each side of this equation we obtain $\frac{1}{x}v = \frac{1}{x} + C$, or equivalently, $v = 1 + Cx$. Since $v = \frac{1}{y}$ we have $\frac{1}{y} = 1 + Cx$, or

$$y = \frac{1}{1 + Cx}.$$

3.6   Modeling with Differential Equations

Exercises 3.6, pp. 151-153

1. Let A represent the amount of money in the entrepreneur's portfolio. Then $\frac{dA}{dt} = 0.1A - 0.25$, because the growth rate of the portfolio is 10% per year and he is withdrawing \$250,000 per year. (Here we are measuring the amount of money in millions of dollars.) The initial condition is $A(0) = 2$. The differential equation is both linear and separable. We solve it as a linear differential equation.

Rewriting the equation as $\frac{dA}{dt} - 0.1A = -0.25$ we see the integrating factor

$$u = e^{\int -0.1\,dt} = e^{-0.1t}.$$

Multiplying each side of the differential equation by u we obtain $\frac{d}{dt}\left(Ae^{-0.1t}\right) = -0.25e^{-0.1t}$. After integrating each side we have $Ae^{-0.1t} = 2.5e^{-0.1t} + C$, or equivalently $A = 2.5 + Ce^{0.1t}$. Now, using the initial condition we see $A(0) = 2 = 2.5 + C$. Therefore $C = -0.5$. The solution to the initial value problem is $A = 2.5 - 0.5e^{0.1t}$.

To find how old he will be when he runs out of money, we first find how long it will take until $A = 0$. We set $2.5 - 0.5e^{0.1t} = 0$ and solve for t. We obtain $t = 10\ln 5 \approx 16$ where t is measured in years. Since the entrepreneur is 32 years old now, he will run out of money when he is 32+16=48 years old.

3. Without the removal of any amoeba, the differential equation modeling the population would be

$$P' = kP \tag{3.21}$$

since the population grows in proportion to the current population. The initial population is $P(0) = 1000$. After one hour we have $P(1) = 2000$. The solution to equation (3.21) is $P = P_0 e^{kt}$, where $P_0 = 1000$ and the growth rate, $k = \ln 2$.

We now consider what happens when we start removing 100 amoeba per hour. The growth rate k remains $k = \ln 2$. The adjusted differential equation then is $P' = (\ln 2)P - 100$ with the initial condition $P(1) = 2000$, since there are 2000 amoeba after 1 hour. The differential equation is both linear and separable. We solve it as a linear differential equation.

Rewriting the equation as $P' - (\ln 2)P = -100$ we see the integrating factor

$$u = e^{\int -\ln 2\,dt} = e^{-(\ln 2)t}.$$

Multiplying each side of the differential equation by u we obtain $\frac{d}{dt}\left(Pe^{-(\ln 2)t}\right) = -100e^{-(\ln 2)t}$. After integrating each side we have $Pe^{-(\ln 2)t} = \frac{100}{\ln 2}e^{-(\ln 2)t} + C$, or equivalently

$$P = \frac{100}{\ln 2} + Ce^{(\ln 2)t}.$$

Now, using the initial condition we see $P(1) = 2000 = \frac{100}{\ln 2} + Ce^{(\ln 2)}$. Solving this equation for C we obtain $C = 1000 - \frac{50}{\ln 2}$. The solution to the initial value problem is

$$P = \frac{100}{\ln 2} + \left(1000 - \frac{50}{\ln 2}\right)e^{(\ln 2)t}.$$

We are asked to find the population at the end of three hours,

$$P(3) = \frac{100}{\ln 2} + \left(1000 - \frac{50}{\ln 2}\right)e^{(\ln 2)3} = 8000 - \frac{300}{\ln 2} \approx 7567 \text{ amoeba.}$$

5. A quantity Q of cesium 137 will decay according to the differential equation $Q' = kQ$. The solution to this differential equation is $Q = Q_0 e^{kt}$. Since the half-life of cesium 137 is approximately 30 years, the decay constant $k = -\frac{\ln 2}{30}$. In the given problem cesium 137 is entering a cesium free pond at the rate of 1 lb/year. Thus the new differential equation is given by $Q' = kQ + 1$, or $Q' = 1 - \frac{\ln 2}{30}Q$. Since the pond is initially cesium free we have $Q(0) = 0$. The differential equation is both linear and separable. We solve it as a linear differential equation.

Rewriting the equation as $Q' + \frac{\ln 2}{30}Q = 1$ we see the integrating factor $u = e^{\int \frac{\ln 2}{30} dt} = e^{\frac{\ln 2}{30}t}$. Multiplying each side of the differential equation by u we obtain $\frac{d}{dt}\left(Qe^{\frac{\ln 2}{30}t}\right) = 1e^{\frac{\ln 2}{30}t}$. After integrating each side we have

$$Qe^{\frac{\ln 2}{30}t} = \frac{30}{\ln 2}e^{\frac{\ln 2}{30}t} + C,$$

or equivalently $Q = \frac{30}{\ln 2} + Ce^{-\frac{\ln 2}{30}t}$. Now, using the initial condition we see $Q(0) = 0 = \frac{30}{\ln 2} + Ce^{-\frac{\ln 2}{30}}$. Solving this equation for C we obtain $C = -\frac{30}{\ln 2}$. The solution to the initial value problem is $Q = \frac{30}{\ln 2}\left(1 - e^{-\frac{\ln 2}{30}t}\right)$. We are asked to find the amount of cesium in the pond after 30 years,

$$Q(30) = \frac{30}{\ln 2}\left(1 - e^{-\ln 2}\right) = \frac{15}{\ln 2} \approx 21.64 \text{ pounds.}$$

7. Because 3 gallons per minute of fluid is entering the tank, but only 2 gallons per minute is being drained off, the volume of fluid in the tank is increasing at 1 gal/min. Therefore the volume of fluid in the tank after t minutes is $(40 + t)$ gallons. The amount of alcohol in the tank is also changing. To determine the differential equation describing the rate of change of alcohol in the tank we consider the separate rates at which alcohol is entering and leaving the tank. The rate at which alcohol is entering the tank is given by the product of the concentration times the flow rate, i.e., $0.5\frac{\text{gal alcohol}}{\text{gal}} \cdot 3\frac{\text{gal}}{\text{min}} = 1.5\frac{\text{gal alcohol}}{\text{min}}$. The rate at which alcohol is leaving the tank is determined in a similar manner, except that the concentration is *not* constant. If we let $A(t)$ denote the amount of alcohol in the tank at time t, then the concentration of alcohol leaving the tank at time t is given by $\frac{A(t)}{40+t}\frac{\text{gal alcohol}}{\text{gal}}$. Thus the rate at which alcohol is leaving the tank is given by

$$\frac{A(t)}{40+t}\frac{\text{gal alcohol}}{\text{gal}} \cdot 2\frac{\text{gal}}{\text{min}} = \frac{2A(t)}{40+t}\frac{\text{gal alcohol}}{\text{min}}.$$

Therefore, the differential equation describing the rate of change of $A(t)$ is

$$A' = 1.5 - \frac{2A}{40+t}.$$

This is a linear differential equation. To solve it, we first rewrite the equation in the form

$$A' + \frac{2A}{40+t} = 1.5.$$

The integrating factor is $u = e^{\int \frac{2}{40+t} dt} = e^{2\ln(40+t)} = (40+t)^2$. Multiplying each side of the differential equation by u we obtain $\frac{d}{dt}\left(A(40+t)^2\right) = 1.5(40+t)^2$. After integrating each side we have $A(40+t)^2 = 0.5(40+t)^3 + C$, or equivalently $A = 0.5(40+t) + \frac{C}{(40+t)^2}$. We now use the initial condition $A(0) = 2$ to solve for C.

Thus, $A(0) = 2 = 20 + \frac{C}{1600}$. Therefore $C = -28800$. The solution to the initial value problem is

$$A = 0.5(40+t) - \frac{28800}{(40+t)^2}.$$

In part (a) we are asked to find the amount alcohol is in the tank after 10 minutes. We compute

$$A(10) = 25 - \frac{28800}{2500} = 13.48 \text{gallons.}$$

For part (b), since we know that the tank has a 100 gallon capacity, initially contains 40 gallons of fluid, and the net amount of 1 gallon per minute is being added, it will take 60 minutes for the tank to overflow. We compute

$$A(60) = 50 - \frac{28800}{10000} = 47.12 \text{ gallons of alcohol}$$

at the time of overflow.

9. Five cubic feet of pure oxygen is entering the room every minute. Well mixed air is escaping from the room at the same rate. To determine the differential equation describing the rate of change of carbon dioxide in the room we consider the separate rates at which the carbon dioxide is entering and leaving the room. However, since pure oxygen is being pumped in, the rate at which carbon dioxide is entering the room in part (a) is 0 cubic feet per minute. If we let $A(t)$ denote the amount of carbon dioxide in the room at time t, then the rate at which carbon dioxide is leaving the room is given by the product of the concentration times the flow rate, i.e.,

$$\frac{A(t)}{10000}\frac{\text{cu ft CO}_2}{\text{cu ft}} \cdot 5\frac{\text{cu ft}}{\text{min}} = \frac{A(t)}{2000}\frac{\text{cu ft CO}_2}{\text{min}}.$$

Therefore, the differential equation describing the rate of change of $A(t)$ is

$$A' = -\frac{A}{2000}. \tag{3.22}$$

This is both a linear and separable differential equation. We solve it as a separable differential equation. We first rewrite the equation in the form $\frac{dA}{A} = -\frac{1}{2000}\,dt$. After integrating each side we have $\ln|A| = -\frac{1}{2000}t + k$. The absolute value is unnecessary since $A > 0$. We solve the equation for $A = Ce^{-\frac{1}{2000}t}$. The initial condition is $A(0) = 2000$, since 20% of the air was initially carbon dioxide. Thus, $A(0) = 2000 = C$. The solution to the initial value problem is $A = 2000e^{-\frac{1}{2000}t}$. We are asked to find how long it will take until the carbon dioxide level has been reduced to 5%. Since 5% of 10000 is 500, we need to solve the equation $500 = 2000e^{-\frac{1}{2000}t}$ for t. We see that $t = 2000\ln 4 \approx 2773$ minutes ≈ 46.2 hours.

In part (b) only the rate at which carbon dioxide is entering the room has changed. We are given that this rate is 1 cubic foot per minute. Thus equation (3.22) is changed to

$$A' = 1 - \frac{1}{2000}A.$$

Again, this is both a linear and separable differential equation. This time we solve it as a linear equation.

To solve it, we first rewrite the equation in the form $A' + \frac{1}{2000} = 1$. The integrating factor is $u = e^{\int \frac{1}{2000}\,dt} = e^{\frac{1}{2000}t}$. Multiplying each side of the differential equation by u we obtain $\frac{d}{dt}\left(Ae^{\frac{1}{2000}t}\right) = e^{\frac{1}{2000}t}$. After integrating each side we have

$$Ae^{\frac{1}{2000}t} = 2000e^{\frac{1}{2000}t} + C,$$

or equivalently $A = 2000 + Ce^{-\frac{1}{2000}t}$. Again, we use the initial condition $A(0) = 2000$ to solve for C.

Thus, $A(0) = 2000 = 2000 + C$. Therefore $C = 0$. We have $A(t) = 2000$. That is, 2000 is an equilibrium solution for this problem. Therefore, there is no solution! Please note, this could also have been discovered by setting $\frac{dA}{dt} = 0$, before trying to solve this new initial value problem.

11. Let θ be the temperature of the can of juice at time t and assume the can has uniform composition. By Newton's law of cooling $\theta' = k(70 - \theta)$. We can solve this equation either as a separable or linear differential equation. As a separable equation $\frac{d\theta}{70-\theta} = kdt$. Integrating we have

$$-\ln(70 - \theta) = kt + c.$$

Solving this equation for θ we obtain $\theta = 70 - Ce^{-kt}$. We now use the conditions $\theta(0) = 20$ and $\theta(35) = 28$ to find the constants C and k. First, $\theta(0) = 20 = 70 - C$. That is, $C = 50$. We now have $\theta = 70 - 50e^{-kt}$. We also know

$$\theta(35) = 28 = 70 - 50e^{-35k}.$$

So, $-k = \frac{\ln\frac{21}{25}}{35}$. Thus the solution to the original problem is $\theta = 70 - 50e^{\frac{\ln\frac{21}{25}}{35}t}$. We are asked to find how long it will take for the juice to reach a temperature of $32°$ F. We set $32 = 70 - 50e^{\frac{\ln\frac{21}{25}}{35}t}$ and solve for t to obtain $t = \frac{35\ln\frac{19}{25}}{\ln\frac{21}{25}} \approx 55.09$ minutes.

13. Let θ be the temperature of the body at time t and assume the body has uniform composition. By Newton's law of cooling $\theta' = k(\theta - 75)$. We can solve this equation either as a separable or linear differential equation. As a separable equation $\frac{d\theta}{\theta-75} = kdt$. Integrating we have

$$\ln(\theta - 75) = kt + c.$$

Solving this equation for θ we obtain $\theta = 75 + Ce^{kt}$. We now use the conditions that the body's temperature when first found was $90°$ F, $\theta(0) = 90$ and then measured to be $85°$ F after one hour, $\theta(1) = 85$, to find the constants C and k. First,

$\theta(0) = 90 = 75 + C$. That is, $C = 15$. We now have $\theta = 75 + 15e^{kt}$. We also know $\theta(1) = 85 = 75 + 15e^k$. So, $k = \ln\frac{2}{3}$. Thus the solution to the original problem is $\theta = 75 + 15e^{\ln\frac{2}{3}t}$. We are asked to approximate the time of death, assuming the person's body temperature was $98.6°$ F at the time of death. We set $98.6 = 75 + 15e^{\ln\frac{2}{3}t}$ and solve for t to obtain $t = \frac{\ln\frac{23.6}{15}}{\ln\frac{2}{3}} \approx -1.12$ hours. That is, the person died approximately 1.12 hours before being found.

15. We use the formula

$$v = \frac{Mg}{k}\left(1 - e^{-\frac{k}{M}t}\right)$$

for velocity derived in section 3.6, with $M = 5$ kg, $g = 10$ m/sec^2 and $k = 10$ kg/sec. Therefore, after 2 seconds,

$$v = \frac{5 \cdot 10}{10}(1 - e^{-\frac{10}{5}2}) = 5(1 - e^{-4}) \approx 4.91 \text{ m/sec.}$$

17. We use the formula

$$x = \frac{v_0^2 R}{2gR - v_0^2}$$

derived in section 3.6 for the maximum height of an object projected from the earth, with $v_0 = 20$ m/sec and $g = 10$ m/sec^2. Thus, $x = \frac{20^2 R}{2 \cdot 10R - 20^2} = \frac{20R}{R - 20}$ meters, where R is the radius of the earth.

19. (a) A circle centered at the origin has equation $x^2 + y^2 = a^2$. Differentiating with respect to x we obtain $2x + 2y\frac{dy}{dx} = 0$. Solving this equation for $\frac{dy}{dx}$ we have $\frac{dy}{dx} = -\frac{x}{y}$. Interpreting this as a slope, the slope of a perpendicular line would be the negative reciprocal. That is, $\frac{dy}{dx} = \frac{y}{x}$.

(b) To solve the separable differential equation above, we first separate the variables $\frac{dy}{y} = \frac{dx}{x}$, and then integrate to obtain $\ln|y| = \ln|x| + C$. Solving this equation for y we obtain $y = mx$. Thus, the orthogonal trajectories are just straight lines through the origin (as we would have expected).

21. We are given the differential equation $\frac{dP}{dt} = \frac{kP(C-P)}{C}$. This is a separable differential equation. Separating the variables we have $\frac{C dP}{P(C-P)} = k\, dt$. We use the method of partial fractions to break up the rational expression on the left and obtain $\left(\frac{1}{P} + \frac{1}{C-P}\right) dP = k\, dt$. Integrating each side we obtain

$$\ln P - \ln(C - P) = kt + \beta,$$

for some constant β. Equivalently, $\ln\frac{P}{C-P} = kt + \beta$. Exponentiating each side, we have $\frac{P}{C-P} = \alpha e^{kt}$, where $\alpha = e^\beta$. Thus,

$$P = \frac{\alpha C e^{kt}}{1 + \alpha e^{kt}}.$$

If the initial population is B, then $P(0) = B = \frac{\alpha C}{1+\alpha}$. Solving this equation for α we have $\alpha = \frac{B}{C-B}$. In this problem $B = 10$ billion and $C = 1$ trillion $= 1000$ billion, thus $\alpha = \frac{1}{99}$, so

$$P = \frac{1000e^{kt}}{99 + e^{kt}}.$$

23. (a) We have the equation $PV = NRT$ where N and R are constants. We differentiate each side of the equation with respect to T to obtain

$$P\frac{dV}{dT} + V\frac{dP}{dT} = NR.$$

Now, writing this in differential form, we have $P\, dV + V\, dP = NR\, dT$.

(b) Since $dU = -P\, dV$ and $dU = Nc_v\, dT$, we have $Nc_v\, dT = -P\, dV$, or equivalently, $dT = -\frac{1}{Nc_v}P\, dV$. Substituting this for dT in differential equation from part (a) we obtain, $P\frac{dV}{dT} + V\frac{dP}{dT} = -\frac{R}{c_v}P\, dV$.

(c) The differential equation above is separable. We may rewrite the equation in the form, $\left(1 + \frac{R}{c_v}\right)\frac{dV}{V} = -\frac{dP}{P}$. Integrating each side we obtain, $\left(1 + \frac{R}{c_v}\right)\ln|V| = -\ln|P| + K$, where K is a constant. However, since both V and P are positive, the absolute values are unnecessary, thus $\left(1 + \frac{R}{c_v}\right)\ln V + \ln P = K$.

3.7   Reduction of Order

Exercises 3.7, pp. 157

1. The differential equation $y'' - 9y' = x$ is second order and does not contain y. We let $v = y'$ and obtain

$$v' - 9v = x. \tag{3.23}$$

This is a linear equation. The integrating factor $u = e^{\int -9\,dx} = e^{-9x}$. Multiplying each side of equation (3.23) by u we obtain, $\frac{d}{dx}(ve^{-9x}) = xe^{-9x}$. Integrating each side of this equation (using integration by parts on the right hand side), we obtain $ve^{-9x} = -\frac{1}{9}xe^{-9x} - \frac{1}{81}e^{-9x} + c$, or equivalently, $v = -\frac{1}{9}x - \frac{1}{81} + ce^{9x}$. Since $y' = v$, integrating y' gives us

$$y = -\frac{1}{18}x^2 - \frac{1}{81}x + \frac{1}{9}ce^{9x} + C_2.$$

Letting $C_1 = \frac{1}{9}c$ we obtain the solution

$$y = -\frac{1}{18}x^2 - \frac{1}{81}x + C_1 e^{9x} + C_2.$$

3. The differential equation $y'' - (y')^3 = 0$ is second order and does not contain y. We let $v = y'$ and obtain $v' - v^3 = 0$. This is a separable equation. Separating the variables we obtain $v^{-3}\,dv = dx$. Integrating we have $-\frac{1}{2}v^{-2} = x + C_1$. Solving this equation for v we obtain $v = \pm(C_1 - 2x)^{-\frac{1}{2}}$. Since $y' = v$, integrating y' gives us $y = \pm(C_1 - 2x)^{\frac{1}{2}} + C_2$.

5. The differential equation $yy'' - 4(y')^2 = 0$ is second order and does not contain x. We make the substitutions $y' = v$ and $y'' = v\frac{dv}{dy}$ to obtain $yv\frac{dv}{dy} - 4v^2 = 0$. This separable equation can be written in the form $\frac{dv}{v} = \frac{4}{y}\,dy$. Integrating we obtain $\ln|v| = 4\ln|y| + c$. Solving for v we have $v = c_1 y^4$. We now resubstitute $y' = v$ to obtain the new separable differential equation $\frac{dy}{dx} = c_1 y^4$, or equivalently, $y^{-4}\,dy = c_1\,dx$. Integrating we have $-\frac{1}{3}y^{-3} = c_1 x + c_2$. Solving this equation for y^3 we obtain $y^3 = \frac{1}{C_1 x + C_2}$, where $C_1 = -3c_1$ and $C_2 = -3c_2$.

7. The differential equation $y'' + 4y = 0$ is second order and does not contain x. We make the substitutions $y' = v$ and $y'' = v\frac{dv}{dy}$ to obtain $v\frac{dv}{dy} + 4y = 0$. This separable equation can be written in the form $2v\,dv = -8y\,dy$. Integrating we obtain $v^2 = -4y^2 + c^2$. Note we let the constant of integration be c^2 here for later convenience. We have, $v = \pm\sqrt{c^2 - 4y^2}$. We now resubstitute $y' = v$ to obtain the new separable differential equation $\frac{dy}{dx} = \pm\sqrt{c^2 - 4y^2}$. This separable equation can be written in the form $\frac{dy}{c^2 - 4y^2} = \pm dx$. Integrating we have $\frac{1}{2}\sin^{-1}\left(\frac{2}{c}y\right) = \pm x + C_2$. Solving this equation for y we obtain $y = \pm\frac{c}{2}\sin(2x + C_2)$. (Recall that sine is an *odd* function.) To simplify slightly we let $C_1 = \pm\frac{c}{2}$ and we have $y = C_1 \sin(2x + C_2)$.

   Note, this answer can be shown to be equivalent to the one in the text if you use the trigonometric identity

$$\sin(A + B) = \sin A \cos B + \cos A \sin B.$$

9. The differential equation $y''y' = 1$ is second order and does not contain y. We let $v = y'$ and obtain $v'v = 1$. This is a separable equation. Separating the variables we have $2v\,dv = 2dx$. Integrating each side of this equation, we obtain $v^2 = 2x + c$. Since $y' = v$, we have $(y')^2 = 2x + c$. We now use the initial condition

$$y'(0) = 1 \tag{3.24}$$

to obtain $c = 1$. Thus,

$$(y')^2 = 2x + 1,$$

or equivalently, $y' = \pm(2x + 1)^{\frac{1}{2}}$. Here, we choose the positive sign because of the initial condition (3.24). Integrating again we obtain $y = \frac{1}{3}(2x + 1)^{\frac{3}{2}} + k$. We use the other initial condition $y(0) = 0$, to see $0 = \frac{1}{3} \cdot 1^{\frac{3}{2}} + k$, that is $k = -\frac{1}{3}$. The solution to the initial value problem is $y = \frac{1}{3}\left((2x + 1)^{\frac{3}{2}} - 1\right)$.

11. The differential equation $x^2 y'' + 2xy' = 1$ is second order and does not contain y. We let $v = y'$ and obtain $x^2 v' + 2xv = 1$. We could write this linear differential equation in the standard form and then find the integrating factor, however, if we just look at the left hand side of this new equation we see that it equals $\frac{d}{dx}(vx^2)$, thus we have $\frac{d}{dx}(vx^2) = 1$. Integrating each side of this equation we have $vx^2 = x + c$, or equivalently $v = \frac{1}{x} + \frac{c}{x^2}$. Since $y' = v$, we have $y' = \frac{1}{x} + \frac{c}{x^2}$. We now use the initial condition $y'(1) = 1$ to obtain $1 = 1 + c$. Thus $c = 0$ and we have $y' = \frac{1}{x}$. Integrating again we have $y = \ln|x| + k$. Because of the initial condition the absolute value is unnecessary. Since $y(1) = 0$, $0 = \ln 1 + k$, thus $k = 0$. The solution to the initial value problem is $y = \ln x$.

13. The differential equation $y'' - 3(y')^2 = 0$ is second order and does not contain x. We make the substitutions $y' = v$ and $y'' = v\frac{dv}{dy}$ to obtain $v\frac{dv}{dy} - 3v^2 = 0$. This separable equation can be written in the form $2v\,dv = 6y^2\,dy$. Integrating we obtain $v^2 = 2y^3 + c$. Solving for v we have $v = \pm(2y^3 + c)^{\frac{1}{2}}$. We now resubstitute $y' = v$ to obtain $y' = \pm(2y^3 + c)^{\frac{1}{2}}$. We now use the initial conditions $y(0) = 2$ and $y'(0) = 4$ to evaluate the constant c. Note that when $y = 2$ we have $y' = 4$, so $4 = (2 \cdot 2^3 + c)^{\frac{1}{2}}$, where we use the positive root. Thus, $c = 0$, and $y' = \sqrt{2}y^{\frac{3}{2}}$. This separable equation can be written $y^{-\frac{3}{2}}\,dy = \sqrt{2}\,dx$. Integrating we have $-2y^{-\frac{1}{2}} = \sqrt{2}x + k$. Again, we use the initial condition $y(0) = 2$ to find $-2 \cdot 2^{-\frac{1}{2}} = k$. That is $k = -\sqrt{2}$. Thus, $-2y^{-\frac{1}{2}} = \sqrt{2}(x - 1)$. Upon solving for y we obtain $y = \frac{2}{(1-x)^2}$.

15. The differential equation $xy''' = y''$ is third order and contains neither y nor y'. We let $v = y''$ and obtain $xv' = v$. This is a separable equation. Separating the variables we have $\frac{dv}{v} = \frac{dx}{x}$. Integrating each side of this equation, we obtain

$$\ln|v| = \ln|x| + c,$$

or equivalently, $v = C_1 x$. Since $y'' = v$, we have $y'' = C_1 x$. We now use the initial condition $y''(1) = 3$ to obtain $C_1 = 3$. Thus, $y'' = 3x$. Integrating again we obtain $y' = \frac{3}{2}x^2 + C_2$. We use the initial condition $y'(1) = 2$, to see $2 = \frac{3}{2} + C_2$, that is, $C_2 = \frac{1}{2}$. We now have $y' = \frac{3}{2}x^2 + \frac{1}{2}$. Integrating one last time we obtain $y = \frac{1}{2}x^3 + \frac{1}{2}x + C_3$. Now using the final initial condition $y(1) = 1$ we see $1 = \frac{1}{2} + \frac{1}{2} + C_3$, or $C_3 = 0$. The solution to the initial value problem is $y = \frac{1}{2}(x^3 + x)$.

17. The differential equation $M\frac{d^2x}{dt^2} = -\frac{MgR^2}{(R+x)^2}$ is second order and does not contain t. We make the substitutions $\frac{dx}{dt} = v$ and $\frac{d^2x}{dt^2} = v\frac{dv}{dx}$ to obtain $Mv\frac{dv}{dx} = -\frac{MgR^2}{(R+x)^2}$. This separable equation can be written in the form $2v\,dv = -2gR^2(R+x)^{-2}\,dx$. Integrating we obtain $v^2 = \frac{2gR^2}{R+x} + c$. We now resubstitute $\frac{dx}{dt} = v$ to obtain

$$\left(\frac{dx}{dt}\right)^2 = \frac{2gR^2}{R+x} + c. \tag{3.25}$$

We now use the initial conditions $v(0) = x'(0) = v_0$ and $x(0) = x_0$ to evaluate the constant c. Note that when $x = x_0$ we have $x' = v_0$, so $v_0^2 = \frac{2gR^2}{R+x_0} + c$. Thus, $c = v_0^2 - \frac{2gR^2}{R+x_0}$. Solving equation (3.25) for $\frac{dx}{dt}$ we obtain

$$\frac{dx}{dt} = \pm\sqrt{\frac{2gR^2}{R+x} + v_0^2 - \frac{2gR^2}{R+x_0}}.$$

If $x_0 = 0$, we would obtain the answer in the book

$$\frac{dx}{dt} = \pm\sqrt{\frac{2gR^2}{R+x} + v_0^2 - 2gR}.$$

3.8   The Theory of First Order Differential Equations

Exercises 3.8, pp. 167-168

1. For the initial value problem $y' = f(x,y)$, $y(x_0) = y_0$ we have the equivalent integral equation

$$y(x) = y_0 + \int_{x_0}^{x} f(s,y(s))\, ds.$$

Thus for $y' = x + 2y$, $y(0) = 0$, we have $y(x) = \int_0^x (s + 2y(s))\, ds$.

3. $y' = 2y$ is a separable differential equation. We separate the variables to obtain $\frac{dy}{y} = 2\, dx$. Integrating we have $\ln|y| = 2x + c$. When we solve for y, we obtain $y = Ce^{2x}$. With the initial condition $y(0) = 1$ we have the solution $y = e^{2x}$.

This function has the Maclaurin series

$$e^{2x} = \sum_{n=0}^{\infty} \frac{(2x)^n}{n!} = 1 + 2x + 2x^2 + \frac{4}{3}x^3 + \frac{2}{3}x^4 + \cdots.$$

The first four Maclaurin polynomials, then, are
$m_1 = 1 + 2x$,
$m_2 = 1 + 2x + 2x^2$,
$m_3 = 1 + 2x + 2x^2 + \frac{4}{3}x^3$ and
$m_4 = 1 + 2x + 2x^2 + \frac{4}{3}x^3 + \frac{2}{3}x^4$.

We now find the Picard iterates. The integral equation for this initial value problem is

$$y(x) = 1 + \int_0^x 2y(s)\, ds.$$

The 0th Picard iterate $p_0(x) = y(0) = 1$.
The first Picard iterate $p_1(x) = 1 + \int_0^x 2 \cdot 1\, ds = 1 + (2s)|_0^x = 1 + 2x$.
The second Picard iterate $p_2(x) = 1 + \int_0^x 2(1 + 2s)\, ds = 1 + \left(2s + 2s^2\right)\big|_0^x = 1 + 2x + 2x^2$.
The third Picard iterate $p_3(x) = 1 + \int_0^x 2(1 + 2s + 2s^2)\, ds = 1 + \left(2s + 2s^2 + \frac{4}{3}s^3\right)\big|_0^x = 1 + 2x + 2x^2 + \frac{4}{3}x^3$.
The fourth Picard iterate

$$\begin{aligned} p_4(x) &= 1 + \int_0^x 2\left(1 + 2s + 2s^2 + \frac{4}{3}s^3\right) ds = 1 + \left(2s + 2s^2 + \frac{4}{3}s^3 + \frac{2}{3}s^4\right)\Big|_0^x \\ &= 1 + 2x + 2x^2 + \frac{4}{3}x^3 + \frac{2}{3}x^4. \end{aligned}$$

We see that the Maclaurin and Picard polynomials are the same.

5. $y' = 2y^2$ is a separable differential equation. We separate the variables to obtain $y^{-2}\, dy = 2\, dx$. Integrating we have $-y^{-1} = 2x + c$. When we solve for y, we obtain $y = -\frac{1}{2x+c}$. With the initial condition $y(0) = 1$ we have the solution $y = \frac{1}{1-2x}$.

This function has the Maclaurin series

$$\frac{1}{1-2x} = \sum_{n=0}^{\infty} (2x)^n = 1 + 2x + 4x^2 + 8x^3 + 16x^4 + \cdots.$$

The first four Maclaurin polynomials, then, are
$m_1 = 1 + 2x$,
$m_2 = 1 + 2x + 4x^2$,
$m_3 = 1 + 2x + 4x^2 + 8x^3$ and
$m_4 = 1 + 2x + 4x^2 + 8x^3 + 16x^4$.

We now find the Picard iterates. The integral equation for this initial value problem is

$$y(x) = 1 + \int_0^x 2y(s)^2\, ds.$$

The 0th Picard iterate $p_0(x) = y(0) = 1$.

The first Picard iterate $p_1(x) = 1 + \int_0^x 2 \cdot 1^2 \, ds = 1 + (2s)|_0^x = 1 + 2x$.

The second Picard iterate

$$
\begin{aligned}
p_2(x) &= 1 + \int_0^x 2(1+2s)^2 \, ds = 1 + \int_0^x (2 + 8s + 8s^2) \, ds = 1 + \left(2s + 4s^2 + \frac{8}{3}s^3\right)\Big|_0^x \\
&= 1 + 2x + 4x^2 + \frac{8}{3}x^3.
\end{aligned}
$$

The third Picard iterate

$$
\begin{aligned}
p_3(x) &= 1 + \int_0^x 2\left(1 + 2s + 4s^2 + \frac{8}{3}s^3\right)^2 \, ds \\
&= 1 + \int_0^x \left(2 + 8s + 24s^2 + \frac{128}{3}s^3 + \frac{160}{3}s^4 + \frac{128}{3}s^5 + \frac{128}{9}s^6\right) \, ds \\
&= 1 + \left(2s + 4s^2 + 8s^3 + \frac{32}{3}s^4 + \frac{32}{3}s^5 + \frac{64}{9}s^6 + \frac{128}{63}s^7\right)\Big|_0^x \\
&= 1 + 2x + 4x^2 + 8x^3 + \frac{32}{3}x^4 + \frac{32}{3}x^5 + \frac{64}{9}x^6 + \frac{128}{63}x^7.
\end{aligned}
$$

The fourth Picard iterate

$$
\begin{aligned}
p_4(x) &= 1 + \int_0^x 2\left(1 + 2s + 4s^2 + 8s^3 + \frac{32}{3}s^4 + \frac{32}{3}s^5 + \frac{64}{9}s^6 + \frac{128}{63}s^7\right)^2 \, ds \\
&= 1 + 2x + 4x^2 + 8x^3 + 16x^4 + \frac{416}{15}x^5 + \frac{128}{3}x^6 + \frac{3712}{63}x^7 + \frac{4544}{63}x^8 + \frac{44032}{567}x^9 \\
&\quad + \frac{22528}{315}x^{10} + \frac{10240}{189}x^{11} + \frac{2048}{63}x^{12} + \frac{8192}{567}x^{13} + \frac{16384}{3969}x^{14} + \frac{32768}{59535}x^{15}.
\end{aligned}
$$

7. The initial value problem $y' = x^2 - y^2$, $y(1) = 0$ has the integral equation

$$
y(x) = 1 + \int_1^x (s^2 - y(s)^2) \, ds.
$$

The 0th Picard iterate $p_0(x) = y(1) = 0$.

The first Picard iterate $p_1(x) = \int_1^x (s^2 - 0^2) \, ds = \frac{1}{3}s^3 \big|_1^x = \frac{1}{3}x^3 - \frac{1}{3}$.

The second Picard iterate

$$
\begin{aligned}
p_2(x) &= \int_1^x \left(s^2 - \left(\frac{1}{3}s^3 - \frac{1}{3}\right)^2\right) ds = \int_1^x \left(s^2 - \frac{1}{9}s^6 + \frac{2}{9}s^3 - \frac{1}{9}\right) ds = \left(-\frac{1}{9}s + \frac{1}{3}s^3 + \frac{1}{18}s^4 - \frac{1}{63}s^7\right)\Big|_1^x \\
&= -\frac{11}{42} - \frac{1}{9}x + \frac{1}{3}x^3 + \frac{1}{18}x^4 - \frac{1}{63}x^7.
\end{aligned}
$$

9. The initial value problem $y' = \cos y$, $y(0) = 0$ has the integral equation

$$
y(x) = \int_0^x \cos y(s) \, ds.
$$

The 0th Picard iterate $p_0(x) = y(0) = 0$.

The first Picard iterate $p_1(x) = \int_0^x \cos 0 \, ds = \int_0^x ds = s|_0^x = x$.

The second Picard iterate $p_2(x) = \int_0^x \cos s \, ds = \sin s|_0^x = \sin x$.

The third Picard iterate $p_3(x) = \int_0^x \cos(\sin s) \, ds$.

11. If $u(x)$ is a solution to the initial value problem $u' = f(x + x_0, u)$, $u(0) = y_0$, then
$u'(x) = f(x + x_0, u)$ and $u(0) = y_0$. Now let $y(x) = u(x - x_0)$. So,

$$y'(x) = u'(x - x_0) = f((x - x_0) + x_0, u(x - x_0)) = f(x, y(x)) = f(x, y).$$

Furthermore, $y(x_0) = u(x_0 - x_0) = u(0) = y_0$. Therefore, $y(x)$ is a solution to the initial value problem

$$y' = f(x, y), \ y(x_0) = y_0.$$

13. Using the approach of Exercise 11, the initial value problem $y' = x - y^2$, $y(1) = 1$ is equivalent to the initial value problem
$u' = (x + 1) - y^2$, $u(0) = 1$. This initial value problem has the integral equation

$$u(x) = 1 + \int_0^x (s + 1 - y(s)^2) \, ds.$$

The 0th Picard iterate for u is $p_0(x) = u(0) = 1$.

The first Picard iterate for u is $p_1(x) = 1 + \int_0^x (s + 1 - 1^2) \, ds = 1 + \left(\frac{1}{2}s^2\right)\big|_0^x = 1 + \frac{1}{2}x^2$.

The second Picard iterate for u is

$$
\begin{aligned}
p_2(x) &= 1 + \int_0^x \left(s + 1 - \left(1 + \frac{1}{2}x^2\right)^2\right) ds \\
&= 1 + \int_0^x \left(s - s^2 - \frac{1}{4}s^4\right) ds \\
&= 1 + \left(\frac{1}{2}s^2 - \frac{1}{3}s^3 - \frac{1}{20}s^5\right)\Big|_0^x \\
&= 1 + \frac{1}{2}x^2 - \frac{1}{3}x^3 - \frac{1}{20}x^5.
\end{aligned}
$$

The third Picard iterate for u is

$$
\begin{aligned}
p_3(x) &= 1 + \int_0^x \left(s + 1 - \left(1 + \frac{1}{2}s^2 - \frac{1}{3}s^3 - \frac{1}{20}s^5\right)^2\right) ds \\
&= 1 + \int_0^x \left(s - s^2 + \frac{2}{3}s^3 - \frac{1}{4}s^4 + \frac{13}{30}s^5 - \frac{1}{9}s^6 + \frac{1}{20}s^7 - \frac{1}{30}s^8 + \frac{1}{400}s^{10}\right) ds \\
&= 1 + \left(\frac{1}{2}s^2 - \frac{1}{3}s^3 + \frac{1}{6}s^4 - \frac{1}{20}s^5 + \frac{13}{180}s^6 - \frac{1}{63}s^7 + \frac{1}{160}s^8 - \frac{1}{270}s^9 - \frac{1}{4400}s^{11}\right)\Big|_0^x \\
&= 1 + \frac{1}{2}x^2 - \frac{1}{3}x^3 + \frac{1}{6}x^4 - \frac{1}{20}x^5 + \frac{13}{180}x^6 - \frac{1}{63}x^7 + \frac{1}{160}x^8 - \frac{1}{270}x^9 - \frac{1}{4400}x^{11}.
\end{aligned}
$$

Therefore the third Picard iterate for y is

$$
\begin{aligned}
p_3(x - 1) &= 1 + \frac{1}{2}(x - 1)^2 - \frac{1}{3}(x - 1)^3 + \frac{1}{6}(x - 1)^4 - \frac{1}{20}(x - 1)^5 + \frac{13}{180}(x - 1)^6 \\
&\quad - \frac{1}{63}(x - 1)^7 + \frac{1}{160}(x - 1)^8 - \frac{1}{270}(x - 1)^9 - \frac{1}{4400}(x - 1)^{11}.
\end{aligned}
$$

To find the third degree Taylor polynomial for y about $x_0 = 1$ we need the first 3 derivatives of y at $x_0 = 1$. Since $y' = x - y^2$
we have $y'' = 1 - 2yy'$ and $y''' = -2(y')^2 - 2yy''$. At $x_0 = 1$ we have

$$y(1) = 1, \ y'(1) = 1 - 1^2 = 0, \ y''(1) = 1 - 2 \cdot 1 \cdot 0 = 1 \text{ and } y'''(1) = -2 \cdot 0^2 - 2 \cdot 1 \cdot 1 = -2.$$

The third degree Taylor polynomial for y about $x_0 = 1$ is given by

$$T_3(x) = y(1) + y'(1)(x - 1) + \frac{y''(1)}{2!}(x - 1)^2 + \frac{y'''(1)}{3!}(x - 1)^3.$$

We have $T_3(x) = 1 + \frac{1}{2}(x - 1)^2 - \frac{1}{3}(x - 1)^3$, which agrees with the first terms of the Picard iteration above.

3.9   Numerical Solutions of Ordinary Differential Equations

Exercises 3.9, pp. 176-177

For problems 1 and 3 we use the Euler method to find approximations to the solution of the initial value problem

$$y' = f(x,y),\ y(x_0) = y_0$$

at the x-values $x_0, x_1, x_2, \ldots, x_N$. The approximations are given by the formula

$$y_{n+1} = y_n + f(x_n, y_n)(x_{n+1} - x_n),\ \text{for } n = 0, 1, 2, \ldots, N-1.$$

1. Here we have $y' = 2y - x$, $y(0) = 1$, $x = 0.1, 0.2, 0.4, 0.5$. Thus, substituting $x_1 = 0.1$,
   $x_2 = 0.2$, $x_3 = 0.4$, $x_4 = 0.5$ and $y_0 = 1$ into $y_{n+1} = y_n + f(x_n, y_n)(x_{n+1} - x_n)$ where $f(x,y) = 2y - x$ we obtain
   $y_1 = 1 + (2 \cdot 1 - 0)(0.1 - 0) = 1 + 0.2 = 1.2$
   $y_2 = 1.2 + (2 \cdot 1.2 - 0.1)(0.2 - 0.1) = 1.2 + 2.3(0.1) = 1.43$
   $y_3 = 1.43 + (2 \cdot 1.43 - 0.2)(0.4 - 0.2) = 1.43 + 2.66(0.2) = 1.962$
   $y_4 = 1.962 + (2 \cdot 1.962 - 0.4)(0.5 - 0.4) = 1.962 + 3.524(0.1) = 2.3144$

   The given differential equation is linear. We rewrite the equation in the form $y' - 2y = -x$ and find the integrating factor $u = e^{\int -2\,dx} = e^{-2x}$. Multiplying each side of the equation by u we see that $\frac{d}{dx}\left(ye^{-2x}\right) = -xe^{-2x}$. Integrating each side of this equation we obtain $ye^{-2x} = \frac{1}{2}xe^{-2x} + \frac{1}{4}e^{-2x} + C$, or equivalently, $y = \frac{1}{2}x + \frac{1}{4} + Ce^{2x}$. Using the initial condition $y(0) = 1 = \frac{1}{4} + C$, we see that $C = \frac{3}{4}$. Thus the solution to the initial value problem is $y = \frac{1}{2}x + \frac{1}{4} + \frac{3}{4}e^{2x}$.
   The exact values at the points $x = 0.1, 0.2, 0.4, 0.5$ would be
   $y(0.1) = 0.05 + 0.25 + 0.75e^{0.2} \approx 1.21605$
   $y(0.2) = 0.1 + 0.25 + 0.75e^{0.4} \approx 1.46887$
   $y(0.4) = 0.2 + 0.25 + 0.75e^{0.8} \approx 2.11916$
   $y(0.5) = 0.25 + 0.25 + 0.75e^{1} \approx 2.53871$

3. Here we have $y' = x^2y^2$, $y(1) = 1$, $x = 1.2, 1.4, 1.6, 1.8$. Thus, substituting $x_1 = 1.2$,
   $x_2 = 1.4$, $x_3 = 1.6$, $x_4 = 1.8$ and $y_0 = 1$ into $y_{n+1} = y_n + f(x_n, y_n)(x_{n+1} - x_n)$ where $f(x,y) = x^2y^2$ we obtain
   $y_1 = 1 + (1^2\, 1^2)(1.2 - 1) = 1 + 0.2 = 1.2$
   $y_2 = 1.2 + (1.2^2\, 1.2^2)(1.4 - 1.2) = 1.61472$
   $y_3 = 1.61472 + (1.4^2\, 1.61472^2)(1.6 - 1.4) = 2.63679$
   $y_4 = 2.63679 + (1.6^2\, 2.63679^2)(1.8 - 1.6) = 6.1966$

   The given differential equation is separable. We rewrite the equation in the form

   $$y^{-2}\,dy = x^2\,dx.$$

   Integrating each side of this equation we obtain $-y^{-1} = \frac{1}{3}x^3 + c$, or equivalently, $y = \frac{3}{C - x^3}$. Using the initial condition $y(1) = 1 = \frac{3}{C-1}$, we see that $C = 4$. Thus the solution to the initial value problem is $y = \frac{3}{4 - x^3}$.
   The exact values at the points $x = 1.2, 1.4, 1.6, 1.8$ would be
   $y(1.2) = \frac{3}{4 - 1.2^3} \approx 1.32042$
   $y(1.4) = \frac{3}{4 - 1.4^3} \approx 2.38854$
   $y(1.6) = \frac{3}{4 - 1.6^3} = -31.25$
   $y(1.8) = \frac{3}{4 - 1.8^3} \approx -1.63755$

5. We again have the initial value problem $y' = 2y - x$, $y(0) = 1$, with $x = 0.1, 0.2, 0.4, 0.5$. For the second order Taylor approximation we need the second derivative of the solution. Since $y' = 2y - x$ this is $y'' = 2y' - 1$. Since

   $$T_2(x) = y(x_0) + y'(x_0)(x - x_0) + \frac{1}{2}y''(x_0)(x - x_0)^2,$$

   we have that

   $$T_2(x) = y(0) + y'(0)(x - 0) + \frac{1}{2}y''(0)(x - 0)^2 = 1 + 2x + \frac{1}{2}3x^2 = 1 + 2x + \frac{3}{2}x^2.$$

Thus,

$$T_2(x_1) = T_2(0.1) = 1 + 2(0.1) + \frac{3}{2}(0.1)^2 = 1.215.$$

We now develop the second order Taylor polynomial at $(x_1, T(x_1)) = (0.1, 1.215)$. Since

$$T_2(x) = y(x_1) + y'(x_1)(x - x_1) + \frac{1}{2}y''(x_1)(x - x_1)^2,$$

we have that

$$
\begin{aligned}
T_2(x_2) &= y(0.1) + y'(0.1)(x_2 - x_1) + \frac{1}{2}y''(0.1)(x_2 - x_1)^2 \\
&= 1.215 + 2.33(0.1) + \frac{1}{2}(3.66)(0.1)^2 \approx 1.4663.
\end{aligned}
$$

Continuing we have

$$T_2(x) = y(x_2) + y'(x_2)(x - x_2) + \frac{1}{2}y''(x_2)(x - x_2)^2,$$

so that

$$
\begin{aligned}
T_2(x_3) &= y(0.2) + y'(0.2)(x_3 - x_2) + \frac{1}{2}y''(0.2)(x_3 - x_2)^2 \\
&\approx 1.4663 + 2.7326(0.2) + \frac{1}{2}(4.4652)(0.2)^2 \approx 2.102124.
\end{aligned}
$$

Finally we have

$$T_2(x) = y(x_3) + y'(x_3)(x - x_3) + \frac{1}{2}y''(x_3)(x - x_3)^2,$$

so that

$$
\begin{aligned}
T_2(x_4)) &= y(0.4) + y'(0.4)(x_4 - x_3) + \frac{1}{2}y''(0.4)(x_4 - x_3)^2 \\
&\approx 2.102124 + 3.804248(0.1) + \frac{1}{2}(6.608496)(0.1)^2 \approx 2.5156.
\end{aligned}
$$

Comparing with Exercise 1, we notice that the second order Taylor polynomial gives a better approximation of $\phi(x)$ at each of the x-values than Euler's method.

7. We again have the initial value problem $y' = x^2 y^2$, $y(1) = 1$, with $x = 1.2$, 1.4, 1.6, 1.8. For the third order Taylor approximation we need the second and third derivatives of the solution. Since $y' = x^2 y^2$ these are

$$y'' = 2xy^2 + 2x^2 yy' = 2xy^2 + 2x^2 yx^2 y^2 = 2xy^2 + 2x^4 y^3 = 2xy^2(1 + x^3 y) \text{ and}$$

$$y''' = 2y^2 + 4xyy' + 8x^3 y^3 + 6x^4 y^2 y' = 2y^2(1 + 6x^3 y + 3x^6 y^2).$$

Since

$$T_3(x) = y(x_0) + y'(x_0)(x - x_0) + \frac{1}{2}y''(x_0)(x - x_0)^2 + \frac{1}{6}y'''(x_0)(x - x_0)^3,$$

we have that

$$
\begin{aligned}
T_3(x_1) &= y(1) + y'(1)(1.2 - 1) + \frac{1}{2}y''(1)(1.2 - 1)^2 + \frac{1}{6}y'''(1)(1.2 - 1)^3 \\
&= 1 + 1(0.2) + \frac{1}{2}4(0.2)^2 + \frac{1}{6}20((0.2)^3 \approx 1.30666.
\end{aligned}
$$

We now develop the third order Taylor polynomial at $(x_1, T(x_1)) = (1.2, 1.30666)$. Since

$$T_3(x) = y(x_1) + y'(x_1)(x - x_1) + \frac{1}{2}y''(x_1)(x - x_1)^2 + \frac{1}{6}y'''(x_1)(x - x_1)^3,$$

we have that

$$T_3(x_2) = y(1.2) + y'(1.2)(1.4 - 1.2) + \frac{1}{2}y''(1.2)(1.4 - 1.2)^2 + y'''(1.2)(1.4 - 1.2)^3 \approx 2.20126.$$

Continuing we have

$$T_3(x) = y(x_2) + y'(x_2)(x - x_2) + \frac{1}{2}y''(x_2)(x - x_2)^2 + \frac{1}{6}y'''(x_2)(x - x_2)^3,$$

so that

$$T_3(x_3) = y(1.4) + y'(1.4)(1.6 - 1.4) + \frac{1}{2}y''(1.4)(1.6 - 1.4)^2 + y'''(1.4)(1.6 - 1.4)^3 \approx 7.90664.$$

Finally we have

$$T_3(x) = y(x_3) + y'(x_3)(x - x_3) + \frac{1}{2}y''(x_3)(x - x_3)^2 + \frac{1}{6}y''(x_3)(x - x_3)^6,$$

so that

$$T_3(x_4) = y(1.6) + y'(1.6)(1.8 - 1.6) + \frac{1}{2}y''(1.6)(1.8 - 1.6)^2 + y'''(1.6)(1.8 - 1.6)^3 \approx 730.5869.$$

Comparing with Exercise 3, we notice that the third order Taylor polynomial gives a better approximation of $\phi(x)$ than Euler's method for $x = 1.2$ and 1.4 and a worse approximation for the other x-values. However, because of the discontinuity of the solution at $\sqrt[3]{4}$, no approximation method will be effective crossing from $x < \sqrt[3]{4}$ to $x > \sqrt[3]{4}$.

9. When $x_{n+1} = x_0 + nh$ the formulæ for the Euler and Taylor methods can be simplified slightly because

$$x_{k+1} - x_k = h.$$

The formula for Euler's method simplifies to

$$y_{n+1} = y_n + f(x_n, y_n)h, \ n = 0, \ 1, \ 2, \ldots, N-1.$$

The formula for Taylor's method simplifies to

$$T(x_{n+1}) = \sum_{k=0}^{n} \frac{y^{(k)}}{k!}h^k, \ n = 0, \ 1, \ 2, \ldots, N-1.$$

11. We again have the initial value problem $y' = 2y - x$, $y(0) = 1$, now with $h = 0.1$ and $N = 5$. We set
$a_0 = f(0,1) = 2 \cdot 1 - 0 = 2$,
$b_0 = f(0 + 0.05, 1 + 0.05 \cdot 2) = f(0.05, 1.1) = 2 \cdot 1.1 - 0.05 = 2.15$,
$c_0 = f(0.05, 1 + 0.05 \cdot 2.15) = f(0.05, 1.1075) = 2 \cdot 1.1075 - 0.05 = 2.165$,
$d_0 = f(0.1, 1 + 0.1 \cdot 2.165) = f(0.1, 1.2165) = 2 \cdot 1.2165 - 0.1 = 2.333$.
Therefore, $y_1 = y_0 + \frac{h}{6}(a_0 + 2b_0 + 2c_0 + d_0) = 1 + \frac{0.1}{6}(2 + 2 \cdot 2.15 + 2 \cdot 2.165 + 2.333) = 1.21605$.

Setting
$a_1 = f(0.1, 1.21605) = 2 \cdot 1.21605 - 0.1 = 2.3321$,
$b_1 = f(0.1 + 0.05, 1.21605 + 0.05 \cdot 2.3321) = f(0.15, 1.33266) = 2.51531$,
$c_1 = f(0.15, 1.21605 + 0.05 \cdot 2.51531) = f(0.15, 1.34182) = 2.53363$,
$d_1 = f(0.2, 1.21605 + 0.1 \cdot 2.53363) = f(0.2, 1.46941) = 2.73883$.

We get

$$
\begin{aligned}
y_2 &= y_1 + \frac{h}{6}(a_1 + 2b_1 + 2c_1 + d_1) \\
&= 1.21605 + \frac{0.1}{6}(2.3321 + 2 \cdot 2.51531 + 2 \cdot 2.53363 + 2.73883) = 1.46886.
\end{aligned}
$$

Setting
$a_2 = f(0.2, 1.46886) = 2.73773$,
$b_2 = f(0.25, 1.46886 + 0.05 \cdot 2.73773) = f(0.25, 1.60575) = 2.96150$,

$c_2 = f(0.25, 1.46886 + 0.05 \cdot 2.96150) = f(0.25, 1.61693) = 2.98388,$
$d_2 = f(0.3, 1.46886 + 0.1 \cdot 2.98388) = f(0.3, 1.76725) = 3.23450.$

We get

$$
\begin{aligned}
y_3 &= y_2 + \frac{h}{6}(a_2 + 2b_2 + 2c_2 + d_2) \\
&= 1.46886 + \frac{0.1}{6}(2.73773 + 2 \cdot 2.96150 + 2 \cdot 2.98388 + 3.23450) = 1.76658.
\end{aligned}
$$

Setting
$a_3 = f(0.3, 1.76658) = 3.23316,$
$b_3 = f(0.35, 1.76658 + 0.05 \cdot 3.23316) = f(0.35, 1.92824) = 3.50648,$
$c_3 = f(0.35, 1.76658 + 0.05 \cdot 3.50648) = f(0.35, 1.94190) = 3.53381,$
$d_3 = f(0.4, 1.76658 + 0.1 \cdot 3.53381) = f(0.4, 2.11996) = 3.83992.$

We get

$$
\begin{aligned}
y_4 &= y_3 + \frac{h}{6}(a_3 + 2b_3 + 2c_3 + d_3) \\
&= 1.76658 + \frac{0.1}{6}(3.23316 + 2 \cdot 3.50648 + 2 \cdot 3.53381 + 3.83992) = 2.11914.
\end{aligned}
$$

Finally setting
$a_4 = f(0.4, 2.11914) = 3.83828,$
$b_4 = f(0.45, 2.11914 + 0.05 \cdot 3.83828) = f(0.45, 2.31105) = 4.17211,$
$c_4 = f(0.45, 2.11914 + 0.05 \cdot 4.17211) = f(0.45, 2.32775) = 4.20549,$
$d_4 = f(0.5, 2.11914 + 0.1 \cdot 4.20549) = f(0.5, 2.53969) = 4.57938.$

We get

$$
\begin{aligned}
y_5 &= y_4 + \frac{h}{6}(a_4 + 2b_4 + 2c_4 + d_4) \\
&= 2.11914 + \frac{0.1}{6}(3.83828 + 2 \cdot 4.17211 + 2 \cdot 4.20549 + 4.57938) = 2.53869.
\end{aligned}
$$

Comparing these approximations with Exercises 1 and 5, we notice that the fourth order Runge-Kutta method gives a better approximation of $\phi(x)$ at each of the x-values than either Euler's method or the second order Taylor polynomial method.

13. We have the initial value problem $y' = x^2 + y^2$, $y(0) = 0$, with $h = 0.1$ and $N = 5$. We set
$a_0 = f(0,0) = 0^2 + 0^2 = 0,$
$b_0 = f(0 + 0.05, 0 + 0.05 \cdot 0) = f(0.05, 0) = 0.05^2 + 0^2 = 0.0025,$
$c_0 = f(0.05, 0 + 0.05 \cdot 0.0025) = f(0.05, 0.000125) = 0.00250,$
$d_0 = f(0.1, 0 + 0.1 \cdot 0.00250) = f(0.1, .00025) = 0.01000.$

Therefore,

$$
\begin{aligned}
y_1 &= y_0 + \frac{h}{6}(a_0 + 2b_0 + 2c_0 + d_0) \\
&= 0 + \frac{0.1}{6}(0 + 2 \cdot 0.0025 + 2 \cdot 0.00250 + 0.01000) = 0.000333.
\end{aligned}
$$

Setting
$a_1 = f(0.1, 0.000333) = 0.01000,$
$b_1 = f(0.1 + 0.05, 0.000333 + 0.05 \cdot 0.01000) = f(0.15, 0.00083) = 0.022501,$
$c_1 = f(0.15, 0.000333 + 0.05 \cdot 0.022501) = f(0.15, 0.001458) = 0.022502,$
$d_1 = f(0.2, 0.000333 + 0.1 \cdot 0.002502) = f(0.2, 0.00258) = 0.040007.$

We get

$$
\begin{aligned}
y_2 &= y_1 + \frac{h}{6}(a_1 + 2b_1 + 2c_1 + d_1) \\
&= 0.000333 + \frac{0.1}{6}(0.01000 + 2 \cdot 0.022501 + 2 \cdot 0.022502 + 0.040007) = 0.0026669.
\end{aligned}
$$

Setting

$a_2 = f(0.2, 0.0026669) = 0.040007,$

$b_2 = f(0.25, 0.0026669 + 0.05 \cdot 0.040007) = f(0.25, 0.004667) = 0.062522,$

$c_2 = f(0.25, 0.0026669 + 0.05 \cdot 0.062522) = f(0.25, 0.005793) = 0.062533,$

$d_2 = f(0.3, 0.0026669 + 0.1 \cdot 0.062533) = f(0.3, 0.008920) = 0.090080.$

We get

$$
\begin{aligned}
y_3 &= y_2 + \frac{h}{6}(a_2 + 2b_2 + 2c_2 + d_2) \\
&= 0.0026669 + \frac{0.1}{6}(0.040007 + 2 \cdot 0.062522 + 2 \cdot 0.062533 + 0.090080) = 0.00900.
\end{aligned}
$$

Setting

$a_3 = f(0.3, 0.00900) = 0.090081,$

$b_3 = f(0.35, 0.00900 + 0.05 \cdot 0.090081) = f(0.35, 0.013504) = 0.122682,$

$c_3 = f(0.35, 0.00900 + 0.05 \cdot 0.122682) = f(0.35, 0.015134) = 0.122729,$

$d_3 = f(0.4, 0.00900 + 0.1 \cdot 0.122729) = f(0.4, 0.021273) = 0.160453.$

We get

$$
\begin{aligned}
y_4 &= y_3 + \frac{h}{6}(a_3 + 2b_3 + 2c_3 + d_3) \\
&= 0.00900 + \frac{0.1}{6}(0.090081 + 2 \cdot 0.122682 + 2 \cdot 0.122729 + 0.160453) = 0.021359.
\end{aligned}
$$

Finally setting

$a_4 = f(0.4, 0.021359) = 0.160456,$

$b_4 = f(0.45, 0.021359 + 0.05 \cdot 0.160456) = f(0.45, 0.029382) = 0.203363,$

$c_4 = f(0.45, 0.021359 + 0.05 \cdot 0.203363) = f(0.45, 0.031527) = 0.203494,$

$d_4 = f(0.5, 0.021359 + 0.1 \cdot 0.203494) = f(0.5, 0.041708) = 0.251740.$

We get

$$
\begin{aligned}
y_5 &= y_4 + \frac{h}{6}(a_4 + 2b_4 + 2c_4 + d_4) \\
&= 0.021359 + \frac{0.1}{6}(0.160456 + 2 \cdot 0.203363 + 2 \cdot 0.203494 + 0.251740) = 0.041791.
\end{aligned}
$$

Chapter 4

Linear Differential Equations

4.1 The Theory of Higher Order Linear Differential Equations

Exercises 4.1, pp. 188-189

1. The given differential equation is linear. It is third order.

3. The given differential equation is linear. It is fourth order.

5. Since each of the functions $q_2(x) = 1$, $q_1(x) = -2x$, $q_0(x) = 4$ and $g(x) = 5x$ are continuous on $(-\infty, \infty)$ and $q_2(x) \neq 0$ on $(-\infty, \infty)$ Theorem 4.1 guarantees a unique solution on $(-\infty, \infty)$.

7. Although the functions $q_3(x) = e^x \ln x$, $q_2(x) = -x$, $q_1(x) = 2$, $q_0(x) = -5$ and $g(x) = 2$ are all simultaneously continuous on the interval $(0, \infty)$, the function $q_3(x) = 0$ at $x_0 = 1$, therefore Theorem 4.1 does not apply to this differential equation.

9. We are given the initial value problem $y'' - y' - 2y = 0$, $y(0) = 1$, $y'(0) = 0$ and the functions $y_1 = e^{-x}$ and $y_2 = e^{2x}$. Therefore, $y_1' = -e^{-x}$, $y_1'' = e^{-x}$, $y_2' = 2e^{2x}$ and $y_2'' = 4e^{2x}$. Thus,

$$y_1'' - y_1' - 2y_1 = e^{-x} - (-e^{-x}) - 2(e^{-x}) = 0 \text{ and}$$

$$y_2'' - y_2' - 2y_2 = 4e^{2x} - 2e^{2x} - 2(e^{2x}) = 0.$$

Therefore, both e^{-x} and e^{2x} are solutions to the differential equation.

The Wronskian, $w(e^{-x}, e^{2x}) = \begin{vmatrix} e^{-x} & e^{2x} \\ -e^{-x} & 2e^{2x} \end{vmatrix} = 2e^x + e^x = 3e^x \neq 0$ for any x. Therefore, e^{-x} and e^{2x} form a fundamental set of solutions.

The general solution to the differential equation is $y = c_1 e^{-x} + c_2 e^{2x}$.

This function has the derivative $y' = -c_1 e^{-x} + 2c_2 e^{2x}$. Considering the initial conditions $y(0) = 1$, $y'(0) = 0$ we obtain the system of equations $y(0) = 1 = c_1 + c_2$, $y'(0) = 0 = -c_1 + 2c_2$, which has solution $c_1 = \frac{2}{3}$, $c_2 = \frac{1}{3}$. Therefore, the given initial value problem has the unique solution $y = \frac{2}{3}e^{-x} + \frac{1}{3}e^{2x}$.

11. We are given the initial value problem $y''' - 7y' + 6y = 0$, $y(0) = 1$, $y'(0) = 0$, $y''(0) = 0$ and the functions $y_1 = e^x$, $y_2 = e^{2x}$ and $y_3 = e^{-3x}$. Therefore,

$$y_1' = e^x, \ y_1'' = e^x, \ y_1''' = e^x,$$

$$y_2' = 2e^{2x}, \ y_2'' = 4e^{2x}, \ y_2''' = 8e^{2x},$$

$$y_3' = -3e^{-3x}, \ y_3'' = 9e^{-3x}, \ y_3''' = -27e^{-3x}.$$

Thus,

$$y_1''' - 7y_1' + 6y_1 = e^x - 7e^x + 6e^x = 0,$$

$$y_2''' - 7y_2' + 6y_2 = 8e^{2x} - 7(2e^{2x}) + 6e^{2x} = 0,$$

$$y_3''' - 7y_3' + 6y_3 = -27e^{2x} - 7(-3e^{-3x}) + 6e^{-3x} = 0.$$

Therefore, e^x, e^{2x} and e^{-3x} are all solutions to the differential equation.

The Wronskian, $w(e^x, e^{2x}, e^{-3x}) = \begin{vmatrix} e^x & e^{2x} & e-3x \\ e^x & 2e^{2x} & -3e-3x \\ e^x & 4e^{2x} & 9e-3x \end{vmatrix} = 20 \neq 0$ for any x. Therefore, e^x, e^{2x} and e^{-3x} form a funda-

mental set of solutions.

The general solution to the differential equation is $y = c_1 e^x + c_2 e^{2x} + c_3 e^{-3x}$.

This function has the derivatives $y' = c_1 e^x + 2c_2 e^{2x} - 3c_3 e^{-3x}$ and $y'' = c_1 e^x + 4c_2 e^{2x} + 9c_3 e^{-3x}$. Considering the initial conditions $y(0) = 1$, $y'(0) = 0$, $y''(0) = 0$ we obtain the system of equations

$$
\begin{aligned}
1 &= c_1 + c_2 + c_3 \\
0 &= c_1 + 2c_2 - 3c_3 \\
0 &= c_1 + 4c_2 + 9c_3
\end{aligned}
$$

which has solution $c_1 = \frac{3}{2}$, $c_2 = -\frac{3}{5}$, $c_3 = \frac{1}{10}$.

The given initial value problem has the unique solution $y = \frac{3}{2}e^x - \frac{3}{5}e^{2x}\frac{1}{10}e^{-3x}$.

13. We are given the initial value problem $y'' - y' - 2y = 4$, $y(0) = 1$, $y'(0) = 0$ and the particular solution $y_P = -2$.

We see that $y_P' = 0$ and $y_P'' = 0$. Thus, $y_P'' - y_P' - 2y_P = 0 - 0 - 2(-2) = 4$, so y_P is a solution to the given nonhomogeneous differential equation. In problem 9 we have already seen that e^{-x} and e^{2x} are solutions to the corresponding homogeneous differential equation, therefore the general solution to the nonhomogeneous differential equation is $y = c_1 e^{-x} + c_2 e^{2x} - 2$.

Considering the initial conditions $y(0) = 1$, $y'(0) = 0$ we now obtain the system of equations

$$
y(0) = 1 = c_1 + c_2 - 2, \quad y'(0) = 0 = -c_1 + 2c_2,
$$

which has solution $c_1 = 2$, $c_2 = 1$. Therefore, the given initial value problem has the unique solution

$$
y = 2e^{-x} + e^{2x} - 2.
$$

15. We are given the initial value problem $y''' - 7y' + 6y = 12e^{-x}$, $y(0) = 1$, $y'(0) = 0$, $y''(0) = 0$ and the particular solution $y_P = e^{-x}$.

We see that $y_P' = -e^{-x}$, $y_P'' = e^{-x}$ and $y_P''' = -e^{-x}$. Thus,

$$
y_P''' - 7y_P' + 6y_P = -e^{-x} - 7(-e^{-x}) + 6e^{-x} = 12e^{-x},
$$

so y_P is a solution to the given nonhomogeneous differential equation. In problem 11 we have already seen that e^x, e^{2x} and e^{-3x} are solutions to the corresponding homogeneous differential equation, therefore the general solution to the nonhomogeneous differential equation is $y = c_1 e^x + c_2 e^{2x} + c_3 e^{-3x} + e^{-x}$.

Considering the initial conditions $y(0) = 1$, $y'(0) = 0$, $y''(0) = 0$ we now obtain the system of equations

$$
\begin{aligned}
1 &= c_1 + c_2 + c_3 + 1 \\
0 &= c_1 + 2c_2 - 3c_3 - 1 \\
0 &= c_1 + 4c_2 + 9c_3 + 1
\end{aligned}
$$

which has solution $c_1 = 0$, $c_2 = \frac{1}{5}$, $c_3 = -\frac{1}{5}$.

Therefore, the given initial value problem has the unique solution $y = \frac{1}{5}e^{2x} - \frac{1}{5}e^{-3x} + e^{-x}$.

17. We are given the initial value problem $\frac{1}{2}x^2 y'' - xy' + y = g(x)$, $y(1) = 1$, $y'(1) = 0$ and the functions $y_1 = x$ and $y_2 = x^2$. Therefore, $y_1' = 1$, $y_1'' = 0$, $y_2' = 2x$ and $y_2'' = 2$. Thus,

$$
\frac{1}{2}x^2 y_1'' - xy_1' + y_1 = \frac{1}{2}x^2 \cdot 0 - x \cdot 1 + x = 0 \text{ and}
$$

$$
\frac{1}{2}x^2 y_2'' - xy_2' + y_2 = \frac{1}{2}x^2 \cdot 2 - x \cdot 2x + x^2 = 0.
$$

Therefore, both x and x^2 are solutions to the corresponding homogeneous differential equation.

The Wronskian, $w(x,x^2) = \begin{vmatrix} x & x^2 \\ 1 & 2x \end{vmatrix} = 2x^2 - x^2 = x^2 \neq 0$ on $(-\infty, 0)$ and $(0, \infty)$. Therefore, x and x^2 form a fundamental set of solutions to the corresponding homogeneous differential equation on either of these intervals.

Now, if $y_P = e^x$, then $y'_P = y''_P = e^x$ and

$$\frac{1}{2}x^2 y''_P - xy'_P + y_P = \frac{1}{2}x^2 \cdot e^x - x \cdot e^x + e^x = e^x \left(\frac{1}{2}x^2 - x + 1 \right).$$

Therefore, if $g(x) = e^x(\frac{1}{2}x^2 - x + 1)$, y_P will be a solution to the given differential equation.

The general solution to this differential equation is $y = c_1 x + c_2 x^2 + e^x$.

This function has the derivative $y' = c_1 + 2c_2 x + e^x$. Considering the initial conditions, we obtain the system of equations $y(1) = 1 = c_1 + c_2 + e$, $y'(1) = 0 = c_1 + 2c_2 + e$, which has solution $c_1 = 2 - e$, $c_2 = -1$. Therefore, the initial value problem has the unique solution $y = (2 - e)x - x^2 + e^x$.

21. By Corollary 4.5 the functions x and x^2 cannot form a fundamental set of solutions to a second order differential equation on $(-\infty, \infty)$ because the Wronskian,

$$w(x,x^2) = \begin{vmatrix} x & x^2 \\ 1 & 2x \end{vmatrix} = 2x^2 - x^2 = x^2 = 0$$

at $x = 0$, which, of course is in $(-\infty, \infty)$.

23. Let y_1, y_2, y_3 be solutions to a homogeneous linear differential equation. The Wronskian,

$$w(y_1(1), y_2(1), y_3(1)) = \begin{vmatrix} 1 & 1 & 1 \\ 1 & 2 & 3 \\ 1 & -1 & 1 \end{vmatrix} = 4 \neq 0.$$

By Theorem 2.15 the three solutions are linearly independent on some interval containing $x = 1$. Therefore, they form a fundamental set of solutions on that interval.

4.2   Homogeneous Constant Coefficient Linear Differential Equations

Exercises 4.2, pp. 201-203

1. The differential equation $y'' + 8y' = 0$ has characteristic equation $\lambda^2 + 8\lambda = 0$. This factors to $\lambda(\lambda + 8) = 0$. The roots are $\lambda = -8, 0$. Thus the solutions are e^{-8x} and 1. The general solution is then $y = c_1 e^{-8x} + c_2$.

3. The differential equation $y'' + 4y' + y = 0$ has characteristic equation $\lambda^2 + 4\lambda + 1 = 0$. Using the quadratic formula we find the roots $\lambda = -2 \pm \sqrt{3}$. Thus the solutions are $e^{(-2+\sqrt{3})x}$ and $e^{-(2+\sqrt{3})x}$. The general solution is then

$$y = c_1 e^{(-2+\sqrt{3})x} + c_2 e^{-(2+\sqrt{3})x}.$$

5. The differential equation $y''' + 4y'' - y' - 4y = 0$ has characteristic equation $\lambda^3 + 4\lambda^2 - \lambda - 4 = (\lambda + 4)(\lambda + 1)(\lambda - 1) = 0$. The roots are $\lambda = -4, -1, 1$. Thus the solutions are e^{-4x}, e^{-x} and e^x. The general solution is then $y = c_1 e^{-4x} + c_2 e^{-x} + c_3 e^x$.

7. The differential equation $y'' + 6y' + 9y = 0$ has characteristic equation $\lambda^2 + 6\lambda + 9 = 0$. This factors to $(\lambda + 3)^2 = 0$. There is a single repeated root $\lambda = -3$. Thus the solutions are e^{-3x} and xe^{-3x}. The general solution is then $y = c_1 e^{-3x} + c_2 xe^{-3x}$.

9. The differential equation $3y''' + 18y'' + 36y' + 24y = 0$ has characteristic equation $3\lambda^3 + 18\lambda^2 + 36\lambda + 24 = 0$. This factors to $3(\lambda + 2)^3 = 0$. There is a single repeated root $\lambda = -2$. Thus the solutions are e^{-2x}, xe^{-2x} and $x^2 e^{-2x}$. The general solution is then $y = c_1 e^{-2x} + c_2 xe^{-2x} c_3 x^2 e^{-2x}$.

11. The differential equation $4y'' + 16y = 0$ has characteristic equation $4\lambda^2 + 16 = 0$, which has imaginary roots $\lambda = \pm 2i$. Thus the general solution is $y = c_1 \cos 2x + c_2 \sin 2x$.

13. The differential equation $y''' - 2y' + 4y = 0$ has characteristic equation $\lambda^3 - 2\lambda + 4$, which has the roots $\lambda = -2$ and $\lambda = 1 \pm i$. Thus the general solution is $y = c_1 e^{-2x} + c_2 e^x \cos x + c_3 e^x \sin x$.

15. (a) The differential equation $y'' - y = 0$ has characteristic equation $\lambda^2 - 1 = 0$, which has the roots $\lambda = \pm 1$. Thus the general solution is $y = c_1 e^x + c_2 e^{-x}$.

    (b) We now use the initial conditions $y(1) = 0$ and $y'(1) = -1$. Since $y = c_1 e^x + c_2 e^{-x}$, we have $y' = c_1 e^x - c_2 e^{-x}$. Thus

$$0 = c_1 e + c_2 \frac{1}{e}$$
$$-1 = c_1 e - c_2 \frac{1}{e}.$$

    This system has the solution $c_1 = -\frac{1}{2e}$, $c_2 = \frac{e}{2}$. Thus, $y = -\frac{1}{2e} e^x + \frac{e}{2} e^{-x}$, or equivalently, $y = -\frac{1}{2} e^{x-1} + \frac{1}{2} e^{1-x}$.

17. (a) The differential equation $y'' + 4y' + 8y = 0$ has characteristic equation $\lambda^2 + 4\lambda + 8 = 0$, which has the roots $\lambda = -2 \pm 2i$. Thus the general solution is $y = c_1 e^{-2x} \cos 2x + c_2 e^{-2x} \sin 2x$.

    (b) We now use the initial conditions $y(0) = 0$ and $y'(0) = -1$. Since $y = c_1 e^{-2x} \cos 2x + c_2 e^{-2x} \sin 2x$, we have

$$y' = -2c_1 e^{-2x} \cos 2x - 2c_1 e^{-2x} \sin 2x - 2c_2 - 2c_2 e^{-2x} \sin 2x + 2c_2 e^{-2x} \cos 2x.$$

    Thus

$$0 = c_1$$
$$-1 = -2c_1 + 2c_2.$$

    This system has the solution $c_1 = 0$, $c_2 = -\frac{1}{2}$. Thus, $y = -\frac{1}{2} e^{-2x} \sin 2x$.

19. (a) The differential equation $4y'' + 4y' + y = 0$ has characteristic equation $4\lambda^2 + 4\lambda + 1 = 0$, which has the repeated root $\lambda = -\frac{1}{2}$. Thus the general solution is $y = (c_1 + c_2 x) e^{-\frac{1}{2}x}$.

    (b) We now use the initial conditions $y(0) = 0$ and $y'(0) = -1$. Since $y = (c_1 + c_2 x) e^{-\frac{1}{2}x}$, we have $y' = -\frac{1}{2}(c_1 + c_2 x) e^{-\frac{1}{2}x} + c_2 e^{-\frac{1}{2}x}$. Thus

$$0 = c_1$$
$$-1 = -\frac{1}{2} c_1 + c_2.$$

    This system has the solution $c_1 = 0$, $c_2 = -1$. Thus, $y = -x e^{-\frac{1}{2}x}$.

21. (a) The differential equation $y''' + 5y'' = 0$ has characteristic equation $\lambda^3 + 5\lambda^2 = \lambda^2(\lambda + 5) = 0$, which has the root $\lambda = -5$ and the repeated root $\lambda = 0$. Thus the general solution is $y = c_1 + c_2 x + c_3 e^{-5x}$.

    (b) We now use the initial conditions $y(-1) = 1$, $y'(-1) = 0$ and $y''(-1) = 2$. Since $y = c_1 + c_2 x + c_3 e^{-5x}$, we have $y' = c_2 - 5c_3 e^{-5x}$ and $y'' = 25 c_3 e^{-5x}$. Thus

$$1 = c_1 - c_2 + c_3 e^5$$
$$0 = c_2 - 5c_3 e^5$$
$$2 = 25 c_3 e^5.$$

    This system has the solution $c_1 = \frac{33}{25}$, $c_2 = \frac{2}{5}$, $c_3 = \frac{2}{25e^5}$. Thus, $y = \frac{33}{25} + \frac{2}{5} x + \frac{2}{25e^5} e^{-5x}$ or equivalently,

$$y = \frac{33}{25} + \frac{2}{5} x + \frac{2}{25} e^{-5x-5}.$$

23. (a) The differential equation $y''' - y'' - 6y' = 0$ has characteristic equation $\lambda^3 - \lambda^2 - 6\lambda = \lambda(\lambda + 2)(\lambda - 3) = 0$. The roots are $\lambda = 0$, -2 and 3. Thus the general solution is $y = c_1 + c_2 e^{-2x} + c_3 e^{3x}$.

(b) We now use the initial conditions $y(0) = 1$, $y'(0) = 0$ and $y''(0) = 2$. Since $y = c_1 + c_2 e^{-2x} + c_3 e^{3x}$, we have

$$y' = -2c_2 e^{-2x} + 3c_3 e^{3x} \text{ and } y'' = 4c_2 e^{-2x} + 9c_3 e^{3x}.$$

Thus

$$
\begin{aligned}
1 &= c_1 + c_2 + c_3 \\
0 &= -2c_2 + 3c_3 \\
2 &= 4c_2 + 9c_3.
\end{aligned}
$$

This system has the solution $c_1 = \frac{2}{3}$, $c_2 = \frac{1}{5}$, $c_3 = \frac{2}{15}$. Thus, $y = \frac{2}{3} + \frac{1}{5}e^{-2x} + \frac{2}{15}e^{3x}$.

25. (a) The differential equation $y''' - 5y'' + 3y' + 9y = 0$ has characteristic equation $\lambda^3 - 5\lambda^2 + 3\lambda + 9 = 0$, which factors to $(\lambda + 1)(\lambda - 3)^2 = 0$. This has the root $\lambda = -1$ and the repeated root $\lambda = 3$. Thus the general solution is

$$y = c_1 e^{-x} + (c_2 + c_3 x)e^{3x}.$$

(b) We now use the initial conditions $y(0) = 0$, $y'(0) = -1$ and $y''(0) = 1$. Since $y = c_1 e^{-x} + (c_2 + c_3 x)e^{3x}$, we have $y' = -c_1 e^{-x} + 3c_2 e^{3x} + 3c_3 x e^{3x} + c_3 e^{3x}$ and $y'' = c_1 e^{-x} + 9c_2 e^{3x} + 9c_3 x e^{3x} + 6c_3 e^{3x}$. Thus

$$
\begin{aligned}
0 &= c_1 + c_2 \\
-1 &= -c_1 + 3c_2 + c_3 \\
1 &= c_1 + 9c_2 + 6c_3.
\end{aligned}
$$

This system has the solution $c_1 = \frac{7}{16}$, $c_2 = -\frac{7}{16}$, $c_3 = \frac{3}{4}$. Thus, $y = \frac{7}{16}e^{-x} + (\frac{3}{4}x - \frac{7}{16})e^{3x}$.

27. The characteristic polynomial $p(\lambda) = (\lambda - 4)^3(\lambda^2 + 4)^2$ has roots $\lambda = \pm 2i$ each with multiplicity 2 and $\lambda = 4$ with multiplicity 3. Therefore the general solution to the homogeneous linear differential equation would be

$$y = (A + Bx)\sin 2x + (C + Dx)\cos 2x + (E + Fx + Gx^2)e^{4x}.$$

29. The characteristic polynomial $p(\lambda) = (2\lambda + 4)^3(\lambda^2 - 9)^2(\lambda^2 + 9)^3$ has roots $\lambda = \pm 3i$ each with multiplicity 3, $\lambda = -2$ with multiplicity 1 and $\lambda = \pm 3$ each with multiplicity 2. Therefore the general solution to the homogeneous linear differential equation would be

$$y = (A + Bx + Cx^2)\cos 3x + (D + Ex + Fx^2)\sin 3x + (E + Fx + Gx^2)e^{4x} + Ge^{-2x} + (H + Ix)e^{-3x} + (J + Kx)e^{3x}.$$

31. Consider the differential equation

$$x^2 y'' - xy' = 0. \tag{4.1}$$

If $y_1 = x^2$ then $y_1' = 2x$ and $y_1'' = 2$. Thus $x^2 y_1'' - xy_1' = x^2 \cdot 2 - x \cdot 2x = 2x^2 - 2x^2 = 0$. Therefore, $y_1 = x^2$ is a solution to the given differential equation.

We now try to find a second linearly independent solution of the form $y_2 = u(x)y_1 = ux^2$. Here

$$y_2' = u'x^2 + 2xu \text{ and } y_2'' = u''x^2 + 4xu' + 2u.$$

Substituting these into equation (4.1) we have $x^2(u''x^2 + 4xu' + 2u) - x(u'x^2 + 2xu) = 0$. Simplifying we obtain

$$u''x^4 + 3x^3 u' = 0 \text{ or equivalently } u''x^4 = -3x^3 u'.$$

This is a separable differential equation. Separating the variables we have $\frac{u''}{u'} = -\frac{3}{x}$. Integrating we obtain

$$\ln|u'| = -3\ln|x| + k, \text{ or equivalently } u' = c_1 x^{-3}.$$

Integrating again to find u we have $u = -\frac{1}{2}c_1 x^{-2} + c_2$. We are free to choose the constants c_1 and c_2 as we wish, as long as $c_1 \neq 0$. Thus we choose $c_1 = -2$, $c_2 = 0$ and have $u = x^{-2}$. Thus, $y_2 = ux^2 = 1$.

33. Consider the differential equation

$$(1+x)y'' + xy' - y = 0. \tag{4.2}$$

If $y_1 = e^{-x}$ then $y_1' = -e^{-x}$ and $y_1'' = e^{-x}$. Thus $(1+x)y'' + xy_1 - y_1 = (1+x)e^{-x} + x(-e^{-x}) - e^{-x} = 0$. Therefore, $y_1 = e^{-x}$ is a solution to the given differential equation.

We now try to find a second linearly independent solution of the form $y_2 = u(x)y_1 = ue^{-x}$. Here $y_2' = u'e^{-x} - ue^{-x}$ and $y_2'' = u''e^{-x} - 2u'e^{-x} + ue^{-x}$. Substituting these into equation (4.2) we have

$$(1+x)(u''e^{-x} - 2u'e^{-x} + ue^{-x}) + x(u'e^{-x} - ue^{-x}) - ue^{-x} = 0.$$

Simplifying we obtain $u'' + xu'' - 2u' - xu' = 0$ or equivalently $u''(1+x) = u'(2+x)$. This is a separable differential equation. Separating the variables we have $\frac{u''}{u'} = \frac{2+x}{1+x}$ or $\frac{u''}{u'} = 1 + \frac{1}{1+x}$. Integrating we obtain $\ln|u'| = x + \ln|1+x| + k$, or equivalently $u' = c_1(1+x)e^x$. Integrating again to find u we have $u = c_1 xe^x + c_2$. We are free to choose the constants c_1 and c_2 as we wish, as long as $c_1 \neq 0$. Thus we choose $c_1 = 1$, $c_2 = 0$ and have $u = xe^x$. Thus, $y_2 = ue^{-x} = x$.

39. 15. $\displaystyle\lim_{x\to\infty}\left(-\frac{1}{2}e^{x-1} + \frac{1}{2}e^{1-x}\right) = -\infty$

  17. $\displaystyle\lim_{x\to\infty} -\frac{1}{2}e^{-2x}\sin 2x = 0$

  19. $\displaystyle\lim_{x\to\infty} -xe^{-\frac{1}{2}x} = 0$

  21. $\displaystyle\lim_{x\to\infty}\left(\frac{33}{25} + \frac{2}{5}x + \frac{2}{25}e^{-5x-5}\right) = \infty$

  23. $\displaystyle\lim_{x\to\infty}\left(\frac{2}{3} + \frac{1}{5}e^{-2x} + \frac{2}{15}e^{3x}\right) = \infty$

  25. $\displaystyle\lim_{x\to\infty}\left(\frac{7}{16}e^{-x} + \left(\frac{3}{4}x - \frac{7}{16}\right)e^{3x}\right) = \infty$

41. If $r_i < 0$ for every $i = 1, \ldots n$ the limit will be 0.

45. For the Euler type equation $x^2 y'' - 3xy' - 5y = 0$, $A = -3$ and $B = -5$. We let $z = \ln x$ and obtain

$$\frac{d^2 y}{dz^2} + (A-1)\frac{dy}{dz} + By = 0.$$

Here we have,

$$\frac{d^2}{dz^2} - 4\frac{dy}{dz} - 5y = 0. \tag{4.3}$$

This differential equation has the characteristic equation $\lambda^2 - 4\lambda + 5 = 0$, which has the roots $\lambda = -1, 5$. Thus, equation (4.3) has the solution $y = c_1 e^{-z} + c_2 e^{5z}$. Therefore the original differential equation has the solution $y = c_1 e^{-\ln x} + c_2 e^{5\ln x}$, or equivalently $y = c_1 \frac{1}{x} + c_2 x^5$.

47. For the Euler type equation $x^2 y'' - xy' + y = 0$, $A = -1$ and $B = 1$. We let $z = \ln x$ and obtain

$$\frac{d^2 y}{dz^2} + (A-1)\frac{dy}{dz} + By = 0.$$

Here we have,

$$\frac{d^2}{dz^2} - 2\frac{dy}{dz} + y = 0. \tag{4.4}$$

This differential equation has the characteristic equation $\lambda^2 - 2\lambda + 1 = 0$, which has the repeated root $\lambda = 1$. Thus, equation (4.4) has the solution $y = (c_1 + c_2 z)e^z$. Therefore the original differential equation has the solution $y = (c_1 + c_2 \ln x)e^{\ln x}$, or equivalently $y = (c_1 + c_2 \ln x)x$.

4.3   The Method of Undetermined Coefficients

Exercises 4.3, pp. 211

1. The differential equation $y'' - y' - 6y = 3e^{2x}$ has the corresponding homogeneous equation $y'' - y' - 6y = 0$. Its characteristic equation is $\lambda^2 - \lambda - 6 = (\lambda + 2)(\lambda - 3) = 0$. Consequently, it has the general solution

$$y_H = c_1 e^{-2x} + c_2 e^{3x}.$$

Since the right-hand side of the original differential equation is $3e^{2x}$ we look for a particular solution of the form $y_P = Ae^{2x}$, where A is a constant. Here, $y_P' = 2Ae^{2x}$ and $y_P'' = 4Ae^{2x}$. Substituting these into the original differential equation we obtain $y_P'' - y_P' - 6y_P = 3e^{2x}$ or equivalently, $4Ae^{2x} - 2Ae^{2x} - 6Ae^{2x} = 3e^{2x}$. More simply, $-4Ae^{2x} = 3e^{2x}$. Hence we obtain a solution if $-4A = 3$ or $A = -\frac{3}{4}$. Thus, $y_P = -\frac{3}{4}e^{2x}$ is a particular solution. The general solution to the nonhomogeneous equation is then $y = c_1 e^{-2x} + c_2 e^{3x} - \frac{3}{4}e^{2x}$.

3. The differential equation $y'' + 25y = x^2 + 25$ has the corresponding homogeneous equation $y'' + 25y = 0$. Its characteristic equation is $\lambda^2 + 25 = 0$, which has roots $\lambda = \pm 5i$. Consequently, it has the general solution $y_H = c_1 \sin 5x + c_2 \cos 5x$.

Since the right-hand side of the original differential equation is $x^2 + 25$ we look for a particular solution of the form

$$y_P = Ax^2 + Bx + C,$$

where A, B and C are constants. Here, $y_P' = 2Ax + B$ and $y_P'' = 2A$. Substituting these into the original differential equation we obtain $y_P'' + 25y_P = x^2 + 25$ or equivalently, $2A + 25(Ax^2 + Bx + C) = x^2 + 25$. For the polynomials on the left and right sides of this equation to be equal, the coefficients of the corresponding terms must be equal. We obtain the system of equations

$$25A = 1, \ 25B = 0, \ 2A + 25C = 25.$$

This system has the solution $A = \frac{1}{25}$, $B = 0$, $C = \frac{623}{625}$. Thus, $y_P = \frac{1}{25}x^2 + \frac{623}{625}$ is a particular solution. The general solution to the nonhomogeneous equation is then $y = c_1 \sin 5x + c_2 \cos 5x + \frac{623}{625} + \frac{1}{25}x^2$.

5. The differential equation $y'' - 4y' + 13y = 4e^x \sin 3x$ has the corresponding homogeneous equation $y'' - 4y' + 13y = 0$. Its characteristic equation is $\lambda^2 - 4\lambda - 13 = 0$, which has roots $\lambda = \frac{4 \pm \sqrt{16-52}}{2} = 2 \pm 3i$. Consequently, it has the general solution $y_H = c_1 e^{2x} \sin 3x + c_2 e^{2x} \cos 3x$. Since the right-hand side of the original differential equation is $4e^x \sin 3x$ we look for a particular solution of the form $y_P = Ae^x \sin 3x + Be^x \cos 3x$, where A and B are constants. Here,

$$\begin{aligned} y_P' &= Ae^x \sin 3x + 3Ae^x \cos 3x + Be^x \cos 3x - 3Be^x \sin 3x \quad \text{and} \\ y_P'' &= -8Ae^x \sin 3x + 6Ae^x \cos 3x - 8Be^x \cos 3x - 6Be^x \sin 3x. \end{aligned}$$

Substituting these into the original differential equation we obtain

$$-8Ae^x \sin 3x + 6Ae^x \cos 3x - 8Be^x \cos 3x - 6Be^x \sin 3x$$

$$-4(Ae^x \sin 3x + 3Ae^x \cos 3x + Be^x \cos 3x - 3Be^x \sin 3x) + 13(Ae^x \sin 3x + Be^x \cos 3x)$$

$$= 4e^x \sin 3x.$$

Equivalently, $(A + 6B)e^x \sin 3x + (-6A + B)e^x \cos 3x = 4e^x \sin 3x$. For the functions on the left and right sides of this equation to be equal, the coefficients of the corresponding terms must be equal. We obtain the system of equations

$$A + 6B = 4, \ -6A + B = 0.$$

This system has the solution $A = \frac{4}{37}$, $B = \frac{24}{37}$.

Thus, $y_P = \frac{4}{37}e^x \sin 3x + \frac{24}{37}e^x \cos 3x$ is a particular solution. The general solution to the nonhomogeneous equation is then $y = c_1 e^{2x} \sin 3x + c_2 e^{2x} \cos 3x + \frac{4}{37}e^x \sin 3x + \frac{24}{37}e^x \cos 3x$.

7. The differential equation $y'' + y' = e^x + \cos 2x - x$ has the corresponding homogeneous equation $y'' + y' = 0$. Its characteristic equation is $\lambda^2 + \lambda = \lambda(\lambda + 1) = 0$. Consequently, it has the general solution $y_H = c_1 e^{-x} + c_2$.

Since the right-hand side of the original differential equation is $e^x + \cos 2x - x$ we look for a particular solution of the form $y_P = Ae^x + B\cos 2x + C\sin 2x + Dx^2 + Ex$, where A, B, C, D and E are constants. Note, the higher powers on the terms $Dx^2 + Ex$ (instead of $Dx + E$) are required because constant functions are solutions to the corresponding homogeneous equation. Here, $y_P' = Ae^x - 2B\cos 2x + 2C\sin 2x + 2Dx + E$ and $y_P'' = Ae^x - 4B\cos 2x - 4C\sin 2x + 2D$. Substituting these into the original differential equation we obtain

(margin note: should be $4B\sin 2x$)

$$Ae^x - 4B\cos 2x - 4C\sin 2x + 2D + Ae^x - 2B\cos 2x + 2C\sin 2x + 2Dx + E = e^x + \cos 2x - x.$$

Equivalently, $2Ae^x + (2C - 4B)\cos 2x - (4C + 2B)\sin 2x + 2Dx + (2D + E) = e^x + \cos 2x - x$. For the functions on the left and right sides of this equation to be equal, the coefficients of the corresponding terms must be equal. We obtain the system of equations

$$2A = 1, \ 2C - 4B = 1, \ -(4C + 2B) = 0, \ 2D = -1, \ 2D + E = 0.$$

This system has the solution $A = \frac{1}{2}$, $B = -\frac{1}{5}$, $C = \frac{1}{10}$, $D = -\frac{1}{2}$, $E = 1$.

Thus, $y_P = \frac{1}{2}e^x - \frac{1}{5}\cos 2x + \frac{1}{10}\sin 2x - \frac{1}{2}x^2 + x$ is a particular solution. The general solution to the nonhomogeneous equation is then $y = c_1 e^{-x} + c_2 + x - \frac{1}{2}x^2 + \frac{1}{2}e^x + \frac{1}{10}\sin 2x - \frac{1}{5}\cos 2x$.

9. The differential equation $4y'' + 16y = 3\cos 2x$ has the corresponding homogeneous equation $y'' + 16y = 0$. Its characteristic equation is $\lambda^2 + 4 = 0$, which has roots $\lambda = \pm 2i$. Consequently, it has the general solution $y_H = c_1 \sin 2x + c_2 \cos 2x$.

Since the right-hand side of the original differential equation is $3\cos 2x$ and $\cos 2x$ is a solution of the corresponding homogeneous equation, we look for a particular solution of the form $y_P = Ax\cos 2x + Bx\sin 2x$, where A and B are constants. Here,

$$\begin{aligned} y_P' &= A\cos 2x - 2Ax\sin 2x + B\sin 2x + 2Bx\cos 2x \quad \text{and} \\ y_P'' &= -2A\sin 2x - 2A\sin 2x - 4Ax\cos 2x + 2B\cos 2x + 2B\cos 2x - 4Bx\sin 2x \\ &= -4A\sin 2x 4Ax\cos 2x + 4B\cos 2x + 2B\cos 2x - 4Bx\sin 2x. \end{aligned}$$

Substituting these into the original differential equation we obtain

$$4(-4A\sin 2x 4Ax\cos 2x + 4B\cos 2x + 2B\cos 2x - 4Bx\sin 2x) + 16(Ax\cos 2x + Bx\sin 2x) = 3\cos 2x.$$

Equivalently, $-16A\sin 2x + 16B\cos 2x = 3\cos 2x$. For the functions on the left and right sides of this equation to be equal, the coefficients of the corresponding terms must be equal. We obtain the system of equations

$$-16A = 0, \ 16B = 3.$$

This system has the solution $A = 0$, $B = \frac{3}{16}$.

Thus, $y_P = \frac{3}{16}x\sin 2x$ is a particular solution. The general solution to the nonhomogeneous equation is then

$$y = c_1 \sin 2x + c_2 \cos 2x + \frac{3}{16}x\sin 2x.$$

11. The differential equation $y'' + 8y' = 2x^2 - 7x + 3$ has the corresponding homogeneous equation $y'' + 8y' = 0$. Its characteristic equation is $\lambda^2 + 8\lambda = \lambda(\lambda + 8) = 0$. Consequently, it has the general solution $y_H = c_1 e^{-8x} + c_2$.

Since the right-hand side of the original differential equation is $2x^2 - 7x + 3$ and the constant function 1 is a solution of the corresponding homogeneous equation, we look for a particular solution of the form $y_P = Ax^3 + Bx^2 + Cx$, where A, B and C are constants. Here, $y_P' = 3Ax^2 + 2Bx + C$ and $y_P'' = 6Ax + 2B$. Substituting these into the original differential equation we obtain $6Ax + 2B + 8(3Ax^2 + 2Bx + C) = 2x^2 - 7x + 3$, or equivalently $24Ax^2 + (6A + 16B)x + (2B + 8C) = 2x^2 - 7x + 3$. For the polynomials on the left and right sides of this equation to be equal, the coefficients of the corresponding terms must be equal. We obtain the system of equations

$$24A = 2, \ 6A + 16B = -7, \ 2B + 8C = 3.$$

This system has the solution $A = \frac{1}{12}$, $B = -\frac{15}{32}$, $C = \frac{63}{128}$. Thus, $y_P = \frac{1}{12}x^3 - \frac{15}{32}x^2 + \frac{63}{128}x$ is a particular solution. The general solution to the nonhomogeneous equation is then

$$y = c_1 e^{-8x} + c_2 + \frac{1}{12}x^3 - \frac{15}{32}x^2 + \frac{63}{128}x.$$

13. Consider the differential equation $y'' - 4y' + 13y = 4e^{2x}\sin 3x$. As we saw in problem 5 the corresponding homogeneous equation has the solution $y_H = c_1 e^{2x}\sin 3x + c_2 e^{2x}\cos 3x$. Since $e^{2x}\sin 3x$ is a solution to the corresponding homogeneous equation, the particular solution will be of the form $y_P = Axe^{2x}\sin 3x + Bxe^{2x}\cos 3x$. Here,

$$
\begin{aligned}
y_P' &= Ae^{2x}\sin 3x + 2Axe^{2x}\sin 3x + 3Axe^{2x}\cos 3x + Be^{2x}\cos 3x + 2Bxe^{2x}\cos 3x - 3Bxe^{2x}\sin 3x \qquad \text{and}\\
y_P'' &= 4Ae^{2x}\sin 3x + 6Ae^{2x}\cos 3x + 4Axe^{2x}\sin 3x + 12Axe^{2x}\cos 3x - 9Axe^{2x}\sin 3x + 4Be^{2x}\cos 3x\\
&\quad -6Be^{2x}\sin 3x + 4Bxe^{2x}\cos 3x - 12Bxe^{2x}\sin 3x - 9Bxe^{2x}\cos 3x.
\end{aligned}
$$

Substituting these into the original differential equation we obtain

$$
4Ae^{2x}\sin 3x + 6Ae^{2x}\cos 3x + 4Axe^{2x}\sin 3x + 12Axe^{2x}\cos 3x - 9Axe^{2x}\sin 3x
$$

$$
+4Be^{2x}\cos 3x - 6Be^{2x}\sin 3x + 4Bxe^{2x}\cos 3x - 12Bxe^{2x}\sin 3x - 9Bxe^{2x}\cos 3x
$$

$$
-4(Ae^{2x}\sin 3x + 2Axe^{2x}\sin 3x + 3Axe^{2x}\cos 3x + Be^{2x}\cos 3x + 2Bxe^{2x}\cos 3x - 3Bxe^{2x}\sin 3x)
$$

$$
+13(Axe^{2x}\sin 3x + Bxe^{2x}\cos 3x) = 4e^{2x}\sin 3x.
$$

Simplifying we obtain $6Ae^{2x}\cos 3x - 6Be^{2x}\sin 3x = 4e^{2x}\sin 3x$. For the functions on the left and right sides of this equation to be equal, the coefficients of the corresponding terms must be equal. We obtain the system of equations

$$
6A = 0, \ -6B = 4.
$$

This system has the solution $A = 0$, $B = -\frac{2}{3}$.

Thus, $y_P = -\frac{2}{3}xe^{2x}\cos 3x$ is a particular solution. The general solution to the nonhomogeneous equation is then

$$
y = c_1 e^{2x}\sin 3x + c_2 e^{2x}\cos 3x - \frac{2}{3}xe^{2x}\cos 3x.
$$

15. The differential equation $y''' - 3y'' - 4y' = x - e^{4x}$ has the corresponding homogeneous equation $y''' - 3y'' - 4y' = 0$. Its characteristic equation is $\lambda^3 - 3\lambda^2 - 4\lambda = (\lambda + 1)(\lambda - 4)\lambda = 0$. Consequently, it has the general solution

$$
y_H = c_1 e^{-x} + c_2 e^{4x} + c_3.
$$

Since the right-hand side of the original differential equation is $x - e^x$ we look for a particular solution of the form

$$
y_P = Ax^2 + Bx + Cxe^{4x},
$$

where A, B and C are constants. Here,

$$
\begin{aligned}
y_P' &= 2Ax + B + Ce^{4x} + 4Cxe^{4x}\\
y_P'' &= 2A + 8Ce^{4x} + 16Cxe^{4x} \qquad \text{and}\\
y_P''' &= 48Ce^{4x} + 64Cxe^{4x}.
\end{aligned}
$$

Substituting these into the original differential equation we obtain

$$
48Ce^{4x} + 64Cxe^{4x} - 3(2A + 8Ce^{4x} + 16Cxe^{4x}) - 4(2Ax + B + Ce^{4x} + 4Cxe^{4x}) = x - e^{4x}.
$$

Simplifying we obtain $20Ce^{4x} - 8Ax - 6A - 4B$. For the functions on the left and right sides of this equation to be equal, the coefficients of the corresponding terms must be equal. We obtain the system of equations

$$
-8A = 1, \ -6A - 4B = 0, \ 20C = -1.
$$

This system has the solution $A = -\frac{1}{8}$, $B = \frac{3}{16}$, $C = -\frac{1}{20}$.

Thus, $y_P = -\frac{1}{8}x^2 + \frac{3}{16}x - \frac{1}{20}xe^{4x}$ is a particular solution. The general solution to the nonhomogeneous equation is then $y = c_1 e^{-x} + c_2 e^{4x} + c_3 - \frac{1}{8}x^2 + \frac{3}{16}x - \frac{1}{20}xe^{4x}$.

17. The differential equation $y'' - 2y' - 8y = x$ has the corresponding homogeneous equation $y'' - 2y' - 8y = 0$. Its characteristic equation is $\lambda^2 - 2\lambda - 8 = (\lambda - 4)(\lambda + 2) = 0$. Consequently, it has the general solution $y_H = c_1 e^{4x} + c_2 e^{-2x}$.

Since the right-hand side of the original differential equation is x we look for a particular solution of the form $y_P = Ax + B$, where A and B are constants. Here, $y_P' = A$ and $y_P'' = 0$. Substituting these into the original differential equation we obtain $-2A - 8(Ax + B) = x$ or equivalently $-8Ax - 2A - 8B = x$. For the polynomials on the left and right sides of this equation to be equal, the coefficients of the corresponding terms must be equal. We obtain the system of equations

$$-8A = 1, \ -2A - 8B = 0.$$

This system has the solution $A = -\frac{1}{8}$, $B = \frac{1}{32}$. Thus, $y_P = -\frac{1}{8}x + \frac{1}{32}$ is a particular solution. The general solution to the nonhomogeneous equation is then

$$y = \frac{1}{32} - \frac{1}{8}x + c_1 e^{4x} + c_2 e^{-2x}. \tag{4.5}$$

We now consider the initial conditions $y(0) = -2$, $y'(0) = 2$. From equation (4.5) we have

$$y' = -\frac{1}{8} + 4c_1 e^{4x} - 2c_2 e^{-2x}.$$

Using the initial conditions we obtain the system of equations $-2 = \frac{1}{32} + c_1 + c_2$ and $2 = -\frac{1}{8} + 4c_1 - 2c_2$ or, more simply,

$$c_1 + c_2 = -\frac{65}{32}, \ 4c_1 - 2c_2 = \frac{17}{8}.$$

This system has the solution $c_1 = -\frac{31}{96}$, $c_2 = -\frac{41}{24}$. Thus the solution to the initial value problem is

$$y = -\frac{1}{32} - \frac{1}{8}x - \frac{31}{96}e^{4x} - \frac{41}{24}e^{-2x}.$$

19. The differential equation $y'' - y = 5e^x$ has the corresponding homogeneous equation $y'' - y = 0$. Its characteristic equation is $\lambda^2 - 1 = (\lambda - 1)(\lambda + 1) = 0$. Consequently, it has the general solution $y_H = c_1 e^x + c_2 e^{-x}$.

Since the right-hand side of the original differential equation is $5e^x$ and e^x is a solution to the corresponding homogeneous equation, we look for a particular solution of the form $y_P = Axe^x$, where A is a constant. Here, $y_P' = Ae^x + Axe^x$ and $y_P'' = 2Ae^x + Axe^x$. Substituting these into the original differential equation we obtain $2Ae^x + Axe^x - Axe^x = 5e^x$ or equivalently $2Ae^x = 5e^x$. For these two equation to be equal we must have $A = \frac{5}{2}$. Thus, $y_P = \frac{5}{2}xe^x$ is a particular solution. The general solution to the nonhomogeneous equation is then

$$y = \frac{5}{2}xe^x + c_1 e^x + c_2 e^{-x}. \tag{4.6}$$

We now consider the initial conditions $y(1) = 0$, $y'(1) = -1$. From equation (4.6) we have

$$y' = \frac{5}{2}e^x + \frac{5}{2}xe^x + c_1 e^x - c_2 e^{-x}.$$

Using the initial conditions we obtain the system of equations $0 = \frac{5}{2}e + c_1 e + c_2 \frac{1}{e}$ and $-1 = 5e + c_1 e - c_2 \frac{1}{e}$ or,

$$c_1 e + c_2 \frac{1}{e} = -\frac{5}{2}e, \ c_1 e - c_2 \frac{1}{e} = -1 - 5e.$$

This system has the solution $c_1 = -\frac{2 + 15e}{4e}$, $c_2 = \frac{5e^2 + 2e}{4}$. Thus the solution to the initial value problem is

$$y = \frac{5}{2}xe^x - \frac{2 + 15e}{4e}e^x + \frac{5e^2 + 2e}{4}e^{-x}.$$

21. The differential equation $y''' - 4y'' - 5y' = 7 - 2\cos x$ has the corresponding homogeneous equation $y''' - 4y'' - 5y' = 0$. Its characteristic equation is $\lambda^3 - 4\lambda^2 - 5\lambda = \lambda(\lambda^2 - 4\lambda - 5) = 0$. The quadratic factor has roots

$$\lambda = \frac{4 \pm \sqrt{16 - 20}}{2} = 2 \pm i.$$

Consequently, it has the general solution $y_H = c_1e^{2x}\cos x + c_2e^{2x}\sin x + c_3$. Since the right-hand side of the original differential equation is $7 - 2\cos x$ we look for a particular solution of the form $y_P = Ax + B\cos x + C\sin x$, where A, B and C are constants. Here, $y_P' = A - B\sin x + C\cos x$, $y_P'' = -B\cos x - C\sin x$ and $y_P''' = B\sin x - C\cos x$. Substituting these into the original differential equation we obtain

$$B\sin x - C\cos x - 4(-B\cos x - C\sin x) - 5A - B\sin x + C\cos x) = 7 - 2\cos x.$$

Simplifying we have $(4C - 4B)\sin x + (4B + 4C)\cos x + 5A = 7 - 2\cos x$. For the functions on the left and right sides of this equation to be equal, the coefficients of the corresponding terms must be equal. We obtain the system of equations

$$5A = 7,\ 4C - 4B = 0,\ 4B + 4C = -2.$$

This system has the solution $A = \frac{7}{5}$, $B = -\frac{1}{4}$, $C = -\frac{1}{4}$.

Thus, $y_P = -\frac{1}{4}\cos x - \frac{1}{4}\sin x + \frac{7}{5}x$ is a particular solution. The general solution to the nonhomogeneous equation is then

$$y = -\frac{1}{4}\cos x - \frac{1}{4}\sin x + \frac{7}{5}x + c_1e^{2x}\cos x + c_2e^{2x}\sin x + c_3. \tag{4.7}$$

We now consider the initial conditions $y(0) = 1$, $y'(0) = 0$, $y''(0) = 2$. From equation (4.7) we have

$$y' = \frac{1}{4}\sin x - \frac{1}{4}\cos x + \frac{7}{5} + 2c_1e^{2x}\cos x - c_1e^{2x}\sin x + 2c_2e^{2x}\sin x + c_2e^{2x}\cos x \qquad \text{and}$$

$$y'' = \frac{1}{4}\cos x + \frac{1}{4}\sin x + 3c_1e^{2x}\cos x - 4c_1e^{2x}\sin x + 3c_2e^{2x}\sin x + 4c_2e^{2x}\cos x.$$

Using the initial conditions we obtain the system of equations $1 = -\frac{1}{4} + c_1 + c_3$, $0 = -\frac{1}{4} + \frac{7}{5} + 2c_1 + c_2$ and $2 = \frac{1}{4} + 3c_1 + 4c_2$ or, more simply,

$$c_1 + c_3 = \frac{5}{4},\ 2c_1 + c_2 = -\frac{23}{20},\ 3c_1 + 4c_2 = \frac{7}{4}.$$

This system has the solution $c_1 = -\frac{127}{100}$, $c_2 = \frac{139}{100}$, $c_3 = \frac{63}{25}$. Thus the solution to the initial value problem is

$$y = -\frac{1}{4}\cos x - \frac{1}{4}\sin x + \frac{7}{5}x - \frac{127}{100}e^{2x}\cos x + \frac{139}{100}e^{2x}\sin x + \frac{63}{25}.$$

23. The right-hand side of the differential equation $y'' + y = 2x^2 + 2x\cos x$ is $2x^2 + 2x\cos x$. Without first solving the corresponding homogeneous equation we see that the particular solution would require terms of the form $Ax^2 + Bx + C$ because of the polynomial term. The particular solution would also require terms of the form $(Dx + E)\sin x + (Fx + G)\cos x$ because of the term $2x\cos x$.

However, we also need to understand the solution of the corresponding homogeneous equation before constructing a correct particular solution. The characteristic equation of the corresponding homogeneous equation is $\lambda^2 + 1 = 0$, which has roots $\lambda = \pm i$. Consequently the corresponding homogeneous equation has the general solution

$$y_H = c_1\cos x + c_2\sin x.$$

Since $\cos x$ and $\sin x$ are solutions, we must multiply each of the proposed trigonmetric terms by x. Thus, the particular solution will be of the form $y_P = Ax^2 + Bx + C + (Dx^2 + Ex)\sin x + (Fx^2 + Gx)\cos x$.

25. The right-hand side of the differential equation

$$y''' + y'' + 9y' + 9y = 4e^{-x} + 3x\sin 3x - 6\cos 3x$$

is $4e^{-x} + 3x\sin 3x - 6\cos 3x$. Without first solving the corresponding homogeneous equation we see that the particular solution would require terms of the form $(Ax + B)\sin 3x + (Cx + D)\cos 3x$ because of the two trigonometric terms. The particular solution would also require a term of the form Ee^{-x} because of the exponential term.

However, we also need to understand the solution of the corresponding homogeneous equation before constructing a correct particular solution. The characteristic equation of the corresponding homogeneous equation is

$$\lambda^3 + \lambda^2 + 9\lambda + 9 = (\lambda + 1)(\lambda^2 + 9) = 0,$$

which has roots $\lambda = -1$ and $\pm 3i$. Consequently the corresponding homogeneous equation has the general solution

$$y_H = c_1 e^{-x} + c_2 \cos x + c_3 \sin x.$$

Since $\cos x$ and $\sin x$ are solutions, we must multiply each of the proposed trigonmetric terms by x. Similarly, since e^{-x} is a solution, we must multiply the proposed exponential term by x. Thus, the particular solution will be of the form $y_P = (Ax^2 + Bx) \sin 3x + (Cx^2 + Dx) \cos 3x + Exe^{-x}$.

4.4   The Method of Variation of Parameters

Exercises 4.4, pp. 217

1. We are given the differential equation

$$y'' - y = 3x^2 - 1 \tag{4.8}$$

and are asked to find a particular solution using both the method of undetermined coefficients and the method of variation of parameters. For either method we first need to solve the corresponding homogeneous equation $y'' - y = 0$. This equation has the solution

$$y_H = c_1 e^x + c_2 e^{-x}.$$

(a) Now using the method of undetermined coefficients we look for a particular solution of the form

$$y_P = Ax^2 + Bx + C.$$

Here, $y'_P = 2Ax + B$ and $y''_P = 2A$. Substituting these in equation (4.8) we have $2A - (Ax^2 + Bx + C) = 3x^2 - 1$. Equating the coefficients of the polynomials on the left and right above, we obtain the system of equations

$$-A = 3, \ -B = 0, \ 2A - C = -1$$

which has the solution $A = -3$, $B = 0$, $C = -5$. Therefore a particular solution is $y_P = -3x^2 - 5$.

(b) We now use the method of variation of parameters. We've already found a fundamental set of solutions

$$y_1 = e^x, \ y_2 = e^{-x}.$$

We have

$$w = \begin{vmatrix} e^x & e^{-x} \\ e^x & -e^{-x} \end{vmatrix} = 2,$$

$$w_1 = \begin{vmatrix} 0 & e^{-x} \\ \frac{3x^2 - 1}{1} & -e^{-x} \end{vmatrix} = -(3x^2 - 1)e^{-x}$$

and

$$w_2 = \begin{vmatrix} e^x & 0 \\ e^x & \frac{3x^2 - 1}{1} \end{vmatrix} = (3x^2 - 1)e^x.$$

Using the formulas for u_1 and u_2 and integration by parts we have,

$$u_1 = \int \frac{w_1}{w} \, dx = \frac{1}{2} \int \left(3x^2 - 1\right) e^{-x} \, dx = \left(-\frac{3}{2}x^2 - 3x - \frac{5}{2}\right) e^{-x}$$

and

$$u_2 = \int \frac{w_2}{w} \, dx = -\frac{1}{2} \int \left(3x^2 - 1\right) e^x \, dx = \left(-\frac{3}{2}x^2 + 3x - \frac{5}{2}\right) e^x.$$

Therefore a particular solution is

$$y_P = \left(\left(-\frac{3}{2}x^2 - 3x - \frac{5}{2}\right) e^{-x}\right) e^x + \left(\left(-\frac{3}{2}x^2 + 3x - \frac{5}{2}\right) e^x\right) e^{-x} = -3x^2 - 5.$$

3. We are given the differential equation $y'' + 6y' + 9y = 3e^{3x}$. There is the single repeated root of the characteristic equation, $\lambda = -3$. We use $y_1 = e^{-3x}$, $y_2 = xe^{-3x}$ for a fundamental set of solutions. We have

$$w = \begin{vmatrix} e^{-3x} & xe^{-3x} \\ -3e^{-3x} & (1-3x)e^{-3x} \end{vmatrix} = e^{-6x},$$

$$w_1 = \begin{vmatrix} 0 & xe^{-3x} \\ 3e^{3x} & (1-3x)e^{-3x} \end{vmatrix} = -3x,$$

and

$$w_2 = \begin{vmatrix} e^{-3x} & 0 \\ -3e^{-3x} & 3e^{3x} \end{vmatrix} = 3.$$

Using the formulas for u_1 and u_2 and integration by parts we have,

$$u_1 = \int \frac{w_1}{w}\, dx = -\int 3xe^{6x}\, dx = \left(-\frac{1}{2}x + \frac{1}{12}\right)e^{6x}$$

and

$$u_2 = \int \frac{w_2}{w}\, dx = \int 3e^{6x}\, dx = \frac{1}{2}e^{6x}.$$

Therefore a particular solution is

$$y_P = \left(\left(-\frac{1}{2}x + \frac{1}{12}\right)e^{6x}\right)e^{-3x} + \left(\frac{1}{2}e^{6x}\right)xe^{-3x} = \frac{1}{12}e^{3x}.$$

Thus, the general solution to the original differential equation is

$$y = c_1 e^{-3x} + c_2 xe^{-3x} + \frac{1}{12}e^{3x}.$$

5. We are given the differential equation $y'' + 4y = 5\csc 2x$. The roots of the characteristic equation are $\pm 2i$. We use

$$y_1 = \cos 2x, \ y_2 = \sin 2x$$

for a fundamental set of solutions. We have

$$w = \begin{vmatrix} \cos 2x & \sin 2x \\ -2\sin 2x & 2\cos 2x \end{vmatrix} = 2,$$

$$w_1 = \begin{vmatrix} 0 & \sin 2x \\ 5\csc 2x & 2\cos 2x \end{vmatrix} = -5,$$

and

$$w_2 = \begin{vmatrix} \cos 2x & 0 \\ -2\sin 2x & 5\csc 2x \end{vmatrix} = 5\cot 2x.$$

Using the formulas for u_1 and u_2,

$$u_1 = \int \frac{w_1}{w}\, dx = -\int \frac{5}{2}\, dx = -\frac{5}{2}x$$

and

$$u_2 = \int \frac{w_2}{w}\, dx = \int \frac{5\cot 2x}{2} = \frac{5}{4}\ln|\sin 2x|.$$

Therefore a particular solution is

$$y_P = \left(-\frac{5}{2}x\right)\cos 2x + \left(\frac{5}{4}\ln|\sin 2x|\right)\sin 2x.$$

Thus, the general solution to the original differential equation is

$$y = c_1 \cos 2x + c_2 \sin 2x - \frac{5}{2}x\cos 2x + \frac{5}{4}\sin 2x \ln|\sin 2x|.$$

7. We are given the differential equation $y'' - y' - 2y = e^x \cos x$. The roots of the characteristic equation are -1 and 2. We use $y_1 = e^{-x}$, $y_2 = e^{2x}$ for a fundamental set of solutions. We have

$$w = \begin{vmatrix} e^{-x} & e^{2x} \\ -e^{-x} & 2e^{2x} \end{vmatrix} = 3e^x,$$

$$w_1 = \begin{vmatrix} 0 & e^{2x} \\ e^x \cos x & 2e^{2x} \end{vmatrix} = -e^{3x} \cos x,$$

and

$$w_2 = \begin{vmatrix} e^{-x} & 0 \\ -e^{-x} & e^x \cos x \end{vmatrix} = \cos x.$$

Using the formulas for u_1 and u_2 and integration by parts we obtain,

$$u_1 = \int \frac{w_1}{w} \, dx = \int \frac{-e^{3x} \cos x}{3e^x} \, dx = -\frac{1}{3} \int e^{2x} \cos x \, dx = -\frac{1}{15} e^{2x} \sin x - \frac{2}{15} e^{2x} \cos x$$

and

$$u_2 = \int \frac{w_2}{w} \, dx = \int \frac{\cos x}{3e^x} = \frac{1}{3} \int e^{-x} \cos x = \frac{1}{6} e^{-x} \sin x - \frac{1}{6} e^{-x} \cos x.$$

Therefore a particular solution is

$$y_P = \left(-\frac{1}{15} e^{2x} \sin x - \frac{2}{15} e^{2x} \cos x \right) e^{-x} + \left(\frac{1}{6} e^{-x} \sin x - frac16 e^{-x} \cos x \right) e^{2x} = \frac{1}{10} e^x \sin x - \frac{3}{10} e^x \cos x.$$

Thus, the general solution to the original differential equation is

$$y = c_1 e^{-x} + c_2 e^{2x} + \frac{1}{10} e^x \sin x - \frac{3}{10} e^x \cos x.$$

9. We are given the differential equation $y''' + 4y'' - y' - 4y = e^{-2x}$. The roots of the characteristic equation are 1, -4 and 1. We use $y_1 = e^{-x}$, $y_2 = e^{-4x}$, $y_3 = e^x$ for a fundamental set of solutions. We have

$$w = \begin{vmatrix} e^{-x} & e^{-4x} & e^x \\ -e^{-x} & -4e^{-4x} & e^x \\ e^{-x} & 16e^{-4x} & e^x \end{vmatrix} = 30e^{-4x},$$

$$w_1 = \begin{vmatrix} 0 & e^{-4x} & e^x \\ 0 & -4e^{-4x} & e^x \\ e^{-2x} & 16e^{-4x} & e^x \end{vmatrix} = 3e^{-7x},$$

$$w_2 = \begin{vmatrix} e^{-x} & 0 & e^x \\ -e^{-x} & 0 & e^x \\ e^{-x} & e^{-2x} & e^x \end{vmatrix} = 2e^{-2x},$$

and

$$w_3 = \begin{vmatrix} e^{-x} & e^{-4x} & 0 \\ -e^{-x} & -4e^{-4x} & 0 \\ e^{-x} & 16e^{-4x} & e^{-2x} \end{vmatrix} = -5e^{-5x}.$$

Using the formulas for u_1, u_2 and u_3,

$$u_1 = \int \frac{w_1}{w} \, dx = \int \frac{3e^{-7x}}{30e^{-4x}} \, dx = \frac{1}{10} \int e^{-3x} \, dx = -\frac{1}{30} e^{-3x},$$

$$u_2 = \int \frac{w_2}{w} \, dx = \int \frac{2e^{-2x}}{30e^{-4x}} \, dx = \frac{1}{15} \int e^{2x} \, dx = \frac{1}{30} e^{2x}$$

and

$$u_3 = \int \frac{w_3}{w} \, dx = \int \frac{-5e^{-5x}}{30e^{-4x}} \, dx = -\frac{1}{6} \int e^{-x} \, dx = \frac{1}{6} e^{-x}.$$

Therefore a particular solution is

$$y_P = \left(-\frac{1}{30}e^{-3x}\right)e^x + \left(\frac{1}{30}e^{2x}\right)e^{-4x} + \left(\frac{1}{6}e^{-x}\right)e^{-x} = \frac{1}{6}e^{-2x}.$$

Thus, the general solution to the original differential equation is

$$y = c_1e^x + c_2e^{-4x} + c_3e^{-x} + \frac{1}{6}e^{-2x}.$$

11. We are given the differential equation $x^2y'' - xy' = x^3$. From Exercise 31 in Section 4.2 we know that $y_1 = x^2$, and $y_2 = 1$ form a fundamental set of solutions. We have

$$w = \begin{vmatrix} x^2 & 1 \\ 2x & 0 \end{vmatrix} = -2x,$$

$$w_1 = \begin{vmatrix} 0 & 1 \\ \frac{x^3}{x^2} & 0 \end{vmatrix} = -x,$$

and

$$w_2 = \begin{vmatrix} x^2 & 0 \\ 2x & \frac{x^3}{x^2} \end{vmatrix} = x^3.$$

Using the formulas for u_1 and u_2,

$$u_1 = \int \frac{w_1}{w}\,dx = \int \frac{-x}{-2x}\,dx = \int \frac{1}{2}\,dx = \frac{1}{2}x$$

and

$$u_2 = \int \frac{w_2}{w}\,dx = \int \frac{x^3}{-2x}\,dx = -\frac{1}{2}\int x^2\,dx = -\frac{1}{6}x^3.$$

Therefore a particular solution is

$$y_P = \left(\frac{1}{2}x\right)x^2 + \left(-\frac{1}{6}x^3\right)1 = \frac{1}{3}x^3.$$

13. We are given the differential equation $(1+x)y'' + xy' - y = (1+x)^2e^x$. From Exercise 33 in Section 4.2 we know that $y_1 = e^{-x}$, $y_2 = x$ form a fundamental set of solutions. We have

$$w = \begin{vmatrix} e^{-x} & x \\ -e^{-x} & 1 \end{vmatrix} = (1+x)e^x,$$

$$w_1 = \begin{vmatrix} 0 & x \\ \frac{(1+x)^2e^x}{1+x} & 1 \end{vmatrix} = -x(1+x)e^x$$

and

$$w_2 = \begin{vmatrix} e^{-x} & 0 \\ -e^{-x} & \frac{(1+x)^2e^x}{1+x} \end{vmatrix} = 1+x.$$

Using the formulas for u_1 and u_2 and integration by parts we have,

$$u_1 = \int \frac{w_1}{w}\,dx = \int \frac{-x(1+x)e^x}{(1+x)e^{-x}}\,dx = -\int xe^{2x}\,dx = e^{2x}\left(\frac{1}{4} - \frac{1}{2}x\right)$$

and

$$u_2 = \int \frac{w_2}{w}\,dx = \int \frac{1+x}{(1+x)e^{-x}}\,dx = \int e^x\,dx = e^x.$$

Therefore a particular solution is

$$y_P = e^{2x}\left(\frac{1}{4} - \frac{1}{2}x\right)e^{-x} + e^x \cdot x = \frac{1}{4}e^x(2x+1).$$

15. We are given the differential equation $x^3y''' + x^2y'' - 2xy' + 2y = -5x^3$, $x > 0$ and the functions $y_1 = \frac{1}{x}$, $y_2 = x$, and $y_3 = x^2$. The corresponding homogeneous equation is

$$x^3y''' + x^2y'' - 2xy' + 2y = 0 \qquad (4.9)$$

Here, $y_1' = -x^{-2}$, $y_1'' = 2x^{-3}$ and $y_1''' = -6x^{-4}$. Substituting these into equation (4.9) we have

$$x^3(-6x^{-4}) + x^2(2x^{-3}) - 2x(-x^{-2}) + 2(x^{-1}) = 0.$$

Thus, y_1 is a solution to equation (4.9).

Next, $y_2' = 1$ and $y_2'' = y_2''' = 0$. Again, substituting these into equation (4.9) we have

$$x^3(0) + x^2(0) - 2x(1) + 2(x) = 0.$$

Thus, y_2 is a solution to equation (4.9).

Finally, $y_3' = 2x$, $y_3'' = 2$ and $y_3''' = 0$. Substituting these into equation (4.9) we have

$$x^3(0) + x^2(2) - 2x(2x) + 2(x^2) = 0,$$

so y_3 is also a solution to equation (4.9).

We next compute the Wronskian.

$$w\left(\frac{1}{x}, x, x^2\right) = \begin{vmatrix} x^{-1} & x & x^2 \\ -x^{-2} & 1 & 2x \\ 2x^{-3} & 0 & 2 \end{vmatrix} = 6x^{-1} \neq 0 \text{ for } x > 0.$$

Therefore, y_1, y_2 and y_3 form a fundamental set of solutions for equation (4.9). We have

$$w_1 = \begin{vmatrix} 0 & x & x^2 \\ 0 & 1 & 2x \\ \frac{-5x^3}{x^3} & 0 & 2 \end{vmatrix} = -5x^2,$$

$$w_2 = \begin{vmatrix} x^{-1} & 0 & x^2 \\ -x^{-2} & 0 & 2x \\ 2x^{-3} & -5 & 2 \end{vmatrix} = 15,$$

and

$$w_3 = \begin{vmatrix} x^{-1} & x & 0 \\ -x^{-2} & 1 & 0 \\ 2x^{-3} & 0 & -5 \end{vmatrix} = -10x^{-1}.$$

Using the formulas for u_1, u_2 and u_3,

$$u_1 = \int \frac{w_1}{w} \, dx = \int \frac{-5x^2}{6x^{-1}} \, dx = -\frac{5}{6} \int x^3 \, dx = -\frac{5}{24}x^4,$$

$$u_2 = \int \frac{w_2}{w} \, dx = \int \frac{15}{6x^{-1}} \, dx = \frac{5}{2} \int x \, dx = \frac{5}{4}x^2$$

and

$$u_3 = \int \frac{w_3}{w} \, dx = \int \frac{-10x^{-1}}{6x^{-1}} \, dx = -\frac{5}{3} \int dx = -\frac{5}{3}x.$$

Therefore a particular solution is

$$y_P = \left(-\frac{5}{24}x^4\right)x^{-1} + \left(\frac{5}{4}x^2\right)x + \left(-\frac{5}{3}x\right)x^2 = -\frac{5}{8}x^3.$$

4.5   Some Applications of Higher Order Differential Equations

Exercises 4.5, pp. 228-229

1. We immediately have that $m = 4$ and $f = 0$. Using $g = 10$ m/sec^2 gives us

$$k = \frac{mg}{l} = \frac{4(10)}{0.1} = 400.$$

Since the spring is stretched and additional 50 cm and it is released from rest, the initial value problem describing the motion of the mass is then

$$4u'' + 400u = 0, \ u(0) = 0.5, \ u'(0) = 0.$$

The characteristic equation $4\lambda^2 + 400 = 0$ has roots $\lambda = \pm 10i$. The general solution is given by

$$u(t) = c_1 \cos 10t + c_2 \sin 10t.$$

We also have $u'(t) = -10c_1 \sin 10t + 10c_2 \cos 10t$. Using the initial conditions, we obtain

$$u(0) = c_1 = 0.5 \text{ and } u'(0) = 10c_2 = 0.$$

Therefore, the position of the object is given by

$$u(t) = 0.5 \cos 10t.$$

The frequency is 10 and the amplitude is 0.5.

3. Since $m = \frac{w}{g}$, here we have $m = \frac{2}{32}$. Since the friction has been eliminated $f = 0$. The spring constant

$$k = \frac{w}{l} = \frac{2}{0.5} = 4.$$

Since the spring is given an initial velocity of 2 ft/sec, the initial value problem describing the motion of the mass is then

$$\frac{1}{16}u'' + 4u = 0, \ u(0) = 0, \ u'(0) = 2.$$

The characteristic equation $\frac{1}{16}\lambda^2 + 4 = 0$ has roots $\lambda = \pm 8i$. The general solution is given by

$$u(t) = c_1 \cos 8t + c_2 \sin 8t.$$

We also have $u'(t) = -8c_1 \sin 8t + 8c_2 \cos 8t$. Using the initial conditions, we obtain

$$u(0) = c_1 = 0 \text{ and } u'(0) = 8c_2 = 2.$$

Therefore, the position of the object is given by

$$u(t) = 0.25 \sin 8t.$$

The frequency is 8 and the amplitude is 0.25.

5. Since $m = \frac{w}{g}$, here we have $m = \frac{8}{32}$. It is given that the friction constant is 1 lb-sec/ft. The spring constant

$$k = \frac{w}{l} = \frac{8}{4} = 2.$$

Since the spring is stretched an additional 6 inches and released from rest, the initial value problem describing the motion of the mass is then

$$\frac{1}{4}u'' + u' + 2u = 0, \ u(0) = 0.5, \ u'(0) = 0.$$

The characteristic equation $\frac{1}{4}\lambda^2 + \lambda + 2 = 0$ has roots $\lambda = -2 \pm 2i$. The general solution is given by

$$u(t) = e^{-2t}(c_1 \cos 2t + c_2 \sin 2t).$$

We also have

$$u'(t) = -2c_1 e^{-2t} \cos 2t - 2c_1 e^{-2t} \sin 2t - 2c_2 e^{-2t} \sin 2t + 2c_2 \cos 2t.$$

Using the initial conditions, we obtain $u(0) = c_1 = 0.5$ and $u'(0) = -2c_1 + 2c_2 = 0$. The solution to this system of equations is $c_1 = c_2 = 0.5$. Therefore, the position of the object is given by

$$u(t) = 0.5e^{-2t}(\cos 2t + \sin 2t).$$

7. We immediately have that $m = 2$ and $f = 14$. Using $g = 10$ m/sec^2 gives us

$$k = \frac{mg}{l} = \frac{2(10)}{1} = 20.$$

Since the spring is stretched an additional 3 m and released from rest, the initial value problem describing the motion of the mass is then

$$2u'' + 14u' + 20u = 0, \; u(0) = 3, \; u'(0) = 0.$$

The characteristic equation $2\lambda^2 + 14\lambda + 20 = 0$ has roots $\lambda = -2$ and $\lambda = -5$. The general solution is given by

$$u(t) = c_1 e^{-2t} + c_2 e^{-5t}.$$

We also have

$$u'(t) = -2c_1 e^{-2t} - 5c_2 e^{-5t}.$$

Using the initial conditions, we obtain $u(0) = c_1 + c_2 = 3$ and $u'(0) = -2c_1 - 5c_2 = 0$. The solution to this system of equations is $c_1 = 5$ and $c_2 = -2$. Therefore, the position of the object is given by

$$u(t) = 5e^{-2t} - 2e^{-5t}.$$

9. From Exercise 1 we know that the homogeneous equation $4u'' + 400u = 0$ has the general solution

$$u(t) = c_1 \cos 10t + c_2 \sin 10t.$$

With an external force of 3 newtons we have the initial value problem

$$4u'' + 400u = 3, \; u(0) = 0.5, \; u'(0) = 0. \tag{4.10}$$

Using the method of undetermined coefficients we look for a particular solution of the form $u_P = A$. Here, $u'_P = u''_P = 0$. Substituting into equation (4.10) we obtain $400A = 3$. Thus we have the particular solution $u_P = \frac{3}{400}$. The general solution to the initial value problem is

$$u(t) = \frac{3}{400} + c_1 \cos 10t + c_2 \sin 10t.$$

From this we see that $u'(t) = -10c_1 \sin 10t + 10c_2 \cos 10t$. Using the initial conditions we obtain the system of equations

$$u(0) = \frac{3}{400} + c_1 = 0.5, \; u'(0) = 10c_2 = 0,$$

which has the solution $c_1 = \frac{197}{400}$ and $c_2 = 0$. Thus the solution to the new initial value problem is

$$u(t) = \frac{3}{400} + \frac{197}{400} \cos 10t.$$

11. From Exercise 5 we know that the homogeneous equation $\frac{1}{4}u'' + u' + 2u = 0$ has the general solution

$$u(t) = e^{-2t}(c_1 \cos 2t + c_2 \sin 2t).$$

With an external force of $2\cos 2t$ lb we have the initial value problem

$$\frac{1}{4}u'' + u' + 2u = 2\cos 2t, \; u(0) = \frac{1}{2}, \; u'(0) = 0. \tag{4.11}$$

Using the method of undetermined coefficients we look for a particular solution of the form $u_P = A\cos 2t + B\sin 2t$. Here, $u'_P = -2A\sin 2t + 2B\cos 2t$ and $u''_P = -4A\cos 2t - 4B\sin 2t$. Substituting into equation (4.11) we obtain

$$\frac{1}{4}(-4A\cos 2t - 4B\sin 2t) + (-2A\sin 2t + 2B\cos 2t) + 2(A\cos 2t + B\sin 2t) = 2\cos 2t,$$

or more simply

$$(A + 2B)\cos 2t + (B - 2A)\sin 2t = 2\cos 2t.$$

We have the system of equations

$$A + 2B = 2, \ B - 2A = 0$$

which has the solution $A = \frac{2}{5}$ and $B = \frac{4}{5}$. Thus we have the particular solution

$$u_P = \frac{2}{5}\cos 2t + \frac{4}{5}\sin 2t.$$

The general solution to the initial value problem is

$$u(t) = \frac{4}{5}\sin 2t + \frac{2}{5}\cos 2t + e^{-2t}(c_1 \cos 2t + c_2 \sin 2t).$$

From this we see that

$$u'(t) = \frac{8}{5}\cos 2t - \frac{4}{5}\cos 2t + -2c_1 e^{-2t}\cos 2t - 2c_1 e^{-2t}\sin 2t - 2c_2 e^{-2t}\sin 2t + 2c_2 \cos 2t.$$

Using the initial conditions we obtain the system of equations

$$u(0) = \frac{2}{5} + c_1 = \frac{1}{2}, \ u'(0) = \frac{8}{5} - 2c_1 + 2c_2 = 0,$$

which has the solution $c_1 = \frac{1}{10}$ and $c_2 = -\frac{7}{10}$. Thus the solution to the new initial value problem is

$$u(t) = \frac{4}{5}\sin 2t + \frac{2}{5}\cos 2t - \frac{7}{10}e^{-2t}\sin 2t + \frac{1}{10}e^{-2t}\cos 2t.$$

13. From Exercise 7 we know that the homogeneous equation $2u'' + 14u' + 20u = 0$ has the general solution

$$u(t) = c_1 e^{-2t} + c_2 e^{-5t}.$$

With an external force of $6e^{-2t}$ newtons we have the initial value problem

$$2u'' + 14u' + 20u = 6e^{-2t}, \ u(0) = 3, \ u'(0) = 0. \tag{4.12}$$

Using the method of undetermined coefficients we look for a particular solution of the form $u_P = Ate^{-2t}$. Here,

$$u_P' = Ae^{-2t} - 2Ate^{-2t} \text{ and } u_P'' = -4Ae^{-2t} + 4Ate^{-2t}.$$

Substituting into equation (4.12) we obtain

$$2(-4Ae^{-2t} + 4Ate^{-2t}) + 14(Ae^{-2t} - 2Ate^{-2t}) + 20(Ate^{-2t}) = 6e^{-2t},$$

or more simply

$$6Ae^{-2t} = 6e^{-2t}.$$

Thus $A = 1$. We have the particular solution $u_P = te^{-2t}$. The general solution to the initial value problem is

$$u(t) = c_1 e^{-2t} + c_2 e^{-5t} + te^{-2t}.$$

From this we see that

$$u'(t) = -2c_1 e^{-2t} - 5c_2 e^{-5t} + e^{-2t} - 2te^{-2t}.$$

Using the initial conditions we obtain the system of equations

$$u(0) = c_1 + c_1 = 3, \ -2c_1 - 5c_2 + 1 = 0,$$

which has the solution $c_1 = \frac{14}{3}$ and $c_2 = -\frac{5}{3}$. Thus the solution to the new initial value problem is

$$u(t) = \frac{14}{3}e^{-2t} - \frac{5}{3}e^{-5t} + te^{-2t}.$$

15. The differential equation $mu'' + fu' + ku = 0$ has the characteristic equation $m\lambda^2 + f\lambda + k = 0$. The roots of this equation are

$$\lambda = \frac{-f \pm \sqrt{f^2 - 4mk}}{2m}.$$

Most of the subsequent answers depend on the sign of the discriminant $f^2 - 4mk$. Note, when $f^2 - 4mk \geq 0$ the roots are real and when $f^2 - 4mk < 0$ the roots are complex.

(a) The critical damping value occurs when $f^2 - 4mk = 0$, i.e., $f = \sqrt{4mk}$.

(b)   i. The system is undamped when the roots are complex and the real parts of the roots are 0. This will occur when $f = 0$.

     ii. The system will be damped and oscillating when $f \neq 0$ and the discriminant is negative, i.e., when $f^2 - 4mk < 0$ or $f^2 < 4mk$.

     iii. The system will be critically damped when there is a single repeated root. This will occur when $f^2 = 4mk$.

     iv. The system will be overdamped when there are two distinct real roots. This occurs when $f^2 > 4mk$.

(c) When $m = 1 = k$

     i. the system will be undamped when $f = 0$.

     ii. the system will be damped and oscillating when $0 < f < 2$.

     iii. the system will be critically damped when $f = 2$.

     iv. the system will be overdamped when $f > 2$.

17. We determine the current in an RLC circuit that has $L = 0.5$, $R = 0.2$, $C = 1$ and $E(t) = 0$, with initial charge 1 and initial current 1.

(Note the change from the problem stated in the text!)

We have the initial value problem

$$0.5I'' + 0.2I' + I = 0, \; I(0) = 1, \; I'(0) = 1.$$

The characteristic polynomial $0.5\lambda^2 + 0.2\lambda + 1 = 0$ has roots $\lambda = \frac{-1 \pm 7i}{5}$. Thus we have the general solution

$$I(t) = e^{-\frac{1}{5}t}\left(c_1 \sin\frac{7}{5}t + c_2 \cos\frac{7}{5}t\right).$$

From this we have $I'(t) = -\frac{1}{5}c_1 e^{-\frac{1}{5}t}\sin\frac{7}{5}t + \frac{7}{5}c_1 e^{-\frac{1}{5}t}\cos\frac{7}{5}t - \frac{1}{5}c_2 e^{-\frac{1}{5}t}\cos\frac{7}{5}t - \frac{7}{5}c_2 e^{-\frac{1}{5}t}\sin\frac{7}{5}t$. Using the initial conditions we obtain the system

$$I(0) = c_2 = 1, \; I'(0) = \frac{7}{5}c_1 - \frac{1}{5}c_2,$$

which has the solution $c_1 = \frac{6}{7}$ and $c_2 = 1$. The current in the RLC circuit is given by

$$I(t) = e^{-\frac{1}{5}t}\left(\frac{6}{7}\sin\frac{7}{5}t + \cos\frac{7}{5}t\right).$$

19. For the voltage $h(t) = \sin\omega t$ to produce resonance in an RLC circuit $\sin\omega t$ must be a solution to the homogeneous equation $LI'' + RI' + \frac{1}{C}I = 0$. The characteristic polynomial here is $L\lambda^2 + R\lambda + \frac{1}{C} = 0$ which has roots

$$\lambda = \frac{-R \pm \sqrt{R^2 - \frac{4L}{C}}}{2L}.$$

For $\sin\omega t$ to be a solution we need $Re(\lambda) = 0$, and $Im(\lambda) = \omega$. This will occur if $R = 0$ and $\omega = \frac{\sqrt{\frac{4L}{C}}}{2L} = \frac{1}{\sqrt{LC}}$.

Chapter 5

Linear Transformations and Eigenvalues and Eigenvectors

5.1 Linear Transformations

Exercises 5.1, pp. 243-245

1. By the definition, T is a linear transformation if, for all vectors u and v in the vector space and for all scalars c, the following two properties are satisfied: (1) $T(u+v) = T(u) + T(v)$, and (2) $T(cv) = cT(v)$. To determine if T is a linear transformation, we check these two properties.

Let $u = \begin{bmatrix} x_1 \\ y_1 \end{bmatrix}$ and $v = \begin{bmatrix} x_2 \\ y_2 \end{bmatrix}$, and let c be a scalar.

To check (1):

$$T(u+v) = T\left(\begin{bmatrix} x_1 \\ y_1 \end{bmatrix} + \begin{bmatrix} x_2 \\ y_2 \end{bmatrix}\right) = T\begin{bmatrix} x_1 + x_2 \\ y_1 + y_2 \end{bmatrix} = \begin{bmatrix} 3(x_1 + x_2) - 2(y_1 + y_2) \\ 5(x_1 + x_2) + 3(y_1 + y_2) \end{bmatrix},$$

and

$$T(u) + T(v) = T\begin{bmatrix} x_1 \\ y_1 \end{bmatrix} + T\begin{bmatrix} x_2 \\ y_2 \end{bmatrix} = \begin{bmatrix} 3x_1 - 2y_1 \\ 5x_1 + 3y_1 \end{bmatrix} + \begin{bmatrix} 3x_2 - 2y_2 \\ 5x_2 + 3y_2 \end{bmatrix} = \begin{bmatrix} 3(x_1 + x_2) - 2(y_1 + y_2) \\ 5(x_1 + x_2) + 3(y_1 + y_2) \end{bmatrix}.$$

So $T(u+v) = T(u) + T(v)$.

To check (2): $T(cv) = T\begin{bmatrix} cx_2 \\ cy_2 \end{bmatrix} = \begin{bmatrix} 3cx_2 - 2cy_2 \\ 5cx_2 + 3cy_2 \end{bmatrix} = c\begin{bmatrix} 3x_2 - 2y_2 \\ 5x_2 + 3y_2 \end{bmatrix} = cT(v)$. So $T(cv) = cT(v)$.

Since both properties are satisfied, T is a linear transformation.

3. By the definition, T is a linear transformation if, for all vectors u and v in the vector space and for all scalars c, the following two properties are satisfied: (1) $T(u+v) = T(u) + T(v)$, and (2) $T(cv) = cT(v)$. To determine if T is a linear transformation, we check these two properties.

Let $u = \begin{bmatrix} x_1 \\ y_1 \\ z_1 \end{bmatrix}$ and $v = \begin{bmatrix} x_2 \\ y_2 \\ z_2 \end{bmatrix}$, and let c be a scalar.

To check (1):

$$T(u+v) = T\left(\begin{bmatrix} x_1 \\ y_1 \\ z_1 \end{bmatrix} + \begin{bmatrix} x_2 \\ y_2 \\ z_2 \end{bmatrix}\right) = T\begin{bmatrix} x_1 + x_2 \\ y_1 + y_2 \\ z_1 + z_2 \end{bmatrix} = \begin{bmatrix} (x_1 + x_2) + (y_1 + y_2) + (z_1 + z_2) \\ (z_1 + z_2) - (y_1 + y_2) - (x_1 + x_2) \\ (x_1 + x_2)(y_1 + y_2)(z_1 + z_2) \end{bmatrix},$$

and

$$T(u) + T(v) = T\begin{bmatrix} x_1 \\ y_1 \\ z_1 \end{bmatrix} + T\begin{bmatrix} x_2 \\ y_2 \\ z_2 \end{bmatrix} = \begin{bmatrix} x_1 + y_1 + z_1 \\ z_1 - y_1 + x_1 \\ x_1 y_1 z_1 \end{bmatrix} + \begin{bmatrix} x_2 + y_2 + z_2 \\ z_2 - y_2 + x_2 \\ x_2 y_2 z_2 \end{bmatrix}$$

$$= \begin{bmatrix} x_1 + x_2 + y_1 + y_2 + z_1 + z_2 \\ z_1 - y_1 + x_1 + z_2 - y_2 + x_2 \\ x_1 y_1 z_1 + x_2 y_2 z_2 \end{bmatrix}.$$

Notice that the third coordinates of $T(u+v)$ and $T(u)+T(v)$ are not the same. So $T(u+v) \neq T(u)+T(v)$, and T is not a linear transformation.

5. By the definition, T is a linear transformation if, for all vectors u and v in the vector space and for all scalars c, the following two properties are satisfied: (1) $T(u+v) = T(u)+T(v)$, and (2) $T(cv) = cT(v)$. To determine if T is a linear transformation, we check these two properties.
Since T operates on ax^2+bx+c by taking its derivative, we know that T is a linear transformation, because we know from calculus that the derivative of a sum is the sum of the derivatives and the derivative of a scalar multiple of a function is the scalar times the derivative of the function.

7. By the definition, T is a linear transformation if, for all vectors u and v in the vector space and for all scalars c, the following two properties are satisfied: (1) $T(u+v) = T(u)+T(v)$, and (2) $T(cv) = cT(v)$. To determine if T is a linear transformation, we check these two properties.
Let $u = a_1x^2 + b_1x + c_1$ and $v = a_2x^2 + b_2x + c_2$, and let d be a scalar. To check (1):

$$T(u+v) = T((a_1+a_2)x^2 + (b_1+b_2)x + (c_1+c_2) = (a_1+a_2)(x+1)^2 + (b_1+b_2)(x+1) + (c_1+c_2)$$

. Multiplying out and collecting terms, we have

$$T(u+v) = (a_1+a_2)x^2 + (2a_1+2a_2+b_1+b_2)x + (a_1+a_2+b_1+b_2+c_1+c_2).$$

$$T(u)+T(v) = (a_1(x+1)^2 + b_1(x+1) + c_1) + (a_2(x+1)^2 + b_2(x+1) + c_2).$$

Multiplying out and collecting terms, we have

$$T(u)+T(v) = (a_1+a_2)x^2 + (2a_1+2a_2+b_1+b_2)x + (a_1+a_2+b_1+b_2+c_1+c_2).$$

So $T(u+v) = T(u)+T(v)$.
To check (2):

$$T(dv) = T(da_2x^2 + db_2x + dc_2) = da_2(x+1)^2 + db_2(x+1) + dc_2 = d(a_2(x+1)^2 + b_2(x+1) + c_2) = dT(v).$$

So $T(dv) = dT(v)$.
Since both properties are satisfied, T is a linear transformation.

9. By the definition, T is a linear transformation if, for all vectors u and v in the vector space and for all scalars c, the following two properties are satisfied: (1) $T(u+v) = T(u)+T(v)$, and (2) $T(cv) = cT(v)$. To determine if T is a linear transformation, we check these two properties. To check (1): Let $u = f(x)$ and $v = g(x)$. Then $T(u+v) = T(f(x)+g(x)) = f(x)+g(x)-2$, and $T(u)+T(v) = f(x)-2+g(x)-2 = f(x)+g(x)-4$. Since $T(u+v) \neq T(u)+T(v)$, T is not a linear transformation.

11. By the definition, T is a linear transformation if, for all vectors u and v in the vector space and for all scalars c, the following two properties are satisfied: (1) $T(u+v) = T(u)+T(v)$, and (2) $T(cv) = cT(v)$. To determine if T is a linear transformation, we check these two properties.
Let $u = A$ and $v = B$, and let c be a scalar. To check (1):

$$T(u+v) = T(A+B) = (A+B)^T = A^T + B^T = T(A)+T(B) = T(u)+T(v).$$

To check (2): $T(cv) = T(cB) = (cB)^T = cB^T = cT(v)$.
Since both properties are satisfied, T is a linear transformation.

13. Proof.

15. The rows of the matrix are the coefficients of the variables x and y in the definition of T: $A = \begin{bmatrix} 3 & 2 \\ 3 & 3 \end{bmatrix}$.

(Try multiplying out $A \begin{bmatrix} x \\ y \end{bmatrix}$ to see why this works.)

17. The rows of the matrix are the coefficients of the variables x, y, and z in the definition of T: $A = \begin{bmatrix} 1 & -1 & 1 \\ 2 & 1 & 2 \\ 3 & 3 & 3 \\ 1 & 2 & 1 \end{bmatrix}$. (Try multiplying out $A \begin{bmatrix} x \\ y \\ z \end{bmatrix}$ to see why this works.) Note: there is a typo in the problem in the textbook: the third row of the column vector giving the definition of T should be $3x + 3y + 3z$, not $3x + 3y + 3x$.

19. Part a: To find $T \begin{bmatrix} 3 \\ 2 \\ 2 \end{bmatrix}$, we need to write $\begin{bmatrix} 3 \\ 2 \\ 2 \end{bmatrix}$ as a linear combination of $\begin{bmatrix} 1 \\ 0 \\ 1 \end{bmatrix}$, $\begin{bmatrix} 1 \\ 1 \\ 0 \end{bmatrix}$, and $\begin{bmatrix} 1 \\ 1 \\ 1 \end{bmatrix}$. This is easy to do by inspection, or use the techniques in Section 2.2.

Since $\begin{bmatrix} 3 \\ 2 \\ 2 \end{bmatrix} = 1\begin{bmatrix} 1 \\ 0 \\ 1 \end{bmatrix} + 1\begin{bmatrix} 1 \\ 1 \\ 0 \end{bmatrix} + 1\begin{bmatrix} 1 \\ 1 \\ 1 \end{bmatrix}$,

$$T \begin{bmatrix} 3 \\ 2 \\ 2 \end{bmatrix} = 1T\begin{bmatrix} 1 \\ 0 \\ 1 \end{bmatrix} + 1T\begin{bmatrix} 1 \\ 1 \\ 0 \end{bmatrix} + 1T\begin{bmatrix} 1 \\ 1 \\ 1 \end{bmatrix} =$$

$$1\begin{bmatrix} 1 \\ 2 \\ 1 \end{bmatrix} + 1\begin{bmatrix} -1 \\ 0 \\ 1 \end{bmatrix} + 1\begin{bmatrix} 0 \\ 1 \\ 1 \end{bmatrix} = \begin{bmatrix} 0 \\ 3 \\ 3 \end{bmatrix}.$$

Part b: To find $T \begin{bmatrix} x \\ y \\ z \end{bmatrix}$, we need to write $\begin{bmatrix} x \\ y \\ z \end{bmatrix}$ as a linear combination of $\begin{bmatrix} 1 \\ 0 \\ 1 \end{bmatrix}$, $\begin{bmatrix} 1 \\ 1 \\ 0 \end{bmatrix}$, and $\begin{bmatrix} 1 \\ 1 \\ 1 \end{bmatrix}$. Use the techniques in Section 2.2 to find the coefficients.

Since $\begin{bmatrix} x \\ y \\ z \end{bmatrix} = (x-y)\begin{bmatrix} 1 \\ 0 \\ 1 \end{bmatrix} + (x-z)\begin{bmatrix} 1 \\ 1 \\ 0 \end{bmatrix} + (-x+y+z)\begin{bmatrix} 1 \\ 1 \\ 1 \end{bmatrix}$,

$$T \begin{bmatrix} x \\ y \\ z \end{bmatrix} = (x-y)T\begin{bmatrix} 1 \\ 0 \\ 1 \end{bmatrix} + (x-z)T\begin{bmatrix} 1 \\ 1 \\ 0 \end{bmatrix} + (-x+y+z)T\begin{bmatrix} 1 \\ 1 \\ 1 \end{bmatrix} =$$

$$(x-y)\begin{bmatrix} 1 \\ 2 \\ 1 \end{bmatrix} + (x-z)\begin{bmatrix} -1 \\ 0 \\ 1 \end{bmatrix} + (-x+y+z)\begin{bmatrix} 0 \\ 1 \\ 1 \end{bmatrix} = \begin{bmatrix} -y+z \\ x-y+z \\ x \end{bmatrix}.$$

21. Part a: To find $T(x)$, we must write x as a linear combination of the vectors $x+1$ and $x-1$. Use the techniques of Section 2.2 to find the coefficients. Since $x = (1/2)(x+1) + (1/2)(x-1)$, we know that

$$T(x) = (1/2)T(x+1) + (1/2)T(x-1) = (1/2)(2x+1) + (1/2)(2x-1) = 2x.$$

Part b: To find $T(ax+b)$, we must write $ax+b$ as a linear combination of the vectors $x+1$ and $x-1$. Use the techniques to Section 2.2 to find the coefficients. Since $ax+b = ((3a-b)/2)(x+1) + ((b-a)/2)(x-1)$, we know that

$$T(x) = ((3a-b)/2)T(x+1) + ((b-a)/2)T(x-1) = ((3a-b)/2)(2x+1) + ((b-a)/2)(2x-1) = 2ax - b.$$

23. To find a basis for the kernel of T, find a basis for the nullspace of the matrix that expresses T. To find a basis for the range of T, find a basis for the column space of the matrix that expresses T.

25. To find the kernel of T, find a basis for the nullspace of the matrix that expresses T. To find a basis for the range of T, find a basis for the column space of the matrix that expresses T.

27. The vectors in the kernel of the differential operator are those functions whose derivative is 0, that is, the constant functions.

5.2   The Algebra of Linear Transformations

Exercises 5.2, pp. 252-253

1. $(S+T)\begin{bmatrix} x \\ y \end{bmatrix} = S\begin{bmatrix} x \\ y \end{bmatrix} + T\begin{bmatrix} x \\ y \end{bmatrix} = \begin{bmatrix} 2x-y \\ x+2y \end{bmatrix} + \begin{bmatrix} x+3y \\ x-y \end{bmatrix} = \begin{bmatrix} 3x+2y \\ 2x+y \end{bmatrix}\begin{bmatrix} x \\ y \end{bmatrix}.$

3. $(2T)\begin{bmatrix} x \\ y \end{bmatrix} = 2T\begin{bmatrix} x \\ y \end{bmatrix} = 2\begin{bmatrix} x+3y \\ x-y \end{bmatrix} = \begin{bmatrix} 2x+6y \\ 2x-2y \end{bmatrix}.$

5. $ST\begin{bmatrix} x \\ y \end{bmatrix} = S\left(T\begin{bmatrix} x \\ y \end{bmatrix}\right) = S\begin{bmatrix} x+3y \\ x-y \end{bmatrix} = \begin{bmatrix} 2(x+3y)-(x-y) \\ (x+3y)+2(x-y) \end{bmatrix} = \begin{bmatrix} x+7y \\ 3x+y \end{bmatrix}.$

7. $(S+3T)(ax+b) = S(ax+b) + 3T(ax+b) = (ax-2b+b) + 3(ax+2a+b) = 4ax+4(a+b).$

9. $T^2(ax+b) = T\left(T(ax+b)\right) = T(ax+2a+b) = ax+2a+2a+b = ax+4a+b.$

11. The kernel of a linear transformation D is the set of all vectors v such that $Dv = 0$. Substitute λ^k for D^k. Solve

$$\lambda^2 - 2\lambda - 3 = (\lambda-3)(\lambda+1) = 0,$$

giving $\lambda = 3$ or $\lambda = -1$. Then the kernel of D is all functions of the form $f(x) = c_1 e^{3x} + c_2 e^{-x}$, and a basis for the kernel of D is e^{3x}, e^{-x}.

13. The kernel of a linear transformation D is the set of all vectors v such that $Dv = 0$. Substitute λ^k for D^k. Solve

$$\lambda^4 + \lambda^2 = (\lambda^2)(\lambda^2+1) = 0,$$

giving $\lambda = 0$, $\lambda = i$, or $\lambda = -i$. Then the kernel of D is all functions of the form $f(x) = c_1 e^0 + c_2 x e^0 + c_3 \cos x + c_4 \sin x$, and a basis for the kernel of D is $1, x, \cos x, \sin x$.

15. Proof.

17. Proof.

19. Proof.

21. Proof.

23. Part a: Since $T\begin{bmatrix} x \\ y \end{bmatrix} = A\begin{bmatrix} x \\ y \end{bmatrix}$, $A = \begin{bmatrix} 1 & 3 \\ 1 & -1 \end{bmatrix}$. (Try multiplying out $A\begin{bmatrix} x \\ y \end{bmatrix}$ to see why.) Similarly, $B = \begin{bmatrix} 2 & -1 \\ 1 & 2 \end{bmatrix}$.

Part b: $ST\begin{bmatrix} x \\ y \end{bmatrix} = S(AX) = BAX$, so $C = BA$.

$$\begin{bmatrix} 2 & -1 \\ 1 & 2 \end{bmatrix}\begin{bmatrix} 1 & 3 \\ 1 & -1 \end{bmatrix} = \begin{bmatrix} 1 & 7 \\ 3 & 1 \end{bmatrix},$$

and $\begin{bmatrix} 1 & 7 \\ 3 & 1 \end{bmatrix}\begin{bmatrix} x \\ y \end{bmatrix} = \begin{bmatrix} x+7y \\ 3x+y \end{bmatrix}$, as we wanted.

Part c: $TS\begin{bmatrix} x \\ y \end{bmatrix} = T(BX) = ABX$, so $D = AB$.

$$\begin{bmatrix} 1 & 3 \\ 1 & -1 \end{bmatrix}\begin{bmatrix} 2 & -1 \\ 1 & 2 \end{bmatrix} = \begin{bmatrix} 5 & 4 \\ 1 & -3 \end{bmatrix},$$

and $\begin{bmatrix} 5 & 4 \\ 1 & -3 \end{bmatrix}\begin{bmatrix} x \\ y \end{bmatrix} = \begin{bmatrix} 5x+4y \\ x-3y \end{bmatrix}$, as we wanted.

5.3   Matrices for Linear Transformations

Exercises 5.3, pp. 267-269

1. Part a: Recall that α is the standard basis consisting of $v_1 = \begin{bmatrix} 1 \\ 0 \end{bmatrix}$ and $v_2 = \begin{bmatrix} 0 \\ 1 \end{bmatrix}$. $T(v_1) = T \begin{bmatrix} 1 \\ 0 \end{bmatrix} = \begin{bmatrix} 1+0 \\ 1-0 \end{bmatrix} = \begin{bmatrix} 1 \\ 1 \end{bmatrix}$

and $T(v_2) = T \begin{bmatrix} 0 \\ 1 \end{bmatrix} = \begin{bmatrix} 0+1 \\ 0-1 \end{bmatrix} = \begin{bmatrix} 1 \\ -1 \end{bmatrix}$. Using $T(v_1)$ and $T(v_2)$ as the columns of $[T]_\alpha^\alpha$, we have $[T]_\alpha^\alpha = \begin{bmatrix} 1 & 1 \\ 1 & -1 \end{bmatrix}$.

Part b: The change of basis matrix P from the standard basis to $\begin{bmatrix} 1 \\ -1 \end{bmatrix}, \begin{bmatrix} -2 \\ 1 \end{bmatrix}$ has these two vectors as its columns from

left to right: $P = \begin{bmatrix} 1 & -1 \\ -2 & 1 \end{bmatrix}$.

Part c: The change of basis matrix from the new basis β to the standard basis α is $P^{-1} = \begin{bmatrix} -1 & -2 \\ -1 & -1 \end{bmatrix}$. (We calculated P in

part b.)

Part d: $[T]_\beta^\beta = P^{-1}[T]_\alpha^\alpha P = \begin{bmatrix} -4 & 7 \\ -2 & 4 \end{bmatrix}$. (We calculated these three matrices in parts a, b, and c.)

Part e: To find $[v]_\beta$, multiply $P^{-1}v = \begin{bmatrix} -1 & -2 \\ -1 & -1 \end{bmatrix} \begin{bmatrix} -2 \\ 3 \end{bmatrix} = \begin{bmatrix} -4 \\ -1 \end{bmatrix}$.

Part f: $[T(v)]_\beta = [T]_\beta^\beta [v]_\beta = \begin{bmatrix} -4 & 7 \\ -2 & 4 \end{bmatrix} \begin{bmatrix} -4 \\ -1 \end{bmatrix} = \begin{bmatrix} 9 \\ 4 \end{bmatrix}$.

Part g: $T(v) = 9 \begin{bmatrix} 1 \\ -1 \end{bmatrix} + 4 \begin{bmatrix} -2 \\ 1 \end{bmatrix} = \begin{bmatrix} 9-8 \\ -9+4 \end{bmatrix} = \begin{bmatrix} 1 \\ -5 \end{bmatrix}$. The coefficients here are the coordinates of the vector we

calculated in part f, and the vectors are the vectors of the basis β.

3. Part a: Recall that α is the standard basis consisting of $v_1 = \begin{bmatrix} 1 \\ 0 \\ 0 \end{bmatrix}$, $v_2 = \begin{bmatrix} 0 \\ 1 \\ 0 \end{bmatrix}$, and $v_3 = \begin{bmatrix} 0 \\ 0 \\ 1 \end{bmatrix}$.

$$T(v_1) = T \begin{bmatrix} 1 \\ 0 \\ 0 \end{bmatrix} = \begin{bmatrix} 17 \cdot 1 - 0 \cdot 0 - 12 \cdot 0 \\ 16 \cdot 1 - 7 \cdot 0 - 12 \cdot 0 \\ 16 \cdot 1 - 8 \cdot 0 - 11 \cdot 0 \end{bmatrix} = \begin{bmatrix} 17 \\ 16 \\ 16 \end{bmatrix}.$$

$T(v_2) = T \begin{bmatrix} 0 \\ 1 \\ 0 \end{bmatrix} = \begin{bmatrix} -8 \\ -7 \\ -8 \end{bmatrix}$, and $T(v_3) = T \begin{bmatrix} 0 \\ 0 \\ 1 \end{bmatrix} = \begin{bmatrix} -12 \\ -12 \\ -11 \end{bmatrix}$. Using $T(v_1)$, $T(v_2)$, and $T(v_3)$ as the columns of $[T]_\alpha^\alpha$,

we have $[T]_\alpha^\alpha = \begin{bmatrix} 17 & -8 & -12 \\ 16 & -7 & -12 \\ 16 & -8 & -11 \end{bmatrix}$.

Part b: The change of basis matrix P from the standard basis α to β is just the three vectors of β used as the columns of P

from left to right: $P = \begin{bmatrix} 1 & 1 & 1 \\ 1 & 2 & -1 \\ 1 & 0 & 2 \end{bmatrix}$.

Part c: The change of basis matrix from the new basis β to the standard basis α is $P^{-1} = \begin{bmatrix} -4 & 2 & 3 \\ 3 & -1 & -2 \\ 2 & -1 & -1 \end{bmatrix}$. (We calculated

P in part b.)

Part d: $[T]_\beta^\beta = P^{-1}[T]_\alpha^\alpha P = \begin{bmatrix} -3 & 0 & 0 \\ 0 & 1 & 0 \\ 0 & 0 & 1 \end{bmatrix}$. (We calculated these three matrices in parts a, b, and c.)

Part e: To find $[v]_\beta$, multiply $P^{-1}v = \begin{bmatrix} -4 & 2 & 3 \\ 3 & -1 & -2 \\ 2 & -1 & -1 \end{bmatrix} \begin{bmatrix} 2 \\ -1 \\ 4 \end{bmatrix} = \begin{bmatrix} 2 \\ -1 \\ 1 \end{bmatrix}$.

Part f: $[T(v)]_\beta = [T]_\beta^\beta [v]_\beta = \begin{bmatrix} -3 & 0 & 0 \\ 0 & 1 & 0 \\ 0 & 0 & 1 \end{bmatrix} \begin{bmatrix} 2 \\ -1 \\ 1 \end{bmatrix} = \begin{bmatrix} -6 \\ -1 \\ 1 \end{bmatrix}$.

Part g: $T(v) = 6 \begin{bmatrix} 1 \\ 1 \\ 1 \end{bmatrix} + -1 \begin{bmatrix} 1 \\ 2 \\ 0 \end{bmatrix} + 1 \begin{bmatrix} 1 \\ -1 \\ 2 \end{bmatrix} = \begin{bmatrix} -6 \\ -9 \\ -4 \end{bmatrix}$. The coefficients here are the coordinates of the vector we calculated in part f, and the vectors are the vectors of the basis β.

5. Part a: Recall that α is the standard basis consisting of $v_1 = x^2 = 1x^2 + 0x + 0 = \begin{bmatrix} 1 \\ 0 \\ 0 \end{bmatrix}$, $v_2 = x = 0x^2 + 1x + 0 = \begin{bmatrix} 0 \\ 1 \\ 0 \end{bmatrix}$,

and $v_3 = 1 = 0x^2 + 0x + 1 = \begin{bmatrix} 0 \\ 0 \\ 1 \end{bmatrix}$. $T(v_1) = T(x^2) = 1(x+1)^2 + 0(x+1) + 0 = x^2 + 2x + 1$, so $T \begin{bmatrix} 1 \\ 0 \\ 0 \end{bmatrix} = \begin{bmatrix} 1 \\ 2 \\ 1 \end{bmatrix}$.

$T(v_2) = T(x) = x + 1$, so $T \begin{bmatrix} 0 \\ 1 \\ 0 \end{bmatrix} = \begin{bmatrix} 0 \\ 1 \\ 1 \end{bmatrix}$, and $T(v_3) = 1$, so $T \begin{bmatrix} 0 \\ 0 \\ 1 \end{bmatrix} = \begin{bmatrix} 0 \\ 0 \\ 1 \end{bmatrix}$. Using $T(v_1)$, $T(v_2)$, and $T(v_3)$ as the

columns of $[T]^\alpha_\alpha$, we have $[T]^\alpha_\alpha = \begin{bmatrix} 1 & 0 & 0 \\ 2 & 1 & 0 \\ 1 & 1 & 1 \end{bmatrix}$.

Part b: View the polynomials as column vectors as we did in part a. The change of basis matrix P from the standard basis α to β is just the three vectors of β used as the columns of P from left to right: $P = \begin{bmatrix} 1 & 0 & -1 \\ 1 & 0 & 1 \\ -1 & 1 & 1 \end{bmatrix}$.

Part c: The change of basis matrix from the new basis β to the standard basis α is $P^{-1} = \begin{bmatrix} 1/2 & 1/2 & -1/2 \\ 1/2 & -1/2 & 1/2 \\ 0 & 1 & 0 \end{bmatrix}$. (We

calculated P in part b.)

Part d: $[T]^\beta_\beta = P^{-1}[T]^\alpha_\alpha P = \begin{bmatrix} 3/2 & 1/2 & -1/2 \\ -1/2 & 1/2 & 1/2 \\ 2 & 2 & 1 \end{bmatrix}$. (We calculated these three matrices in parts a, b, and c.)

Part e: To find $[v]_\beta$, multiply $P^{-1}v = \begin{bmatrix} 1/2 & 1/2 & -1/2 \\ 1/2 & -1/2 & 1/2 \\ 0 & 1 & 0 \end{bmatrix} \begin{bmatrix} 1 \\ 1 \\ 1 \end{bmatrix} = \begin{bmatrix} 1/2 \\ 1/2 \\ 1 \end{bmatrix}$.

Part f: $[T(v)]_\beta = [T]^\beta_\beta [v]_\beta = \begin{bmatrix} 3/2 & 1/2 & -1/2 \\ -1/2 & 1/2 & 1/2 \\ 2 & 2 & 1 \end{bmatrix} \begin{bmatrix} 1/2 \\ 1/2 \\ 1 \end{bmatrix} = \begin{bmatrix} 1/2 \\ 1/2 \\ 3 \end{bmatrix}$.

Part g: $T(v) = 1/2 \begin{bmatrix} 1 \\ 0 \\ -1 \end{bmatrix} + 1/2 \begin{bmatrix} 1 \\ 0 \\ 1 \end{bmatrix} + 3 \begin{bmatrix} 0 \\ 1 \\ 1 \end{bmatrix} = \begin{bmatrix} 1 \\ 3 \\ 3 \end{bmatrix}$. The coefficients here are the coordinates of the vector we calculated in part f, and the vectors are the vectors of the basis β.

7. Part a: D is the differential operator, so Dv is the derivative of v. Recall that α is the standard basis consisting of

$$v_1 = 1 \sin x + 0 \cos x = \begin{bmatrix} 1 \\ 0 \end{bmatrix} \text{ and } v_2 = 0 \sin x + 1 \cos x = \begin{bmatrix} 0 \\ 1 \end{bmatrix}.$$

Since the derivative of $\sin x$ is $\cos x = 0 \sin x + 1 \cos x$,

$$D(v_1) = D \begin{bmatrix} 1 \\ 0 \end{bmatrix} = \begin{bmatrix} 1+0 \\ 1-0 \end{bmatrix} = \begin{bmatrix} 0 \\ 1 \end{bmatrix} \text{ and } D(v_2) = D \begin{bmatrix} 0 \\ 1 \end{bmatrix} = \begin{bmatrix} -1 \\ 0 \end{bmatrix},$$

since the derivative of $\cos x$ is $-\sin x = -1 \sin x + 0 \cos x$. Using $D(v_1)$ and $D(v_2)$ as the columns of $[T]^\alpha_\alpha$, we have

$$[T]^\alpha_\alpha = \begin{bmatrix} 0 & -1 \\ 1 & 0 \end{bmatrix}.$$

Part b: Since $\sin x + \cos x = 1 \sin x + 1 \cos x$, and since $\sin x - \cos x = 1 \sin x - 1 \cos x$, the change of basis matrix P from the standard basis to β is $P = \begin{bmatrix} 1 & 1 \\ 1 & -1 \end{bmatrix}$. (The columns of P are the vectors of β from left to right.)

Part c: The change of basis matrix from the new basis β to the standard basis α is $P^{-1} = \begin{bmatrix} 1/2 & 1/2 \\ 1/2 & -1/2 \end{bmatrix}$.

(We calculated P in part b.)

Part d: $[T]_\beta^\beta = P^{-1}[T]_\alpha^\alpha P = \begin{bmatrix} 0 & 1 \\ -1 & 0 \end{bmatrix}$. (We calculated these three matrices in parts a, b, and c.)

Part e: For $v = 3\cos x - 2\sin x$, whose vector in terms of α is $\begin{bmatrix} -2 \\ 3 \end{bmatrix}$, to find $[v]_\beta$, multiply

$P^{-1}v = \begin{bmatrix} 1/2 & 1/2 \\ 1/2 & -1/2 \end{bmatrix}\begin{bmatrix} -2 \\ 3 \end{bmatrix} = \begin{bmatrix} 1/2 \\ -5/2 \end{bmatrix}$.

Part f: $[T(v)]_\beta = [T]_\beta^\beta[v]_\beta = \begin{bmatrix} 0 & 1 \\ -1 & 0 \end{bmatrix}\begin{bmatrix} 1/2 \\ -5/2 \end{bmatrix} = \begin{bmatrix} -5/2 \\ 1/2 \end{bmatrix}$.

Part g: $T(v) = -5/2\begin{bmatrix} 1 \\ 1 \end{bmatrix} - 1/2\begin{bmatrix} 1 \\ -1 \end{bmatrix} = \begin{bmatrix} -3 \\ 2 \end{bmatrix} = -3\sin x - 2\cos x$. The coefficients here are the coordinates of the vector we calculated in part f, and the vectors are the vectors of the basis β.

9. Part a: Since $T(v_1) = v_1 - v_2 = \begin{bmatrix} 1 \\ -1 \\ 0 \end{bmatrix}$, $T(v_2) = v_2 - v_3 = \begin{bmatrix} 0 \\ 1 \\ -1 \end{bmatrix}$, and $T(v_3) = v_3 - v_1 = \begin{bmatrix} -1 \\ 0 \\ 1 \end{bmatrix}$,

$$[T]_\alpha^\alpha = \begin{bmatrix} 1 & 0 & -1 \\ -1 & 1 & 0 \\ 0 & -1 & 1 \end{bmatrix}.$$

Part b: $[T(v)]_\alpha = [T]_\alpha^\alpha[v]_\alpha = \begin{bmatrix} 1 & 0 & -1 \\ -1 & 1 & 0 \\ 0 & -1 & 1 \end{bmatrix}\begin{bmatrix} 1 \\ -2 \\ 3 \end{bmatrix} = \begin{bmatrix} -2 \\ -3 \\ 5 \end{bmatrix}$.

Part c: Use the coordinates of the vector we calculated in part b. $T(v) = -2v_1 - 3v_2 + 5v_3$.

5.4 Eigenvalues and Eigenvectors of Matrices

Exercises 5.4, pp. 277-278

1. To find the eigenvalues of a matrix A, set the determinant of $\lambda I - A$ equal to zero and solve for λ.

$$\det(\lambda I - A) = \det\left(\begin{bmatrix} \lambda - 3 & 0 \\ -8 & \lambda + 1 \end{bmatrix}\right) = (\lambda - 3)(\lambda + 1) = 0,$$

so, solving for λ, we see that the eigenvalues of A are 3 and -1.
To find a basis for the eigenspace of the eigenvalue 3, we find a basis for the nullspace of the matrix $3I - A$. First we find the reduced row-echelon form of $3I - A$:

$$3I - A = \begin{bmatrix} 0 & 0 \\ -8 & 2 \end{bmatrix} \xrightarrow{\text{row reduction}} \begin{bmatrix} 1 & -1/2 \\ 0 & 0 \end{bmatrix}.$$

The eigenspace of the eigenvector 3 consists of all vectors with the property that $x - y/2 = 0$; that is, all vectors

$$\begin{bmatrix} t/2 \\ t \end{bmatrix} = t\begin{bmatrix} 1/2 \\ 1 \end{bmatrix},$$

so a basis for the eigenspace of the eigenvalue 3 is $\begin{bmatrix} 1/2 \\ 1 \end{bmatrix}$. To find a basis for the eigenspace of the eigenvalue -1, we find a basis for the nullspace of the matrix $-1I - A$. First we find the reduced row-echelon form of $-1I - A$:

$$-1I - A = \begin{bmatrix} -4 & 0 \\ -8 & 0 \end{bmatrix} \xrightarrow{\text{row reduction}} \begin{bmatrix} 1 & 0 \\ 0 & 0 \end{bmatrix}.$$

The eigenspace of the eigenvector -1 consists of all vectors with the property that $x = 0$; that is, all vectors $\begin{bmatrix} 0 \\ t \end{bmatrix} = t\begin{bmatrix} 0 \\ 1 \end{bmatrix}$,

so a basis for the eigenspace of the eigenvalue -1 is $\begin{bmatrix} 0 \\ 1 \end{bmatrix}$.

3. To find the eigenvalues of a matrix A, set the determinant of $\lambda I - A$ equal to zero and solve for λ.

$$\det(\lambda I - A) = \det\left(\begin{bmatrix} \lambda - 10 & -9 \\ -4 & \lambda + 2 \end{bmatrix}\right) = \lambda^2 - 8\lambda + 16 = 0,$$

so, solving for λ, we see that there is one eigenvalues of A, which is 4.

To find a basis for the eigenspace of the eigenvalue 4, we find a basis for the nullspace of the matrix $4I - A$. First we find the reduced row-echelon form of $4I - A$:

$$4I - A = \begin{bmatrix} -6 & 9 \\ -4 & 6 \end{bmatrix} \xrightarrow{\text{row reduction}} \begin{bmatrix} 1 & -3/2 \\ 0 & 0 \end{bmatrix}.$$

The eigenspace of the eigenvector 4 consists of all vectors with the property that $x - 3y/2 = 0$; that is, all vectors

$$\begin{bmatrix} 3t/2 \\ t \end{bmatrix} = t\begin{bmatrix} 3/2 \\ 1 \end{bmatrix},$$

so a basis for the eigenspace of the eigenvalue 4 is $\begin{bmatrix} 3/2 \\ 1 \end{bmatrix}$.

5. To find the eigenvalues of a matrix A, set the determinant of $\lambda I - A$ equal to zero and solve for λ.

$$\det(\lambda I - A) = \det\left(\begin{bmatrix} \lambda & -4 \\ -4 & \lambda \end{bmatrix}\right) = \lambda^2 - 12 = 0,$$

so, solving for λ, we see that the eigenvalues of A are $2\sqrt{3}$ and $-2\sqrt{3}$.

To find a basis for the eigenspace of the eigenvalue $2\sqrt{3}$, we find a basis for the nullspace of the matrix $2\sqrt{3}I - A$. First we find the reduced row-echelon form of $2\sqrt{3}I - A$:

$$2\sqrt{3}I - A = \begin{bmatrix} 2\sqrt{3} & -3 \\ -4 & 2\sqrt{3} \end{bmatrix} \xrightarrow{\text{row reduction}} \begin{bmatrix} 1 & (-1/2)\sqrt{3} \\ 0 & 0 \end{bmatrix}.$$

The eigenspace of the eigenvector $2\sqrt{3}$ consists of all vectors with the property that $x - \sqrt{3}y/2 = 0$; that is, all vectors $\begin{bmatrix} \sqrt{3}t/2 \\ t \end{bmatrix} = t\begin{bmatrix} \sqrt{3}/2 \\ 1 \end{bmatrix}$, so a basis for the eigenspace of the eigenvalue $2\sqrt{3}$ is $\begin{bmatrix} \sqrt{3}/2 \\ 1 \end{bmatrix}$. To find a basis for the eigenspace of the eigenvalue $-2\sqrt{3}$, we find a basis for the nullspace of the matrix $-2\sqrt{3}I - A$. First we find the reduced row-echelon form of $-2\sqrt{3}I - A$:

$$2\sqrt{3}I - A = \begin{bmatrix} -2\sqrt{3} & -3 \\ -4 & -2\sqrt{3} \end{bmatrix} \xrightarrow{\text{row reduction}} \begin{bmatrix} 1 & (1/2)\sqrt{3} \\ 0 & 0 \end{bmatrix}.$$

The eigenspace of the eigenvector $2\sqrt{3}$ consists of all vectors with the property that $x + \sqrt{3}y/2 = 0$; that is, all vectors $\begin{bmatrix} -\sqrt{3}t/2 \\ t \end{bmatrix} = t\begin{bmatrix} -\sqrt{3}/2 \\ 1 \end{bmatrix}$, so a basis for the eigenspace of the eigenvalue $2\sqrt{3}$ is $\begin{bmatrix} -\sqrt{3}/2 \\ 1 \end{bmatrix}$.

7. To find the eigenvalues of a matrix A, set the determinant of $\lambda I - A$ equal to zero and solve for λ.

$$\det(\lambda I - A) = \det\left(\begin{bmatrix} \lambda - 4 & 0 & -1 \\ 2 & \lambda - 1 & 0 \\ 2 & 0 & \lambda - 1 \end{bmatrix}\right) = \lambda^3 - 6\lambda^2 + 11\lambda - 6 = (\lambda - 1)(\lambda - 2)(\lambda - 3) = 0,$$

so, solving for λ, we see that the eigenvalues of A are 1, 2, and 3.

To find a basis for the eigenspace of the eigenvalue 1, we find a basis for the nullspace of the matrix $1I - A$. First we find the reduced row-echelon form of $1I - A$:

$$1I - A = \begin{bmatrix} -3 & 0 & -1 \\ 2 & 0 & 0 \\ 2 & 0 & 0 \end{bmatrix} \xrightarrow{\text{row reduction}} \begin{bmatrix} 1 & 0 & 0 \\ 0 & 0 & 0 \\ 0 & 0 & 1 \end{bmatrix}.$$

The eigenspace of the eigenvector 1 consists of all vectors with the property that $x = z = 0$; that is, all vectors

$$\begin{bmatrix} 0 \\ t \\ 0 \end{bmatrix} = t \begin{bmatrix} 0 \\ 1 \\ 0 \end{bmatrix},$$

so a basis for the eigenspace of the eigenvalue 1 is $\begin{bmatrix} 0 \\ 1 \\ 0 \end{bmatrix}$. To find a basis for the eigenspace of the eigenvalue 2, we find a

basis for the nullspace of the matrix $2I - A$. First we find the reduced row-echelon form of $2I - A$:

$$2I - A = \begin{bmatrix} -2 & 0 & -1 \\ 2 & 1 & 0 \\ 2 & 0 & 1 \end{bmatrix} \xrightarrow{\text{row reduction}} \begin{bmatrix} 1 & 0 & 1/2 \\ 0 & 1 & -1 \\ 0 & 0 & 0 \end{bmatrix}.$$

The eigenspace of the eigenvector 2 consists of all vectors with the property that $x + z/2 = 0$; that is, all vectors

$$\begin{bmatrix} -t/2 \\ 0 \\ t \end{bmatrix} = t \begin{bmatrix} -1/2 \\ 0 \\ 1 \end{bmatrix},$$

so a basis for the eigenspace of the eigenvalue 2 is $\begin{bmatrix} -1/2 \\ 0 \\ 1 \end{bmatrix}$. To find a basis for the eigenspace of the eigenvalue 3, we

find a basis for the nullspace of the matrix $3I - A$. First we find the reduced row-echelon form of $3I - A$:

$$3I - A = \begin{bmatrix} -1 & 0 & -1 \\ 2 & 2 & 0 \\ 2 & 0 & 2 \end{bmatrix} \xrightarrow{\text{row reduction}} \begin{bmatrix} 1 & 0 & 1 \\ 0 & 1 & -1 \\ 0 & 0 & 0 \end{bmatrix}.$$

The eigenspace of the eigenvector 3 consists of all vectors with the property that $x + z = 0$ and $y - z = 0$; that is, all vectors
$\begin{bmatrix} -t \\ t \\ t \end{bmatrix} = t \begin{bmatrix} -1 \\ 1 \\ 1 \end{bmatrix}$, so a basis for the eigenspace of the eigenvalue 3 is $\begin{bmatrix} -1 \\ 1 \\ 1 \end{bmatrix}$.

9. To find the eigenvalues of a matrix A, set the determinant of $\lambda I - A$ equal to zero and solve for λ. Since A is upper triangular with fours on the diagonal, the determinant of $\lambda I - A$ is $(\lambda - 4)^3$, and the only eigenvalue of A is 4. To find a basis for the eigenspace of the eigenvalue 4, we find a basis for the nullspace of the matrix $4I - A$. First we find the reduced row-echelon form of $4I - A$:

$$4I - A = \begin{bmatrix} 0 & 0 & 0 \\ -1 & 0 & 0 \\ 0 & -1 & 0 \end{bmatrix} \xrightarrow{\text{row reduction}} \begin{bmatrix} 1 & 0 & 0 \\ 0 & 1 & 0 \\ 0 & 0 & 0 \end{bmatrix}.$$

The eigenspace of the eigenvector 4 consists of all vectors with the property that $x = y = 0$; that is, all vectors

$$\begin{bmatrix} 0 \\ 0 \\ t \end{bmatrix} = t \begin{bmatrix} 0 \\ 0 \\ 1 \end{bmatrix},$$

so a basis for the eigenspace of the eigenvalue 4 is $\begin{bmatrix} 0 \\ 0 \\ 1 \end{bmatrix}$.

11. To find the eigenvalues of a matrix A, set the determinant of $\lambda I - A$ equal to zero and solve for λ.

$$\det(\lambda I - A) = \det\left(\begin{bmatrix} \lambda - 5 & -6 & -2 \\ 0 & \lambda + 1 & 8 \\ -1 & 0 & \lambda + 2 \end{bmatrix}\right) = \lambda^3 - 2\lambda^2 - 15\lambda + 36 = (\lambda - 3)^2(\lambda + 4) = 0,$$

so, solving for λ, we see that the eigenvalues of A are 3 and -4.

To find a basis for the eigenspace of the eigenvalue 3, we find a basis for the nullspace of the matrix $3I - A$. First we find the reduced row-echelon form of $3I - A$:

$$3I - A = \begin{bmatrix} 2 & -6 & -2 \\ 0 & 4 & 8 \\ -1 & 0 & 5 \end{bmatrix} \quad \xrightarrow{\text{row reduction}} \quad \begin{bmatrix} 1 & 0 & -5 \\ 0 & 1 & 2 \\ 0 & 0 & 0 \end{bmatrix}.$$

The eigenspace of the eigenvector 3 consists of all vectors with the property that $x - 5z = 0$ and $y + 2z = 0$; that is, all vectors $\begin{bmatrix} 5t \\ -2t \\ t \end{bmatrix} = t \begin{bmatrix} 5 \\ -2 \\ 1 \end{bmatrix}$, so a basis for the eigenspace of the eigenvalue 3 is $\begin{bmatrix} 5 \\ -2 \\ 1 \end{bmatrix}$. To find a basis for the eigenspace of the eigenvalue -4, we find a basis for the nullspace of the matrix $-4I - A$. First we find the reduced row-echelon form of $-4I - A$:

$$-4I - A = \begin{bmatrix} 9 & -6 & 2 \\ 0 & -3 & 8 \\ -1 & 0 & -2 \end{bmatrix} \quad \xrightarrow{\text{row reduction}} \quad \begin{bmatrix} 1 & 0 & 2 \\ 0 & 1 & -8/3 \\ 0 & 0 & 0 \end{bmatrix}.$$

The eigenspace of the eigenvector -4 consists of all vectors with the property that $x + 2z = 0$ and $y - 8z/3 = 0$; that is, all vectors $\begin{bmatrix} -2t \\ 8t/3 \\ t \end{bmatrix} = t \begin{bmatrix} -2 \\ 8/3 \\ 1 \end{bmatrix}$, so a basis for the eigenspace of the eigenvalue -4 is $\begin{bmatrix} -2 \\ 8/3 \\ 1 \end{bmatrix}$.

13. To find the eigenvalues of a matrix A, set the determinant of $\lambda I - A$ equal to zero and solve for λ.

$$\det(\lambda I - A) = \det\left(\begin{bmatrix} \lambda & -2 & -2 \\ -2 & \lambda & -2 \\ -2 & -2 & \lambda \end{bmatrix} \right) = \lambda^3 - 12\lambda - 16 = (\lambda + 2)^2(\lambda - 4) = 0,$$

so, solving for λ, we see that the eigenvalues of A are -2 and 4.

To find a basis for the eigenspace of the eigenvalue -2, we find a basis for the nullspace of the matrix $-2I - A$. First we find the reduced row-echelon form of $-2I - A$:

$$-2I - A = \begin{bmatrix} -2 & -2 & -2 \\ -2 & -2 & -2 \\ -2 & -2 & -2 \end{bmatrix} \quad \xrightarrow{\text{row reduction}} \quad \begin{bmatrix} 1 & 1 & 1 \\ 0 & 0 & 0 \\ 0 & 0 & 0 \end{bmatrix}.$$

The eigenspace of the eigenvector -2 consists of all vectors with the property that $x + y + z = 0$; that is, all vectors of the form $\begin{bmatrix} -s - t \\ s \\ t \end{bmatrix}$, which can be written as $s \begin{bmatrix} -1 \\ 1 \\ 0 \end{bmatrix} + t \begin{bmatrix} -1 \\ 0 \\ 1 \end{bmatrix}$, so a basis for the eigenspace of the eigenvalue -2 is given by the two vectors $\begin{bmatrix} -1 \\ 1 \\ 0 \end{bmatrix}$ and $\begin{bmatrix} -1 \\ 0 \\ 1 \end{bmatrix}$. To find a basis for the eigenspace of the eigenvalue 4, we find a basis for the nullspace of the matrix $4I - A$. First we find the reduced row-echelon form of $4I - A$:

$$4I - A = \begin{bmatrix} 4 & -2 & -2 \\ -2 & 4 & -2 \\ -2 & -2 & 4 \end{bmatrix} \quad \xrightarrow{\text{row reduction}} \quad \begin{bmatrix} 1 & 0 & -1 \\ 0 & 1 & -1 \\ 0 & 0 & 0 \end{bmatrix}.$$

The eigenspace of the eigenvector 4 consists of all vectors with the property that $x - z = 0$ and $y - z = 0$; that is, all vectors $\begin{bmatrix} t \\ t \\ t \end{bmatrix} = t \begin{bmatrix} 1 \\ 1 \\ 1 \end{bmatrix}$, so a basis for the eigenspace of the eigenvalue 4 is $\begin{bmatrix} 1 \\ 1 \\ 1 \end{bmatrix}$.

15. To find the eigenvalues of a matrix A, set the determinant of $\lambda I - A$ equal to zero and solve for λ.

$$\det(\lambda I - A) = \det\left(\begin{bmatrix} \lambda - 3 & 2 \\ -4 & \lambda + 1 \end{bmatrix} \right) = \lambda^2 - 2\lambda + 5 = 0,$$

so, solving for λ (use the quadratic formula), we see that the eigenvalues of A are $1 + 2i$ and $1 - 2i$.
To find a basis for the eigenspace of the eigenvalue $1 + 2i$, we find a basis for the nullspace of the matrix $(1 + 2i)I - A$. First we find the reduced row-echelon form of $(1 + 2i)I - A$:

$$(1 + 2i)I - A = \begin{bmatrix} -2 + 2i & 2 \\ -4 & 2 + 2i \end{bmatrix} \xrightarrow[\rightarrow]{\text{row reduction}} \begin{bmatrix} 1 & (-1 - i)/2 \\ 0 & 0 \end{bmatrix}.$$

The eigenspace of the eigenvector $1 + 2i$ consists of all vectors with the property that $x - (1 + i)y/2 = 0$; that is, all vectors $\begin{bmatrix} (1 + i)t/2 \\ t \end{bmatrix} = t \begin{bmatrix} (1 + i)/2 \\ 1 \end{bmatrix}$, so a basis for the eigenspace of the eigenvalue $1 + 2i$ is $\begin{bmatrix} 1/2 \\ 1 \end{bmatrix}$. To find a basis for the eigenspace of the eigenvalue $1 - 2i$, we find a basis for the nullspace of the matrix $(1 - 2i)I - A$. First we find the reduced row-echelon form of $(1 - 2i)I - A$:

$$(1 - 2i)I - A = \begin{bmatrix} -2 - 2i & 2 \\ -4 & 2 - 2i \end{bmatrix} \xrightarrow[\rightarrow]{\text{row reduction}} \begin{bmatrix} 1 & (-1 - 2i) \\ 0 & 0 \end{bmatrix}.$$

So a basis for the eigenspace of the eigenvalue $1 - 2i$ is $\begin{bmatrix} 1/2 - i/2 \\ 1 \end{bmatrix}$.

17. To find the eigenvalues of a matrix A, set the determinant of $\lambda I - A$ equal to zero and solve for λ.

$$\det(\lambda I - A) = \det\left(\begin{bmatrix} \lambda - 1 & 0 & 0 \\ 0 & \lambda & -1 \\ 0 & 1 & \lambda \end{bmatrix}\right) = (\lambda^2 + 1)(\lambda - 1) = 0,$$

so, solving for λ, we see that the eigenvalues of A are $i, -i$, and 1.
To find a basis for the eigenspace of the eigenvalue i, we find a basis for the nullspace of the matrix $iI - A$. First we find the reduced row-echelon form of $iI - A$:

$$iI - A = \begin{bmatrix} -1 + i & 0 & 0 \\ 0 & i & -1 \\ 0 & 1 & i \end{bmatrix} \xrightarrow[\rightarrow]{\text{row reduction}} \begin{bmatrix} 1 & 0 & 0 \\ 0 & 1 & i \\ 0 & 0 & 0 \end{bmatrix}.$$

The eigenspace of the eigenvector i consists of all vectors with the property that $x = 0$ and $y + iz = 0$; that is, all vectors of the form $\begin{bmatrix} 0 \\ -it \\ t \end{bmatrix}$, which can be written as $s \begin{bmatrix} 0 \\ -i \\ 1 \end{bmatrix}$, so a basis for the eigenspace of the eigenvalue i is given by the vector $\begin{bmatrix} 0 \\ -i \\ 1 \end{bmatrix}$. Use similar calculations to find the eigenspaces of $-i$ and 1.

5.5   Similar Matrices, Diagonalization, and Jordan Canonical Form

Exercises 5.5, pp. 286-287

1. By Theorem 5.20, we can tell if an $n \times n$ matrix A is diagonalizable by adding the dimensions of the eigenspaces of A. If they add up to n, then A is diagonalizable. On the corresponding diagonal matrix D, the eigenvalues of A will appear in their correct multiplicities on the diagonal. Construct the matrix P with the property that $P^{-1}AP = D$ by using the basis vectors of the eigenspaces of A as P's columns. (Arrange the columns so the order of the eigenvector columns matches the order of their eigenvalues on the diagonal of D.) If you don't have your worked-out problems from Section 5.4, use the Section 5.4 answers to get the information you need to work this problem.
   Since $\dim(E_{-1}) + \dim(E_3) = 1 + 1 = 2$, this matrix is diagonalizable. For $P = \begin{bmatrix} 0 & 1/2 \\ 1 & 1 \end{bmatrix}$, $P^{-1}AP = \begin{bmatrix} -1 & 0 \\ 0 & 3 \end{bmatrix}$.

3. By Theorem 5.20, we can tell if an $n \times n$ matrix A is diagonalizable by adding the dimensions of the eigenspaces of A. If they add up to n, then A is diagonalizable. If you don't have your worked-out problems from Section 5.4, use the Section 5.4 answers to get the information you need to work this problem. Since $\dim(E_4) = 1 \neq 2$, this matrix is not diagonalizable.

5. By Theorem 5.20, we can tell if an $n \times n$ matrix A is diagonalizable by adding the dimensions of the eigenspaces of A. If they add up to n, then A is diagonalizable. On the corresponding diagonal matrix D, the eigenvalues of A will appear in their correct multiplicities on the diagonal. Construct the matrix P with the property that $P^{-1}AP = D$ by using the basis vectors of the eigenspaces of A as P's columns. (Arrange the columns so the order of the eigenvector columns matches the order of their eigenvalues on the diagonal of D.) If you don't have your worked-out problems from Section 5.4, use the Section 5.4 answers to get the information you need to work this problem.
Since $\dim(E_{2\sqrt{3}}) + \dim(E_{2-\sqrt{3}}) = 1 + 1 = 2$, this matrix is diagonalizable.
For $P = \begin{bmatrix} \sqrt{3}/2 & -\sqrt{3}/2 \\ 1 & 1 \end{bmatrix}$, $P^{-1}AP = \begin{bmatrix} 2\sqrt{3} & 0 \\ 0 & -2\sqrt{3} \end{bmatrix}$.

7. By Theorem 5.20, we can tell if an $n \times n$ matrix A is diagonalizable by adding the dimensions of the eigenspaces of A. If they add up to n, then A is diagonalizable. On the corresponding diagonal matrix D, the eigenvalues of A will appear in their correct multiplicities on the diagonal. Construct the matrix P with the property that $P^{-1}AP = D$ by using the basis vectors of the eigenspaces of A as P's columns. (Arrange the columns so the order of the eigenvector columns matches the order of their eigenvalues on the diagonal of D.) If you don't have your worked-out problems from Section 5.4, use the Section 5.4 answers to get the information you need to work this problem.
Since $\dim(E_2) + \dim(E_3) + \dim(E_1) = 1 + 1 + 1 = 3$, this matrix is diagonalizable. For

$$P = \begin{bmatrix} -1/2 & 1 & 0 \\ 1 & 1 & 1 \\ 1 & 1 & 0 \end{bmatrix}, \; P^{-1}AP = \begin{bmatrix} 2 & 0 & 0 \\ 0 & 3 & 0 \\ 0 & 0 & 1 \end{bmatrix}.$$

9. By Theorem 5.20, we can tell if an $n \times n$ matrix A is diagonalizable by adding the dimensions of the eigenspaces of A. If they add up to n, then A is diagonalizable. If you don't have your worked-out problems from Section 5.4, use the Section 5.4 answers to get the information you need to work this problem. Since $\dim(E_4) = 1 \neq 3$, this matrix is not diagonalizable.

11. By Theorem 5.20, we can tell if an $n \times n$ matrix A is diagonalizable by adding the dimensions of the eigenspaces of A. If they add up to n, then A is diagonalizable. If you don't have your worked-out problems from Section 5.4, use the Section 5.4 answers to get the information you need to work this problem. Since $\dim(E_3 + \dim(E_4) = 1 + 1 = 2 \neq 3$, this matrix is not diagonalizable.

13. By Theorem 5.20, we can tell if an $n \times n$ matrix A is diagonalizable by adding the dimensions of the eigenspaces of A. If they add up to n, then A is diagonalizable. On the corresponding diagonal matrix D, the eigenvalues of A will appear in their correct multiplicities on the diagonal. Construct the matrix P with the property that $P^{-1}AP = D$ by using the basis vectors of the eigenspaces of A as P's columns. (Arrange the columns so the order of the eigenvector columns matches the order of their eigenvalues on the diagonal of D.) If you don't have your worked-out problems from Section 5.4, use the Section 5.4 answers to get the information you need to work this problem.
Since $\dim(E_4) + \dim(E_{-2}) = 1 + 2 = 3$, this matrix is diagonalizable. For

$$P = \begin{bmatrix} 1 & -1 & -1 \\ 1 & 1 & 0 \\ 1 & 0 & 1 \end{bmatrix}, \; P^{-1}AP = \begin{bmatrix} 4 & 0 & 0 \\ 0 & -2 & 0 \\ 0 & 0 & -2 \end{bmatrix}.$$

15. By Theorem 5.20, we can tell if an $n \times n$ matrix A is diagonalizable by adding the dimensions of the eigenspaces of A. If they add up to n, then A is diagonalizable. On the corresponding diagonal matrix D, the eigenvalues of A will appear in their correct multiplicities on the diagonal. Construct the matrix P with the property that $P^{-1}AP = D$ by using the basis vectors of the eigenspaces of A as P's columns. (Arrange the columns so the order of the eigenvector columns matches the order of their eigenvalues on the diagonal of D.) If you don't have your worked-out problems from Section 5.4, use the Section 5.4 answers to get the information you need to work this problem.
Since $\dim(E_{1+2i}) + \dim(E_{1-2i}) = 1 + 1 = 2$, this matrix is diagonalizable.
For $P = \begin{bmatrix} 1/2 + i/2 & 1/2 - i/2 \\ 1 & 1 \end{bmatrix}$, $P^{-1}AP = \begin{bmatrix} 1 + 2i & 0 \\ 0 & 1 - 2i \end{bmatrix}$.

17. By Theorem 5.20, we can tell if an $n \times n$ matrix A is diagonalizable by adding the dimensions of the eigenspaces of A. If they add up to n, then A is diagonalizable. On the corresponding diagonal matrix D, the eigenvalues of A will appear in their correct multiplicities on the diagonal. Construct the matrix P with the property that $P^{-1}AP = D$ by using the basis vectors of the eigenspaces of A as P's columns. (Arrange the columns so the order of the eigenvector columns matches the order of

their eigenvalues on the diagonal of D.) If you don't have your worked-out problems from Section 5.4, use the Section 5.4 answers to get the information you need to work this problem.

Since $\dim(E_1) + \dim(E_i) + \dim(E_{-i}) = 1 + 1 + 1 = 3$, this matrix is diagonalizable. For

$$P = \begin{bmatrix} 1 & 0 & 0 \\ 0 & -i & i \\ 0 & 1 & 1 \end{bmatrix}, \; P^{-1}AP = \begin{bmatrix} 1 & 0 & 0 \\ 0 & i & 0 \\ 0 & 0 & -i \end{bmatrix}.$$

19. Factor $\lambda^2 - 3\lambda + 2 = (\lambda - 1)(\lambda - 2)$. The roots of this polynomial are the eigenvalues of the matrix. Since each eigenvalue appears with multiplicity 1, there is only one possible Jordan canonical form: $\begin{bmatrix} 1 & 0 \\ 0 & 2 \end{bmatrix}$.

21. Factor $\lambda^3 - \lambda^2 = \lambda^2(\lambda - 1)$. The roots of this polynomial are the eigenvalues: 0 (with multiplicity 2) and 1 (with multiplicity 1). Since the eigenvalue 1 has multiplicity 1, its only possible Jordan block is [1]. The possible forms for B_0 are $\begin{bmatrix} 0 & 0 \\ 0 & 0 \end{bmatrix}$ or $\begin{bmatrix} 0 & 1 \\ 0 & 0 \end{bmatrix}$. So the possible Jordan canonical forms for this matrix are $\begin{bmatrix} 0 & 0 & 0 \\ 0 & 0 & 0 \\ 0 & 0 & 1 \end{bmatrix}$ or $\begin{bmatrix} 0 & 1 & 0 \\ 0 & 0 & 0 \\ 0 & 0 & 1 \end{bmatrix}$.

23. For $(\lambda - 5)(\lambda - 1)^2(\lambda + 2)^3$, we know that $B_5 = [5]$, B_1 is either $\begin{bmatrix} 1 & 0 \\ 0 & 1 \end{bmatrix}$ or $\begin{bmatrix} 1 & 1 \\ 0 & 1 \end{bmatrix}$, and B_{-2} is one of

$$\begin{bmatrix} -2 & 0 & 0 \\ 0 & -2 & 0 \\ 0 & 0 & -2 \end{bmatrix}, \begin{bmatrix} -2 & 1 & 0 \\ 0 & -2 & 0 \\ 0 & 0 & -2 \end{bmatrix} \text{ or } \begin{bmatrix} -2 & 1 & 0 \\ 0 & -2 & 1 \\ 0 & 0 & -2 \end{bmatrix}.$$

So, there are $1 \times 2 \times 3 = 6$ possible Jordan canonical forms, each consisting of a matrix with one choice of B_5, one choice of B_1, and one choice of B_{-2} on the diagonal, and zeros elsewhere. See the answers in the back of your textbook for a complete listing.

25. Since the matrix is diagonalizable (see Problem 7), its Jordan canonical form is the corresponding diagonal matrix:

$$\begin{bmatrix} 2 & 0 & 0 \\ 0 & 3 & 0 \\ 0 & 0 & 1 \end{bmatrix}.$$

27. Since this matrix is not diagonalizable (see Problem 3), its Jordan canonical form must be the only other possibility: $\begin{bmatrix} 4 & 1 \\ 0 & 4 \end{bmatrix}$.

29. Since this matrix is not diagonalizable (see Problem 11), its Jordan canonical form must be one of the other possibilities. There is only one form for B_{-4}, the 1×1 matrix $[-4]$. There are two possible forms for B_3, of which one is diagonal and must be rejected, since that would give us a diagonal matrix for the Jordan canonical form. Therefore, B_3 must be $\begin{bmatrix} 3 & 1 \\ 0 & 3 \end{bmatrix}$ and the Jordan canonical form must be $\begin{bmatrix} 3 & 1 & 0 \\ 0 & 3 & 0 \\ 0 & 0 & -4 \end{bmatrix}$.

5.6   Eigenvectors and Eigenvalues of Linear Transformations

Exercises 5.6, pp. 291-292

1. The matrix that expresses T is $A = \begin{bmatrix} 7 & 0 & -5 \\ 15 & 2 & -15 \\ 10 & 0 & -8 \end{bmatrix}$. Now that we have the matrix A, we can use the techniques of Section 5.5 to find that A has eigenvalues 2 (with basis for its eigenspace V_2 given by $\begin{bmatrix} 1 \\ 0 \\ 1 \end{bmatrix}$ and $\begin{bmatrix} 0 \\ 1 \\ 0 \end{bmatrix}$) and -3 (with its eigenspace

V_{-3} having basis $\begin{bmatrix} 1/2 \\ 3/2 \\ 1 \end{bmatrix}$). Since $\dim(V_2) + \dim(V_{-3}) = 2 + 1 = 3$, A (and hence T) is diagonalizable, and its Jordan

canonical form is its diagonal matrix, $\begin{bmatrix} 2 & 0 & 0 \\ 0 & 2 & 0 \\ 0 & 0 & -3 \end{bmatrix}$.

3. Write T in terms of the standard basis α for P_1: x, 1. Since $T(x) = T(1x+0) = 2(0)x+1+0 = 1 = 0x+1$ and

$$T(1) = T(0x+1) = 2(1)x+0+1 = 2x+1,$$

we know that the matrix for T in terms of this basis is $A = \begin{bmatrix} 0 & 2 \\ 1 & 1 \end{bmatrix}$. (The columns of A are the vectors to which T sends the basis vectors.) Now we can use the techniques of Section 5.5 on matrix A to find that A has eigenvalues -1 (with basis for its eigenspace V_{-1} given by $\begin{bmatrix} -2 \\ 1 \end{bmatrix} = -2x+1$) and 2 (with its eigenspace V_2 having basis $\begin{bmatrix} 1 \\ 1 \end{bmatrix} = x+1$). Since $\dim(V_{-1}) + \dim(V_2) = 1 + 1 = 2$ (which is the dimension of P_1), A (and hence T) is diagonalizable, and its Jordan canonical form is its diagonal matrix, $\begin{bmatrix} -1 & 0 \\ 0 & 2 \end{bmatrix}$.

5. Solve this linear constant coefficient differential equation by solving $\lambda^2 - 1 = 0$. So $\lambda = 1$ or $\lambda = -1$. Thus V is the set of all functions $y = c_1 e^x + c_2 e^{-x}$, with basis e^x, e^{-x}. D sends $e^x = 1e^x + 0e^{-x}$ to its derivative $e^x = 1e^x + 0e^{-x}$ and sends $e^{-x} = 0e^x + 1e^{-x}$ to its derivative $-e^{-x} = 0e^x - 1e^{-x}$. So in terms of this basis, D is given by the matrix $A = \begin{bmatrix} 1 & 0 \\ 0 & -1 \end{bmatrix}$. (The columns of A are the vectors to which D sends the basis vectors.)

Now we can use the techniques of Section 5.5 on matrix A to find that A has eigenvalues -1 (with basis for its eigenspace V_{-1} given by $\begin{bmatrix} 0 \\ 1 \end{bmatrix} = e^{-x}$) and 1 (with its eigenspace V_1 having basis $\begin{bmatrix} 1 \\ 0 \end{bmatrix} = e^x$). Since $\dim(V_{-1}) + \dim(V_1) = 1 + 1 = 2$ (which is the dimension of V), A (and hence D) is diagonalizable, and its Jordan canonical form is its diagonal matrix, $\begin{bmatrix} 1 & 0 \\ 0 & -1 \end{bmatrix}$.

7. Hint: To show this, use change of basis matrices (which are invertible), and the fact that $\det(AB) = \det(A)\det(B)$.

9. Simply find the determinant of the matrix that we found in Problem 3. It is -2.

Chapter 6

Systems of Differential Equations

6.1   The Theory of Systems of Linear Differential Equations

Exercises 6.1, pp. 301-302

1. Let $Y = \begin{bmatrix} c_1 e^{2x} + c^2 e^{3x} \\ 2c_1 e^{2x} + c^2 e^{3x} \end{bmatrix}$. Then $Y' = \begin{bmatrix} 2c_1 e^{2x} + 3c^2 e^{3x} \\ 4c_1 e^{2x} + 3c^2 e^{3x} \end{bmatrix}$.

   We also have $\begin{bmatrix} 4 & -1 \\ 2 & 1 \end{bmatrix} \begin{bmatrix} c_1 e^{2x} + c^2 e^{3x} \\ 2c_1 e^{2x} + c^2 e^{3x} \end{bmatrix} = \begin{bmatrix} 2c_1 e^{2x} + 3c^2 e^{3x} \\ 4c_1 e^{2x} + 3c^2 e^{3x} \end{bmatrix}$.

   Thus Y is a solution to the given system of equations.

3. The determinant $\begin{vmatrix} e^{2x} \cos 3x & e^{2x} \sin 3x \\ -e^{2x} \sin 3x & e^{2x} \cos 3x \end{vmatrix} = e^{4x} \cos^2 3x + e^{4x} \sin^2 3x = e^{4x} \neq 0$ for any $x \in (-\infty, \infty)$. Therefore the given columns of functions are linearly independent on $(-\infty, \infty)$.

5. The system of differential equation $Y' = AY$ is equivalent to the system

$$\begin{aligned} y_1' &= y_1 \\ y_2' &= -2y_2 \end{aligned}$$

   which has the general solution

$$y_1 = c_1 e^x, \; y_2 = c_2 e^{-2x}.$$

   If we let $Y_1 = \begin{bmatrix} e^x \\ 0 \end{bmatrix}$, and $Y_2 = \begin{bmatrix} 0 \\ e^{-2x} \end{bmatrix}$, then $M = \begin{bmatrix} Y_1 & Y_2 \end{bmatrix} = \begin{bmatrix} e^x & 0 \\ 0 & e^{-2x} \end{bmatrix}$ is a matrix of fundamental solutions.

7. The system of differential equation $Y' = AY$ is equivalent to the system

$$\begin{aligned} y_1' &= -y_1 \\ y_2' &= 0 \\ y_3' &= 4y_3 \end{aligned}$$

   which has the general solution

$$y_1 = c_1 e^{-x}, \; y_2 = c_2, \; y_3 = c_3 e^{4x}.$$

   If we let $Y_1 = \begin{bmatrix} e^{-x} \\ 0 \\ 0 \end{bmatrix}$, $Y_2 = \begin{bmatrix} 0 \\ 1 \\ 0 \end{bmatrix}$ and $Y_3 = \begin{bmatrix} 0 \\ 0 \\ e^{4x} \end{bmatrix}$, then

$$M = \begin{bmatrix} Y_1 & Y_2 & Y3 \end{bmatrix} = \begin{bmatrix} e^{-x} & 0 & 0 \\ 0 & 1 & 0 \\ 0 & 0 & e^{4x} \end{bmatrix}$$

   is a matrix of fundamental solutions.

9. From Exercise 5 we have the general solution $Y = \begin{bmatrix} c_1 e^x \\ c_2 e^{-2x} \end{bmatrix}$. Using the initial condition we obtain

$$Y(0) = \begin{bmatrix} 2 \\ 1 \end{bmatrix} = \begin{bmatrix} c_1 \\ c_2 \end{bmatrix}.$$

Thus the solution to the initial value problem is

$$Y = \begin{bmatrix} 2e^x \\ e^{-2x} \end{bmatrix},$$

or equivalently $y_1 = 2e^x$, $y_2 = e^{-2x}$.

11. From Exercise 7 we have the general solution $Y = \begin{bmatrix} c_1 e^{-x} \\ c_2 \\ c_3 e^{4x} \end{bmatrix}$. Using the initial condition we obtain

$$Y(0) = \begin{bmatrix} 2 \\ 1 \\ 0 \end{bmatrix} = \begin{bmatrix} c_1 \\ c_2 \\ c_3 \end{bmatrix}.$$

Thus the solution to the initial value problem is

$$Y = \begin{bmatrix} 2e^{-x} \\ 1 \\ 0 \end{bmatrix},$$

or equivalently $y_1 = 2e^{-x}$, $y_2 = 1$, $y_3 = 0$.

13. From Exercise 5 we already have the solutions to the corresponding homogeneous equations. These are $y_1 = c_1 e^x$ and $y_2 = c_2 e^{-2x}$. We now look for particular solutions to each of the following equations

$$y_1' = y_1 + 2 \text{ and } y_2' = -2y_2 + x.$$

For the first equation we look for a particular solution of the form $y_{1P} = A$. Here, $y_{1P}' = 0$, so we have $0 = A + 2$. Therefore, $A = -2$. The particular solution is $y_{1P} = -2$.

For the second equation, we look for a particular solution of the form $y_{2P} = Bx + C$. Now, $y_{2P}' = B$, so from the second equation we have $B = -2(Bx + C) + x$. From this equation we obtain the system

$$-2B + 1 = 0, \ 1 = -2C$$

which has the solution $B = \frac{1}{4}$, $C = -\frac{1}{2}$. Thus, the particular solution for this equation is $y_{2P} = \frac{1}{4}x - \frac{1}{2}$. Therefore the solution to the nonhomogeneous system is

$$y_1 = c_1 e^x - 2, \ y_2 = c_2 e^{-2x} + \frac{1}{4}x - \frac{1}{2}.$$

15. From Exercise 7 we already have the solutions to the corresponding homogeneous equations. These are

$$y_1 = c_1 e^{-x}, \ y_2 = c_2 \text{ and } y_3 = c_3 e^{4x}.$$

We now look for particular solutions to each of the following equations

$$y_1' = -y_1 + 1 - 2x, \ y_2' = xe^{-x} \text{ and } y_3' = 4y_4 + \cos 2x.$$

For the first equation we look for a particular solution of the form $y_{1P} = Ax + B$. Here, $y_{1P}' = A$, so we have

$$A = -(Ax + B) + 1 - 2x.$$

From this equation we obtain the system

$$0 = -A - 2, \ A = -B + 1$$

which has the solution $A = -2$, $B = 3$. The particular solution is $y_{1P} = -2A + 3$.

For the second equation, we look for a particular solution of the form $y_{2P} = Cxe^{-x} + De^{-x}$. Now,

$$y'_{2P} = Ce^{-x} - Cxe^{-x} - De^{-x}.$$

From the second equation we have

$$Ce^{-x} - Cxe^{-x} - De^{-x} = xe^{-x}.$$

From this equation we obtain the system

$$C - D = 0, \ -C = 1$$

which has the solution $C = D = -1$. Thus, the particular solution for this equation is $y_{2P} = -e^{-x}(x+1)$.

Finally, for the third equation we look for a particular solution of the form

$$y_{3P} = E\cos 2x + F\sin 2x.$$

Here, $y'_{3P} = -2E\sin 2x + 2F\cos 2x$. We have

$$-2E\sin 2x + 2F\cos 2x = 4(E\cos 2x + F\sin 2x) + \cos 2x.$$

We obtain the system

$$-2E = 4F, \ 2F = 4E + 1$$

which has the solution $E = -\frac{1}{5}$, $F = \frac{1}{10}$. Thus, the particular solution for the third equation is

$$y_{3P} = -\frac{1}{5}\cos 2x + \frac{1}{10}\sin 2x.$$

Therefore the solution to the nonhomogeneous system is

$$y_1 = c_1 e^{-x} - 2A + 3, \ y_2 = c_2 - e^{-x}(x+1), \ y_3 = c_3 e^{4x} - \frac{1}{5}\cos 2x + \frac{1}{10}\sin 2x.$$

25. Since $A(x) = \begin{bmatrix} 4x & -e^x \\ 2e^{-x} & x \end{bmatrix}$ we have $A'(x) = \begin{bmatrix} 4 & -e^x \\ -2e^{-x} & 1 \end{bmatrix}$.

Thus $(3A)' = \begin{bmatrix} 12 & -3e^x \\ -6e^{-x} & 3 \end{bmatrix}$.

27. Here $A(x) = \begin{bmatrix} 4x & -e^x \\ 2e^{-x} & x \end{bmatrix}$ and $B(x) = \begin{bmatrix} e^{2x} & -x \\ 2x & e^{3x} \end{bmatrix}$.

Thus $AB = \begin{bmatrix} 4xe^{2x} - 2xe^x & -4x^2 - e^{4x} \\ 2x^2 + 2e^x & -2xe^{-x} + xe^{3x} \end{bmatrix}$

and $(AB)' = \begin{bmatrix} 4e^{2x} + 8xe^{2x} - 2e^x - 2xe^x & -8x - 4e^{4x} \\ 4x + 2e^x & 2xe^{-x} - 2e^{-x} + e^{3x} + 3xe^{3x} \end{bmatrix}$.

6.2   Homogeneous Systems with Constant Coefficients: The Diagonalizable Case

Exercises 6.2, pp. 311

1. We have the system $Y' = \begin{bmatrix} 3 & 0 \\ 8 & -1 \end{bmatrix} Y$. From Exercise 1 of Section 5.4 we know that the eigenvalues of matrix A are $\lambda = 3$ and $\lambda = -1$. The eigenspace E_3 has $\begin{bmatrix} \frac{1}{2} \\ 1 \end{bmatrix}$ as a basis vector. The eigenspace E_{-1} has $\begin{bmatrix} 0 \\ 1 \end{bmatrix}$ as a basis vector. Matrix A is similar to the diagonal matrix $D = P^{-1}AP = \begin{bmatrix} 3 & 0 \\ 0 & -1 \end{bmatrix}$ where $P = \begin{bmatrix} \frac{1}{2} & 0 \\ 1 & 1 \end{bmatrix}$. The columns of P are the basis vectors for

the eigenspaces. We also know that $\begin{bmatrix} c_1e^{3x} \\ c_2e^{-x} \end{bmatrix}$ is the general solution to $Y' = DY$. Therefore the general solution to $Y' = AY$ is

$$P\begin{bmatrix} c_1e^{3x} \\ c_2e^{-x} \end{bmatrix} = \begin{bmatrix} \frac{1}{2} & 0 \\ 1 & 1 \end{bmatrix}\begin{bmatrix} c_1e^{3x} \\ c_2e^{-x} \end{bmatrix} = \begin{bmatrix} \frac{1}{2}c_1e^{3x} \\ c_1e^{3x} + c_2e^{-x} \end{bmatrix}.$$

Thus, $y_1 = \frac{1}{2}c_1e^{3x}$, $y_2 = c_1e^{3x} + c_2e^{-x}$. Incorporating the fraction $\frac{1}{2}$ into the constant c_1 this system is equivalent to the answer in the text

$$y_1 = c_1e^{3x}, \ y_2 = 2c_1e^{3x} + c_2e^{-x}.$$

3. We have the system $Y' = \begin{bmatrix} 0 & 3 \\ 4 & 0 \end{bmatrix}Y$. From Exercise 5 of Section 5.4 we know that the eigenvalues of matrix A are $\lambda = 2\sqrt{3}$ and $\lambda = -2\sqrt{3}$. The eigenspace $E_{2\sqrt{3}}$ has $\begin{bmatrix} \frac{\sqrt{3}}{2} \\ 1 \end{bmatrix}$ as a basis vector. The eigenspace $E_{-2\sqrt{3}}$ has $\begin{bmatrix} -\frac{\sqrt{3}}{2} \\ 1 \end{bmatrix}$ as a basis vector.

Matrix A is similar to the diagonal matrix $D = P^{-1}AP = \begin{bmatrix} 2\sqrt{3} & 0 \\ 0 & -2\sqrt{3} \end{bmatrix}$ where $P = \begin{bmatrix} \frac{\sqrt{3}}{2} & -\frac{\sqrt{3}}{2} \\ 1 & 1 \end{bmatrix}$. The columns of P are the basis vectors for the eigenspaces. We also know that $\begin{bmatrix} c_1e^{2\sqrt{3}x} \\ c_2e^{-2\sqrt{3}x} \end{bmatrix}$ is the general solution to $Y' = DY$. Therefore the general solution to $Y' = AY$ is

$$P\begin{bmatrix} c_1e^{2\sqrt{3}x} \\ c_2e^{-2\sqrt{3}x} \end{bmatrix} = \begin{bmatrix} \frac{\sqrt{3}}{2} & -\frac{\sqrt{3}}{2} \\ 1 & 1 \end{bmatrix}\begin{bmatrix} c_1e^{2\sqrt{3}x} \\ c_2e^{-2\sqrt{3}x} \end{bmatrix} = \begin{bmatrix} \frac{\sqrt{3}}{2}c_1e^{2\sqrt{3}x} - \frac{\sqrt{3}}{2}c_2e^{-2\sqrt{3}x} \\ c_1e^{2\sqrt{3}x} + c_2e^{-2\sqrt{3}x} \end{bmatrix}.$$

Thus, $y_1 = \frac{\sqrt{3}}{2}c_1e^{2\sqrt{3}x} - \frac{\sqrt{3}}{2}c_2e^{-2\sqrt{3}x}$, $y_2 = c_1e^{2\sqrt{3}x} + c_2e^{-2\sqrt{3}x}$.

Note: Although the form is slightly different, this answer is equivalent to the answer in the text.

5. We have the system $Y' = \begin{bmatrix} 4 & 0 & 1 \\ -2 & 1 & 0 \\ -2 & 0 & 1 \end{bmatrix}Y$. From Exercise 7 of Section 5.4 we know that the eigenvalues of matrix A are $\lambda = 1$, $\lambda = 2$ and $\lambda = 3$. The eigenspace E_1 has $\begin{bmatrix} 0 \\ 1 \\ 0 \end{bmatrix}$ as a basis vector. The eigenspace E_2 has $\begin{bmatrix} -\frac{1}{2} \\ 1 \\ 1 \end{bmatrix}$ as a basis vector.

The eigenspace E_3 has $\begin{bmatrix} -1 \\ 1 \\ 1 \end{bmatrix}$ as a basis vector. Matrix A is similar to the diagonal matrix $D = P^{-1}AP = \begin{bmatrix} 1 & 0 & 0 \\ 0 & 2 & 0 \\ 0 & 0 & 3 \end{bmatrix}$ where $P = \begin{bmatrix} 0 & -\frac{1}{2} & -1 \\ 1 & 1 & 1 \\ 0 & 1 & 1 \end{bmatrix}$. The columns of P are the basis vectors for the eigenspaces. We also know that $\begin{bmatrix} c_1e^x \\ c_2e^{2x} \\ c_3e^{3x} \end{bmatrix}$ is the general solution to $Y' = DY$. Therefore the general solution to $Y' = AY$ is

$$P\begin{bmatrix} c_1e^x \\ c_2e^{2x} \\ c_3e^{3x} \end{bmatrix} = \begin{bmatrix} 0 & -\frac{1}{2} & -1 \\ 1 & 1 & 1 \\ 0 & 1 & 1 \end{bmatrix}\begin{bmatrix} c_1e^x \\ c_2e^{2x} \\ c_3e^{3x} \end{bmatrix} = \begin{bmatrix} -\frac{1}{2}c_2e^{2x} - c_3e^{3x} \\ c_1e^x + c_2e^{2x} + c_3e^{3x} \\ c_2e^{2x} + c_3e^{3x} \end{bmatrix}.$$

Thus, $y_1 = -\frac{1}{2}c_2e^{2x} - c_3e^{3x}$, $y_2 = c_1e^x + c_2e^{2x} + c_3e^{3x}$, $y_3 = c_2e^{2x} + c_3e^{3x}$.

Incorporating the fraction $-\frac{1}{2}$ into the constant c_2 this system is equivalent to

$$y_1 = c_2e^{2x} - c_3e^{3x}, \ y_2 = c_1e^x - 2c_2e^{2x} + c_3e^{3x}, \ y_3 = -2c_2e^{2x} + c_3e^{3x}.$$

7. We have the system $Y' = \begin{bmatrix} 0 & 2 & 2 \\ 2 & 0 & 2 \\ 2 & 2 & 0 \end{bmatrix}Y$. From Exercise 13 of Section 5.4 we know that the eigenvalues of matrix A are $\lambda = -2$ and $\lambda = 4$. The eigenspace E_{-2} has $\begin{bmatrix} -1 \\ 1 \\ 0 \end{bmatrix}$ and $\begin{bmatrix} -1 \\ 0 \\ 1 \end{bmatrix}$ as basis vectors. The eigenspace E_4 has $\begin{bmatrix} 1 \\ 1 \\ 1 \end{bmatrix}$ as a

basis vector. Matrix A is similar to the diagonal matrix $D = P^{-1}AP = \begin{bmatrix} -2 & 0 & 0 \\ 0 & -2 & 0 \\ 0 & 0 & 4 \end{bmatrix}$ where $P = \begin{bmatrix} -1 & -1 & 1 \\ 1 & 0 & 1 \\ 0 & 1 & 1 \end{bmatrix}$. The

columns of P are the basis vectors for the eigenspaces. We also know that $\begin{bmatrix} c_1 e^{-2x} \\ c_2 e^{-2x} \\ c_3 e^{4x} \end{bmatrix}$ is the general solution to $Y' = DY$.

Therefore the general solution to $Y' = AY$ is

$$P \begin{bmatrix} c_1 e^{-2x} \\ c_2 e^{-2x} \\ c_3 e^{4x} \end{bmatrix} = \begin{bmatrix} -1 & -1 & 1 \\ 1 & 0 & 1 \\ 0 & 1 & 1 \end{bmatrix} \begin{bmatrix} c_1 e^{-2x} \\ c_2 e^{-2x} \\ c_3 e^{4x} \end{bmatrix} = \begin{bmatrix} -c_1 e^{-2x} - c_2 e^{-2x} + c_3 e^{4x} \\ c_1 e^{-2x} + c_3 e^{4x} \\ c_2 e^{-2x} + c_3 e^{4x} \end{bmatrix}.$$

Thus, $y_1 = -c_1 e^{-2x} - c_2 e^{-2x} + c_3 e^{4x}$, $y_2 = c_1 e^{-2x} + c_3 e^{4x}$, $y_3 = c_2 e^{-2x} + c_3 e^{4x}$.

9. We have the system $Y' = \begin{bmatrix} 3 & -2 \\ 4 & -1 \end{bmatrix} Y$. From Exercise 15 of Section 5.4 we know that the eigenvalues of matrix A are

$\lambda = 1 \pm 2i$. The eigenspace E_{1+2i} has $\begin{bmatrix} \frac{1}{2} + \frac{1}{2}i \\ 1 \end{bmatrix}$ as a basis vector. The eigenspace E_{1-2i} has $\begin{bmatrix} \frac{1}{2} - \frac{1}{2}i \\ 1 \end{bmatrix}$ as a basis vector.

Matrix A is similar to the diagonal matrix $D = P^{-1}AP = \begin{bmatrix} 1+2i & 0 \\ 0 & 1-2i \end{bmatrix}$ where $P = \begin{bmatrix} \frac{1}{2} + \frac{1}{2}i & \frac{1}{2} - \frac{1}{2}i \\ 1 & 1 \end{bmatrix}$. The columns

of P are the basis vectors for the eigenspaces. One complex valued solution to $Y' = DY$ is

$$Z = \begin{bmatrix} e^{(1+2i)x} \\ 0 \end{bmatrix} = \begin{bmatrix} e^x \cos 2x + i e^x \sin 2x \\ 0 \end{bmatrix}.$$

By the first part of Theorem 6.7,

$$\begin{aligned} PZ &= \begin{bmatrix} \frac{1}{2} + \frac{1}{2}i & \frac{1}{2} - \frac{1}{2}i \\ 1 & 1 \end{bmatrix} \begin{bmatrix} e^x \cos 2x + i e^x \sin 2x \\ 0 \end{bmatrix} \\ &= \begin{bmatrix} \left(\frac{1}{2}e^x \cos 2x - \frac{1}{2}e^x \sin 2x \right) + i\left(\frac{1}{2}e^x \sin 2x + \frac{1}{2}e^x \cos 2x \right) \\ e^x \cos 2x + i e^x \sin 2x \end{bmatrix} \\ &= \begin{bmatrix} \frac{1}{2}e^x \cos 2x - \frac{1}{2}e^x \sin 2x \\ e^x \cos 2x \end{bmatrix} + i \begin{bmatrix} \frac{1}{2}e^x \sin 2x + \frac{1}{2}e^x \cos 2x \\ e^x \sin 2x \end{bmatrix} \end{aligned}$$

is a complex valued solution to $Y' = AY$. By Theorem 6.8

$$\begin{bmatrix} \frac{1}{2}e^x \cos 2x - \frac{1}{2}e^x \sin 2x \\ e^x \cos 2x \end{bmatrix} \quad \text{and} \quad \begin{bmatrix} \frac{1}{2}e^x \sin 2x + \frac{1}{2}e^x \cos 2x \\ e^x \sin 2x \end{bmatrix}$$

are real valued solutions to $Y' = AY$.

The general solution to $Y' = AY$ is then

$$c_1 \begin{bmatrix} \frac{1}{2}e^x \cos 2x - \frac{1}{2}e^x \sin 2x \\ e^x \cos 2x \end{bmatrix} + c_2 \begin{bmatrix} \frac{1}{2}e^x \sin 2x + \frac{1}{2}e^x \cos 2x) \\ e^x \sin 2x \end{bmatrix} =$$
$$\begin{bmatrix} c_1 \left(\frac{1}{2}e^x \cos 2x - \frac{1}{2}e^x \sin 2x \right) + c_2 \left(\frac{1}{2}e^x \sin 2x + \frac{1}{2}e^x \cos 2x \right) \\ c_1 e^x \cos 2x + c_2 e^x \sin 2x \end{bmatrix}.$$

Thus, $y_1 = c_1 \left(\frac{1}{2}e^x \cos 2x - \frac{1}{2}e^x \sin 2x \right) + c_2 \left(\frac{1}{2}e^x \sin 2x + \frac{1}{2}e^x \cos 2x \right)$, $y_2 = c_1 e^x \cos 2x + c_2 e^x \sin 2x$. Incorporating the fraction $\frac{1}{2}$ into the constants c_1 and c_2 this system is equivalent to

$$y_1 = c_1 \left(e^x \cos 2x - e^x \sin 2x \right) + c_2 \left(e^x \sin 2x + e^x \cos 2x \right), \ y_2 = 2c_1 e^x \cos 2x + 2c_2 e^x \sin 2x.$$

Please note the answer to this problem in the back of the text is incorrect.

11. We have the system $Y' = \begin{bmatrix} 1 & 0 & 0 \\ 0 & 0 & 1 \\ 0 & -1 & 0 \end{bmatrix} Y$. From Exercise 15 of Section 5.4 we know that the eigenvalues of matrix A are

$\lambda = 1$ and $\lambda = \pm i$. The eigenspace E_1 has $\begin{bmatrix} 1 \\ 0 \\ 0 \end{bmatrix}$ as a basis vector. The eigenspace E_i has $\begin{bmatrix} 0 \\ -i \\ 1 \end{bmatrix}$ as a basis vector. The

eigenspace E_{-i} has $\begin{bmatrix} 0 \\ i \\ 1 \end{bmatrix}$ as a basis vector. Matrix A is similar to the diagonal matrix $D = P^{-1}AP = \begin{bmatrix} 1 & 0 & 0 \\ 0 & i & 0 \\ 0 & 0 & -i \end{bmatrix}$ where

$P = \begin{bmatrix} 1 & 0 & 0 \\ 0 & -i & i \\ 0 & 1 & 1 \end{bmatrix}$. The columns of P are the basis vectors for the eigenspaces.

From the diagonal entry 1 of matrix D, we obtain one solution to $Y' = DY$: $\begin{bmatrix} e^x \\ 0 \\ 0 \end{bmatrix}$. This gives us the solution

$$P \begin{bmatrix} e^x \\ 0 \\ 0 \end{bmatrix} = \begin{bmatrix} 1 & 0 & 0 \\ 0 & -i & i \\ 0 & 1 & 1 \end{bmatrix} \begin{bmatrix} e^x \\ 0 \\ 0 \end{bmatrix} = \begin{bmatrix} e^x \\ 0 \\ 0 \end{bmatrix}$$

to $Y' = AY$.

From the diagonal entry i of D we obtain the complex valued solution to $Y' = DY$,

$$Z = \begin{bmatrix} 0 \\ e^{ix} \\ 0 \end{bmatrix} = \begin{bmatrix} 0 \\ \cos x + i \sin x \\ 0 \end{bmatrix}.$$

By the first part of Theorem 6.7,

$$
\begin{aligned}
PZ &= \begin{bmatrix} 1 & 0 & 0 \\ 0 & -i & i \\ 0 & 1 & 1 \end{bmatrix} \begin{bmatrix} 0 \\ \cos x + i \sin x \\ 0 \end{bmatrix} \\
&= \begin{bmatrix} 0 \\ \sin x - i \cos x \\ \cos x + i \sin x \end{bmatrix} \\
&= \begin{bmatrix} 0 \\ \sin x \\ \cos x \end{bmatrix} + i \begin{bmatrix} 0 \\ -\cos x \\ \sin x \end{bmatrix}
\end{aligned}
$$

is a complex valued solution to $Y' = AY$. By Theorem 6.8 $\begin{bmatrix} 0 \\ \sin x \\ \cos x \end{bmatrix}$ and $\begin{bmatrix} 0 \\ -\cos x \\ \sin x \end{bmatrix}$ are real valued solutions to $Y' = AY$.

The general solution to $Y' = AY$ is then

$$c_1 \begin{bmatrix} e^x \\ 0 \\ 0 \end{bmatrix} + c_2 \begin{bmatrix} 0 \\ \sin x \\ \cos x \end{bmatrix} + c_3 \begin{bmatrix} 0 \\ -\cos x \\ \sin x \end{bmatrix} = \begin{bmatrix} c_1 e^x \\ c_2 \sin x - c_3 \cos x \\ c_2 \cos x + c_3 \sin x \end{bmatrix}.$$

Thus, $y_1 = c_1 e^x$, $y_2 = c_2 \sin x - c_3 \cos x$, $y_3 = c_2 \cos x + c_3 \sin x$.

Note: Although the form is slightly different, this answer is equivalent to the answer in the text.

13. From Exercise 1 we have the general solution $y_1 = \frac{1}{2}c_1 e^{3x}$, $y_2 = c_1 e^{3x} + c_2 e^{-x}$. We are now given the initial values

$$Y(0) = \begin{bmatrix} 2 \\ 1 \end{bmatrix}.$$

We have the system of equations

$$y_1(0) = \frac{1}{2}c_1 = 2, \; y_2(0) = c_1 + c_2 = 1.$$

This system has the solution $c_1 = 4$, $c_2 = -3$. Thus the solution to the initial value problem is

$$y_1 = 2e^{3x}, \; y_2 = 4e^{3x} - 3e^{-x}.$$

15. From Exercise 5 we have the general solution

$$y_1 = -\frac{1}{2}c_2 e^{2x} - c_3 e^{3x}, \; y_2 = c_1 e^x + c_2 e^{2x} + c_3 e^{3x}, \; y_3 = c_2 e^{2x} + c_3 e^{3x}.$$

We are now given the initial values $Y(0) = \begin{bmatrix} 0 \\ 1 \\ -1 \end{bmatrix}$. We have the system of equations

$$y_1(0) = \frac{1}{2}c_2 - c_3 = 0, \; y_2(0) = c_1 + c_2 + c_3 = 1, \; y_3(0) = c_2 + c_3 = -1.$$

This system has the solution $c_1 = 2$, $c_2 = -2$, $c_3 = 1$. Thus the solution to the initial value problem is

$$y_1 = e^{2x} - e^{3x}, \; y_2 = 2e^x - 2e^{2x} + e^{3x}, \; y_3 = -2e^{2x} + e^{3x}.$$

17. From Exercise 9 we have the general solution

$$y_1 = c_1 \left(\frac{1}{2}e^x \cos 2x - \frac{1}{2}e^x \sin 2x \right) + c_2 \left(\frac{1}{2}e^x \sin 2x + \frac{1}{2}e^x \cos 2x \right), \; y_2 = c_1 e^x \cos 2x + c_2 e^x \sin 2x.$$

We are now given the initial values $Y(0) = \begin{bmatrix} 2 \\ 1 \end{bmatrix}$. We have the system of equations

$$y_1(0) = \frac{1}{2}c_1 + \frac{1}{2}c_2 = 2, \; y_2(0) = c_1 = 1.$$

This system has the solution $c_1 = 1$, $c_2 = 3$. Thus the solution to the initial value problem is

$$y_1 = 1\left(\frac{1}{2}e^x \cos 2x - \frac{1}{2}e^x \sin 2x \right) + 3\left(\frac{1}{2}e^x \sin 2x + \frac{1}{2}e^x \cos 2x \right), \; y_2 = 1e^x \cos 2x + 3e^x \sin 2x.$$

More simply,

$$y_1 = e^x(2\cos 2x + \sin 2x), \; y_2 = e^x(\cos 2x + 3\sin 2x).$$

19. We have the system $Y' = AY$, Where $A = \begin{bmatrix} 8 & -10 \\ 5 & -7 \end{bmatrix}$. The characteristic equation of A is

$$\det(\lambda I - A) = \begin{vmatrix} \lambda - 8 & 10 \\ -5 & \lambda + 7 \end{vmatrix}$$

$$= (\lambda - 8)(\lambda + 7) + 50 = \lambda^2 - \lambda - 6 = (\lambda + 2)(\lambda - 3) = 0.$$

Therefore, the eigenvalues of A are $\lambda = -2$ and $\lambda = 3$. We now wish to find bases for the corresponding eigenspaces.

For $\lambda = -2$ we need to find a basis for the null space of $-2I - A = \begin{bmatrix} -10 & 10 \\ -5 & 5 \end{bmatrix}$. Reducing the matrix for the associated

homogeneous system, $\begin{bmatrix} -10 & 10 & \vdots & 0 \\ -5 & 5 & \vdots & 0 \end{bmatrix} \rightarrow \begin{bmatrix} 1 & -1 & \vdots & 0 \\ 0 & 0 & \vdots & 0 \end{bmatrix}$, we see our solutions are

$$\begin{bmatrix} x \\ y \end{bmatrix} = \begin{bmatrix} y \\ y \end{bmatrix} = y \begin{bmatrix} 1 \\ 1 \end{bmatrix}$$

and hence the column vector $\begin{bmatrix} 1 \\ 1 \end{bmatrix}$ forms a basis for E_{-2}.

For $\lambda = 3$ we need to find a basis for the null space of $3I - A = \begin{bmatrix} -5 & 10 \\ -5 & 10 \end{bmatrix}$. Reducing the matrix for the associated

homogeneous system, $\begin{bmatrix} -5 & 10 & \vdots & 0 \\ -5 & 10 & \vdots & 0 \end{bmatrix} \rightarrow \begin{bmatrix} 1 & -2 & \vdots & 0 \\ 0 & 0 & \vdots & 0 \end{bmatrix}$, we see our solutions are

$$\begin{bmatrix} x \\ y \end{bmatrix} = \begin{bmatrix} 2y \\ y \end{bmatrix} = y \begin{bmatrix} 2 \\ 1 \end{bmatrix}$$

and hence the column vector $\begin{bmatrix} 2 \\ 1 \end{bmatrix}$ forms a basis for E_3.

Matrix A is similar to the diagonal matrix $D = P^{-1}AP = \begin{bmatrix} -2 & 0 \\ 0 & 3 \end{bmatrix}$ where $P = \begin{bmatrix} 1 & 2 \\ 1 & 1 \end{bmatrix}$. The columns of P are the basis

vectors for the eigenspaces. We also know that $\begin{bmatrix} c_1 e^{-2x} \\ c_2 e^{3x} \end{bmatrix}$ is the general solution to $Y' = DY$. Therefore the general solution

to $Y' = AY$ is

$$P \begin{bmatrix} c_1 e^{-2x} \\ c_2 e^{3x} \end{bmatrix} = \begin{bmatrix} 1 & 2 \\ 1 & 1 \end{bmatrix} \begin{bmatrix} c_1 e^{-2x} \\ c_2 e^{3x} \end{bmatrix} = \begin{bmatrix} c_1 e^{-2x} + 2c_2 e^{3x} \\ c_1 e^{-2x} + c_2 e^{3x} \end{bmatrix}.$$

Thus, $y_1 = c_1 e^{-2x} + 2c_2 e^{3x}$, $y_2 = c_1 e^{-2x} + c_2 e^{3x}$.

21. We have the system $Y' = AY$, Where $A = \begin{bmatrix} 3 & 1 \\ 9 & 1 \end{bmatrix}$. The characteristic equation of A is

$$\begin{aligned} \det(\lambda I - A) &= \begin{vmatrix} \lambda - 3 & -1 \\ -9 & \lambda - 1 \end{vmatrix} \\ &= (\lambda - 3)(\lambda - 1) - 9 = \lambda^2 - 4\lambda - 6 = 0. \end{aligned}$$

Therefore, the eigenvalues of A are $\lambda = \frac{4 \pm \sqrt{16+24}}{2} = 2 \pm \sqrt{10}$. We now wish to find bases for the corresponding eigenspaces.

For $\lambda = 2 + \sqrt{10}$ we need to find a basis for the null space of $(2 + \sqrt{10}) I - A = \begin{bmatrix} -1 + \sqrt{10} & -1 \\ -9 & 1 + \sqrt{10} \end{bmatrix}$. Reducing the

matrix for the associated homogeneous system, $\begin{bmatrix} -1 + \sqrt{10} & -1 & \vdots & 0 \\ -9 & 1 + \sqrt{10} & \vdots & 0 \end{bmatrix} \rightarrow \begin{bmatrix} -1 + \sqrt{10} & -1 & \vdots & 0 \\ 0 & 0 & \vdots & 0 \end{bmatrix}$, we see

our solutions are

$$\begin{bmatrix} x \\ y \end{bmatrix} = \begin{bmatrix} x \\ (-1 + \sqrt{10}) x \end{bmatrix} = x \begin{bmatrix} 1 \\ -1 + \sqrt{10} \end{bmatrix}$$

and hence the column vector $\begin{bmatrix} 1 \\ -1 + \sqrt{10} \end{bmatrix}$ forms a basis for $E_{2+\sqrt{10}}$.

For $\lambda = 2 - \sqrt{10}$ we need to find a basis for the null space of $(2 - \sqrt{10}) I - A = \begin{bmatrix} -1 - \sqrt{10} & -1 \\ -9 & 1 - \sqrt{10} \end{bmatrix}$. Reducing the

matrix for the associated homogeneous system, $\begin{bmatrix} -1 - \sqrt{10} & -1 & \vdots & 0 \\ -9 & 1 - \sqrt{10} & \vdots & 0 \end{bmatrix} \rightarrow \begin{bmatrix} -1 - \sqrt{10} & -1 & \vdots & 0 \\ 0 & 0 & \vdots & 0 \end{bmatrix}$, we see

our solutions are

$$\begin{bmatrix} x \\ y \end{bmatrix} = \begin{bmatrix} x \\ (-1 - \sqrt{10}) x \end{bmatrix} = x \begin{bmatrix} 1 \\ -1 - \sqrt{10} \end{bmatrix}$$

and hence the column vector $\begin{bmatrix} 1 \\ -1 - \sqrt{10} \end{bmatrix}$ forms a basis for $E_{2-\sqrt{10}}$.

Matrix A is similar to the diagonal matrix

$$D = P^{-1}AP = \begin{bmatrix} 2 + \sqrt{10} & 0 \\ 0 & 2 - \sqrt{10} \end{bmatrix} \text{ where } P = \begin{bmatrix} 1 & 1 \\ -1 + \sqrt{10} & -1 - \sqrt{10} \end{bmatrix}.$$

The columns of P are the basis vectors for the eigenspaces. We also know that $\begin{bmatrix} c_1 e^{(2+\sqrt{10})x} \\ c_2 e^{(2-\sqrt{10})x} \end{bmatrix}$ is the general solution to $Y' = DY$. Therefore the general solution to $Y' = AY$ is

$$P \begin{bmatrix} c_1 e^{(2+\sqrt{10})x} \\ c_2 e^{(2-\sqrt{10})x} \end{bmatrix} = \begin{bmatrix} 1 & 1 \\ -1+\sqrt{10} & -1-\sqrt{10} \end{bmatrix} \begin{bmatrix} c_1 e^{(2+\sqrt{10})x} \\ c_2 e^{(2-\sqrt{10})x} \end{bmatrix}$$

$$= \begin{bmatrix} c_1 e^{(2+\sqrt{10})x} + c_2 e^{(2-\sqrt{10})x} \\ \left(-1+\sqrt{10}\right) c_1 e^{(2+\sqrt{10})x} - \left(1+\sqrt{10}\right) c_2 e^{(2-\sqrt{10})x} \end{bmatrix}.$$

Thus, $y_1 = c_1 e^{(2+\sqrt{10})x} + c_2 e^{(2-\sqrt{10})x}$, $y_2 = \left(-1+\sqrt{10}\right) c_1 e^{(2+\sqrt{10})x} - \left(1+\sqrt{10}\right) c_2 e^{(2-\sqrt{10})x}$.

23. We have the system $Y' = AY$, Where $A = \begin{bmatrix} 5 & -3 & 6 \\ 6 & -4 & 24 \\ 0 & 0 & 5 \end{bmatrix}$. The characteristic equation of A is

$$\det(\lambda I - A) = \begin{vmatrix} \lambda-5 & 3 & -6 \\ -6 & \lambda+4 & -24 \\ 0 & 0 & \lambda-5 \end{vmatrix}$$

$$= (\lambda-5)\left((\lambda-5)(\lambda+4)+18\right) = (\lambda-5)(\lambda^2-\lambda-2) = (\lambda-5)(\lambda-2)(\lambda+1) = 0.$$

Therefore, the eigenvalues of A are $\lambda = 5$, $\lambda = 2$ and $\lambda = -1$. We now wish to find bases for the corresponding eigenspaces. For $\lambda = 5$ we need to find a basis for the null space of $5I - A = \begin{bmatrix} 0 & 3 & -6 \\ -6 & 9 & -24 \\ 0 & 0 & 0 \end{bmatrix}$. Reducing the matrix for the associated homogeneous system,

$$\begin{bmatrix} 0 & 3 & -6 & : & 0 \\ -6 & 9 & -24 & : & 0 \\ 0 & 0 & 0 & : & 0 \end{bmatrix} \rightarrow \begin{bmatrix} 2 & -3 & 8 & : & 0 \\ 0 & 3 & -6 & : & 0 \\ 0 & 0 & 0 & : & 0 \end{bmatrix} \rightarrow \begin{bmatrix} 2 & 0 & 2 & : & 0 \\ 0 & 1 & -2 & : & 0 \\ 0 & 0 & 0 & : & 0 \end{bmatrix} \rightarrow \begin{bmatrix} 1 & 0 & 1 & : & 0 \\ 0 & 1 & -2 & : & 0 \\ 0 & 0 & 0 & : & 0 \end{bmatrix}$$

we see our solutions are

$$\begin{bmatrix} x \\ y \\ z \end{bmatrix} = \begin{bmatrix} -z \\ 2z \\ z \end{bmatrix} = z \begin{bmatrix} -1 \\ 2 \\ 1 \end{bmatrix}$$

and hence the column vector $\begin{bmatrix} -1 \\ 2 \\ 1 \end{bmatrix}$ forms a basis for E_5.

For $\lambda = 2$ we need to find a basis for the null space of $2I - A = \begin{bmatrix} -3 & 3 & -6 \\ -6 & 6 & -24 \\ 0 & 0 & -3 \end{bmatrix}$. Reducing the matrix for the associated homogeneous system,

$$\begin{bmatrix} -3 & 3 & -6 & : & 0 \\ -6 & 6 & -24 & : & 0 \\ 0 & 0 & -3 & : & 0 \end{bmatrix} \rightarrow \begin{bmatrix} 1 & -1 & 2 & : & 0 \\ 1 & -1 & 4 & : & 0 \\ 0 & 0 & 1 & : & 0 \end{bmatrix} \rightarrow \begin{bmatrix} 1 & -1 & 0 & : & 0 \\ 0 & 0 & 1 & : & 0 \\ 0 & 0 & 0 & : & 0 \end{bmatrix}$$

we see our solutions are

$$\begin{bmatrix} x \\ y \\ z \end{bmatrix} = \begin{bmatrix} x \\ x \\ 0 \end{bmatrix} = x \begin{bmatrix} 1 \\ 1 \\ 0 \end{bmatrix}$$

and hence the column vector $\begin{bmatrix} 1 \\ 1 \\ 0 \end{bmatrix}$ forms a basis for E_2.

For $\lambda = -1$ we need to find a basis for the null space of $-I - A = \begin{bmatrix} -6 & 3 & -6 \\ -6 & 3 & -24 \\ 0 & 0 & -6 \end{bmatrix}$. Reducing the matrix for the associated

homogeneous system,

$$\begin{bmatrix} -6 & 3 & -6 & \vdots & 0 \\ -6 & 3 & -24 & \vdots & 0 \\ 0 & 0 & -6 & \vdots & 0 \end{bmatrix} \rightarrow \begin{bmatrix} -6 & 3 & 0 & \vdots & 0 \\ 0 & 0 & 1 & \vdots & 0 \\ 0 & 0 & 0 & \vdots & 0 \end{bmatrix} \rightarrow \begin{bmatrix} 2 & -1 & 0 & \vdots & 0 \\ 0 & 0 & 1 & \vdots & 0 \\ 0 & 0 & 0 & \vdots & 0 \end{bmatrix}$$

we see our solutions are

$$\begin{bmatrix} x \\ y \\ z \end{bmatrix} = \begin{bmatrix} x \\ 2x \\ 0 \end{bmatrix} = x \begin{bmatrix} 1 \\ 2 \\ 0 \end{bmatrix}$$

and hence the column vector $\begin{bmatrix} 1 \\ 2 \\ 0 \end{bmatrix}$ forms a basis for E_{-1}.

Matrix A is similar to the diagonal matrix

$$D = P^{-1}AP = \begin{bmatrix} 5 & 0 & 0 \\ 0 & 2 & 0 \\ 0 & 0 & -1 \end{bmatrix} \text{ where } P = \begin{bmatrix} -1 & 1 & 1 \\ 2 & 1 & 2 \\ 1 & 0 & 0 \end{bmatrix}.$$

The columns of P are the basis vectors for the eigenspaces. We also know that $\begin{bmatrix} c_1 e^{5t} \\ c_2 e^{2t} \\ c_3 e^{-t} \end{bmatrix}$ is the general solution to $Y' = DY$.

Therefore the general solution to $Y' = AY$ is

$$P \begin{bmatrix} c_1 e^{5t} \\ c_2 e^{2t} \\ c_3 e^{-t} \end{bmatrix} = \begin{bmatrix} -1 & 1 & 1 \\ 2 & 1 & 2 \\ 1 & 0 & 0 \end{bmatrix} \begin{bmatrix} c_1 e^{5t} \\ c_2 e^{2t} \\ c_3 e^{-t} \end{bmatrix} = \begin{bmatrix} -c_1 e^{5t} + c_2 e^{2t} + c_3 e^{-t} \\ 2c_1 e^{5t} + c_2 e^{2t} + 2c_3 e^{-t} \\ c_1 e^{5t} \end{bmatrix}.$$

Thus, $y_1 = -c_1 e^{5t} + c_2 e^{2t} + c_3 e^{-t}$, $y_2 = 2c_1 e^{5t} + c_2 e^{2t} + 2c_3 e^{-t}$, $y_3 = c_1 e^{5t}$.

25. We have the system $Y' = AY$, Where $A = \begin{bmatrix} 1 & -3 \\ 3 & 1 \end{bmatrix}$. The characteristic equation of A is

$$\begin{aligned} \det(\lambda I - A) &= \begin{vmatrix} \lambda - 1 & 3 \\ -3 & \lambda - 1 \end{vmatrix} \\ &= (\lambda - 1)^2 + 9 = \lambda^2 - 2\lambda + 10 = 0. \end{aligned}$$

Therefore, the eigenvalues of A are $\lambda = \frac{2 \pm \sqrt{4-40}}{2} = 1 \pm 3i$. We now wish to find bases for the corresponding eigenspaces.

For $\lambda = 1 + 3i$ we need to find a basis for the null space of $(1 + 3i)I - A = \begin{bmatrix} 3i & 3 \\ -3 & 3i \end{bmatrix}$. Reducing the matrix for the

associated homogeneous system, $\begin{bmatrix} 3i & 3 & \vdots & 0 \\ -3 & 3i & \vdots & 0 \end{bmatrix} \rightarrow \begin{bmatrix} i & 1 & \vdots & 0 \\ 0 & 0 & \vdots & 0 \end{bmatrix}$, we see our solutions are

$$\begin{bmatrix} x \\ y \end{bmatrix} = \begin{bmatrix} x \\ -ix \end{bmatrix} = x \begin{bmatrix} 1 \\ -i \end{bmatrix}$$

and hence the column vector $\begin{bmatrix} 1 \\ -i \end{bmatrix}$ forms a basis for E_{1+3i}.

For $\lambda = 1 - 3i$ we need to find a basis for the null space of $(1 - 3i)I - A = \begin{bmatrix} -3i & 3 \\ -3 & -3i \end{bmatrix}$. Reducing the matrix for the

associated homogeneous system, $\begin{bmatrix} -3i & 3 & \vdots & 0 \\ -3 & -3i & \vdots & 0 \end{bmatrix} \rightarrow \begin{bmatrix} -i & 1 & \vdots & 0 \\ 0 & 0 & \vdots & 0 \end{bmatrix}$, we see our solutions are

$$\begin{bmatrix} x \\ y \end{bmatrix} = \begin{bmatrix} x \\ ix \end{bmatrix} = x \begin{bmatrix} 1 \\ i \end{bmatrix}$$

and hence the column vector $\begin{bmatrix} 1 \\ i \end{bmatrix}$ forms a basis for E_{1-3i}.

Matrix A is similar to the diagonal matrix

$$D = P^{-1}AP = \begin{bmatrix} 1 + 3i & 0 \\ 0 & 1 - 3i \end{bmatrix} \text{ where } P = \begin{bmatrix} 1 & 1 \\ -i & i \end{bmatrix}.$$

The columns of P are the basis vectors for the eigenspaces. One complex valued solution to $Y' = DY$ is

$$Z = \begin{bmatrix} e^{(1+3i)x} \\ 0 \end{bmatrix} = \begin{bmatrix} e^x \cos 3x + ie^x \sin 3x \\ 0 \end{bmatrix}.$$

By the first part of Theorem 6.7,

$$\begin{aligned} PZ &= \begin{bmatrix} 1 & 1 \\ -i & i \end{bmatrix} \begin{bmatrix} e^x \cos 3x + ie^x \sin 3x \\ 0 \end{bmatrix} \\ &= \begin{bmatrix} e^x \cos 3x + ie^x \sin 3x \\ e^x \sin 3x - ie^x \cos 3x \end{bmatrix} \\ &= \begin{bmatrix} e^x \cos 3x \\ e^x \sin 3x \end{bmatrix} + i \begin{bmatrix} e^x \sin 3x \\ -e^x \cos 3x \end{bmatrix} \end{aligned}$$

is a complex valued solution to $Y' = AY$. By Theorem 6.8

$$\begin{bmatrix} e^x \cos 3x \\ e^x \sin 3x \end{bmatrix} \text{ and } \begin{bmatrix} e^x \sin 3x \\ -e^x \cos 3x \end{bmatrix}$$

are real valued solutions to $Y' = AY$.

The general solution to $Y' = AY$ is then

$$c_1 \begin{bmatrix} e^x \cos 3x \\ e^x \sin 3x \end{bmatrix} + c_2 \begin{bmatrix} e^x \sin 3x \\ -e^x \cos 3x \end{bmatrix} = \\ \begin{bmatrix} e^x(c_1 \cos 3x + c_2 \sin 3x) \\ e^x(c_1 \sin 3x - c_2 \cos 3x) \end{bmatrix}.$$

Thus, $y_1 = e^x(c_1 \cos 3x + c_2 \sin 3x)$, $y_2 = e^x(c_1 \sin 3x - c_2 \cos 3x)$.

27. We have the system $Y' = AY$, Where $A = \begin{bmatrix} 2 & -5 \\ 1 & -2 \end{bmatrix}$, along with the initial conditions $Y(0) = \begin{bmatrix} 1 \\ 1 \end{bmatrix}$. The characteristic equation of A is

$$\det(\lambda I - A) = \begin{vmatrix} \lambda - 2 & 5 \\ -1 & \lambda + 2 \end{vmatrix} = (\lambda - 2)(\lambda + 2) + 5 = \lambda^2 + 1 = 0.$$

Therefore, the eigenvalues of A are $\lambda = \pm i$. We now wish to find bases for the corresponding eigenspaces.

For $\lambda = i$ we need to find a basis for the null space of $iI - A = \begin{bmatrix} -2 + i & 5 \\ -1 & 2 + i \end{bmatrix}$. Reducing the matrix for the associated

homogeneous system, $\begin{bmatrix} -2 + i & 5 & \vdots & 0 \\ -1 & 2 + i & \vdots & 0 \end{bmatrix} \rightarrow \begin{bmatrix} -2 + i & 5 & \vdots & 0 \\ 0 & 0 & \vdots & 0 \end{bmatrix}$, we see our solutions are

$$\begin{bmatrix} x \\ y \end{bmatrix} = \begin{bmatrix} x \\ \frac{2-i}{5}x \end{bmatrix} = x \begin{bmatrix} 1 \\ \frac{2-i}{5} \end{bmatrix}$$

and hence the column vector $\begin{bmatrix} 1 \\ \frac{2-i}{5} \end{bmatrix}$ forms a basis for E_i.

For $\lambda = -i$ we need to find a basis for the null space of $-iI - A = \begin{bmatrix} -2-i & 5 \\ -1 & 2-i \end{bmatrix}$. Reducing the matrix for the associated

homogeneous system, $\begin{bmatrix} -2-i & 5 & \vdots & 0 \\ -1 & 2-i & \vdots & 0 \end{bmatrix} \rightarrow \begin{bmatrix} -2-i & 5 & \vdots & 0 \\ 0 & 0 & \vdots & 0 \end{bmatrix}$, we see our solutions are

$$\begin{bmatrix} x \\ y \end{bmatrix} = \begin{bmatrix} x \\ \frac{2+i}{5}x \end{bmatrix} = x \begin{bmatrix} 1 \\ \frac{2+i}{5} \end{bmatrix}$$

and hence the column vector $\begin{bmatrix} 1 \\ \frac{2+i}{5} \end{bmatrix}$ forms a basis for E_{-i}.

Matrix A is similar to the diagonal matrix

$$D = P^{-1}AP = \begin{bmatrix} i & 0 \\ 0 & -i \end{bmatrix} \text{ where } P = \begin{bmatrix} 1 & 1 \\ \frac{2-i}{5} & \frac{2+i}{5} \end{bmatrix}.$$

The columns of P are the basis vectors for the eigenspaces. One complex valued solution to $Y' = DY$ is

$$Z = \begin{bmatrix} e^{it} \\ 0 \end{bmatrix} = \begin{bmatrix} \cos t + i \sin t \\ 0 \end{bmatrix}.$$

By the first part of Theorem 6.7,

$$\begin{aligned} PZ &= \begin{bmatrix} 1 & 1 \\ \frac{2-i}{5} & \frac{2+i}{5} \end{bmatrix} \begin{bmatrix} \cos t + i \sin t \\ 0 \end{bmatrix} \\ &= \begin{bmatrix} \cos t + i \sin t \\ \frac{2}{5}\cos t + \frac{1}{5}\sin t + i\left(\frac{2}{5}\sin t - \frac{1}{5}\cos t\right) \end{bmatrix} \\ &= \begin{bmatrix} \cos t \\ \frac{2}{5}\cos t + \frac{1}{5}\sin t \end{bmatrix} + i \begin{bmatrix} \sin t \\ \frac{2}{5}\sin t - \frac{1}{5}\cos t \end{bmatrix} \end{aligned}$$

is a complex valued solution to $Y' = AY$. By Theorem 6.8

$$\begin{bmatrix} \cos t \\ \frac{2}{5}\cos t + \frac{1}{5}\sin t \end{bmatrix} \text{ and } \begin{bmatrix} \sin t \\ \frac{2}{5}\sin t - \frac{1}{5}\cos t \end{bmatrix}$$

are real valued solutions to $Y' = AY$.

The general solution to $Y' = AY$ is then

$$c_1 \begin{bmatrix} \cos t \\ \frac{2}{5}\cos t + \frac{1}{5}\sin t \end{bmatrix} + c_2 \begin{bmatrix} \sin t \\ \frac{2}{5}\sin t - \frac{1}{5}\cos t \end{bmatrix} = \\ \begin{bmatrix} c_1 \cos t + c_2 \sin t \\ c_1 \left(\frac{2}{5}\cos t + \frac{1}{5}\sin t\right) + c_2 \left(\frac{2}{5}\sin t - \frac{1}{5}\cos t\right) \end{bmatrix}.$$

We now use the initial condition $Y(0) = \begin{bmatrix} 1 \\ 1 \end{bmatrix} = \begin{bmatrix} c_1 \\ \frac{2}{5}c_1 - \frac{1}{5}c_2 \end{bmatrix}$. This system has the solution $c_1 = 1$, $c_2 = -3$.

Thus we have, $Y(t) = \begin{bmatrix} \cos t - 3\sin t \\ \left(\frac{2}{5}\cos t + \frac{1}{5}\sin t\right) - 3\left(\frac{2}{5}\sin t - \frac{1}{5}\cos t\right) \end{bmatrix}$, or equivalently

$$x = \cos t - 3\sin t, \ y = \cos t - \sin t.$$

6.3   Homogeneous Systems with Constant Coefficients: The Nondiagonalizable Case

Exercises 6.3, pp. 314-315

1. We have the system $Y' = \begin{bmatrix} 10 & -9 \\ 4 & -2 \end{bmatrix} Y$. From Exercise 3 of Section 5.4 we know that the sole eigenvalue of matrix A is $\lambda = 4$. We use Maple to find a Jordan canonical form, B, for matrix A and a matrix P such that $P^{-1}AP = B$. To do this we use the commands:

```
A:=matrix(2,2,[10,-9,4,-2]);
B:=jordan(A,'P');
print(P);
```

Doing this we find $B = \begin{bmatrix} 4 & 1 \\ 0 & 4 \end{bmatrix}$ and $P = \begin{bmatrix} 6 & 1 \\ 4 & 0 \end{bmatrix}$. We now solve the triangular system $Y' = BY$, which is

$$y_1' = 4y_1 + y_2, \; y_2' = 4y_2.$$

The second equation has the general solution $y_2 = c_2 e^{4x}$. Substituting this into the first equation we have $y_1' = 4y_1 + c_2 e^{4x}$. The related homogeneous equation $y_1' - 4y_1 = 0$ has the solution $y_{1H} = c_1 e^{4x}$. We may then use the method of undetermined coefficients to find the particular solution of the form $y_{1P} = Axe^{4x}$. Here, $y_{1P}' = Ae^{4x} + 4Axe^{4x}$. Substituting these into the equation $y_1' = 4y_1 + c_2 e^{4x}$ we obtain

$$Ae^{4x} + 4Axe^{4x} = 4(Axe^{4x}) + c_2 e^{4x}.$$

Simplifying this we see that $Ae^{4x} = c_2 e^{4x}$. Thus $A = c_2$. Adding the particular solution to the general solution of the related homogeneous equation we have $y_1 = c_1 e^{4x} + c_2 xe^{4x}$. Thus the general solution to the triangular system $Y' = BY$ is
$Z = \begin{bmatrix} c_1 e^{4x} + c_2 xe^{4x} \\ c_2 e^{4x} \end{bmatrix}$.

Therefore the general solution to the original system $Y' = AY$ is

$$PZ = \begin{bmatrix} 6 & 1 \\ 4 & 0 \end{bmatrix} \begin{bmatrix} c_1 e^{4x} + c_2 xe^{4x} \\ c_2 e^{4x} \end{bmatrix} = \begin{bmatrix} 6c_1 e^{4x} + c_2 e^{4x} + 6c_2 xe^{4x} \\ 4c_1 e^{4x} + 4c_2 xe^{4x} \end{bmatrix}.$$

Thus, $y_1 = (6c_1 + c_2 + 6c_2 x)e^{4x}$, $y_2 = (4c_1 + 4c_2 x)e^{4x}$.

3. We have the system $Y' = \begin{bmatrix} 4 & 0 & 0 \\ 1 & 4 & 0 \\ 0 & 1 & 4 \end{bmatrix} Y$. From Exercise 9 of Section 5.4 we know that the sole eigenvalue of matrix A is $\lambda = 4$. We use Maple to find a Jordan canonical form, B, for matrix A and a matrix P such that $P^{-1}AP = B$. To do this we use the commands:

```
A:=matrix(3,3,[4,0,0,1,4,0,0,1,4]);
B:=jordan(A,'P');
print(P);
```

Doing this we find $B = \begin{bmatrix} 4 & 1 & 0 \\ 0 & 4 & 1 \\ 0 & 0 & 4 \end{bmatrix}$ and $P = \begin{bmatrix} 0 & 0 & 1 \\ 0 & 1 & 0 \\ 1 & 0 & 0 \end{bmatrix}$. We now solve the triangular system $Y' = BY$, which is

$y_1' = 4y_1 + y_2$, $y_2' = 4y_2 + y_3$, $y_3' = 4y_3$. The third equation has the general solution $y_3 = c_3 e^{4x}$.

Substituting this into the second equation we have $y_2' = 4y_2 + c_3 e^{4x}$. The related homogeneous equation $y_2' - 4y_2 = 0$ has the solution $y_{2H} = c_2 e^{4x}$. We may then use the method of undetermined coefficients to find the particular solution of the form $y_{2P} = Axe^{4x}$. Here, $y_{2P}' = Ae^{4x} + 4Axe^{4x}$. Substituting these into the equation $y_2' = 4y_2 + c_3 e^{4x}$ we obtain

$$Ae^{4x} + 4Axe^{4x} = 4(Axe^{4x}) + c_3 e^{4x}.$$

Simplifying this we see that $Ae^{4x} = c_3 e^{4x}$. Thus $A = c_3$. Adding the particular solution to the general solution of the related homogeneous equation we have $y_2 = c_2 e^{4x} + c_3 xe^{4x}$.

We now substitute this into the first equation to obtain $y_1' = 4y_1 + c_2 e^{4x} + c_3 xe^{4x}$. The related homogeneous equation

$$y_1' - 4y_1 = 0$$

has the solution $y_{1H} = c_1 e^{4x}$. We may then use the method of undetermined coefficients to find the particular solution of the form $y_{1P} = Bxe^{4x} + Cx^2 e^{4x}$. Here, $y'_{1P} = Be^{4x} + 4Bxe^{4x} + 2Cxe^{4x} + 4Cx^2 e^{4x}$. Substituting these into the equation $y'_1 = 4y_1 + c_2 e^{4x} + c_3 xe^{4x}$ we obtain

$$Be^{4x} + 4Bxe^{4x} + 2Cxe^{4x} + 4Cx^2 e^{4x} = 4(Bxe^{4x} + Cx^2 e^{4x}) + c_2 e^{4x} + c_3 xe^{4x}.$$

This simplifies to

$$Be^{4x} + 2Cxe^{4x} = c_2 e^{4x} + c_3 xe^{4x}.$$

Thus $B = c_2$ and $C = \frac{1}{2}c_3$. Adding the particular solution to the general solution of the related homogeneous equation we have $y_1 = c_1 e^{4x} + c_2 xe^{4x} + \frac{1}{2}c_3 x^2 e^{4x}$. Thus the general solution to the triangular system $Y' = BY$ is

$$Z = \begin{bmatrix} c_1 e^{4x} + c_2 xe^{4x} + \frac{1}{2}c_3 x^2 e^{4x} \\ c_2 e^{4x} + c_3 xe^{4x} \\ c_3 e^{4x} \end{bmatrix}.$$

Therefore the general solution to the original system $Y' = AY$ is

$$PZ = \begin{bmatrix} 0 & 0 & 1 \\ 0 & 1 & 0 \\ 1 & 0 & 0 \end{bmatrix} \begin{bmatrix} c_1 e^{4x} + c_2 xe^{4x} + \frac{1}{2}c_3 x^2 e^{4x} \\ c_2 e^{4x} + c_3 xe^{4x} \\ c_3 e^{4x} \end{bmatrix} = \begin{bmatrix} c_3 e^{4x} \\ c_2 e^{4x} + c_3 xe^{4x} \\ c_1 e^{4x} + c_2 xe^{4x} + \frac{1}{2}c_3 x^2 e^{4x} \end{bmatrix}.$$

Thus, $y_1 = c_3 e^{4x}$, $y_2 = (c_2 + c_3 x)e^{4x}$, $y_3 = (c_1 + c_2 x + \frac{1}{2}c_3 x^2)e^{4x}$.

5. We have the system $Y' = \begin{bmatrix} 5 & 6 & 2 \\ 0 & -1 & -8 \\ 1 & 0 & -2 \end{bmatrix} Y$. We use Maple to find a Jordan canonical form, B, for matrix A and a matrix P such that $P^{-1}AP = B$. To do this we use the commands:

```
A:=matrix(3,3,[5,6,2,0,-1,-8,1,0,2]);
B:=jordan(A,'P');
print(P);
```

Doing this we find

$$B = \begin{bmatrix} -4 & 0 & 0 \\ 0 & 3 & 1 \\ 0 & 0 & 3 \end{bmatrix} \text{ and } P = \begin{bmatrix} \frac{6}{49} & \frac{20}{7} & \frac{43}{49} \\ -\frac{8}{49} & -\frac{8}{7} & \frac{8}{49} \\ -\frac{3}{49} & \frac{4}{7} & \frac{3}{49} \end{bmatrix}.$$

We now solve the triangular system $Y' = BY$, which is $y'_1 = -4y_1$, $y'_2 = 3y_2 + y_3$, $y'_3 = 3y_3$. The third equation has the general solution $y_3 = c_3 e^{3x}$.

Substituting this into the second equation we have $y'_2 = 3y_2 + c_3 e^{3x}$. The related homogeneous equation $y'_2 - 3y_2 = 0$ has the solution $y_{2H} = c_2 e^{3x}$. We may then use the method of undetermined coefficients to find the particular solution of the form $y_{2P} = Axe^{3x}$. Here, $y'_{2P} = Ae^{3x} + 3Axe^{3x}$. Substituting these into the equation $y'_2 = 3y_2 + c_3 e^{3x}$ we obtain

$$Ae^{3x} + 3Axe^{3x} = 3(Axe^{3x}) + c_3 e^{3x}.$$

Simplifying this we see that $Ae^{3x} = c_3 e^{3x}$. Thus $A = c_3$. Adding the particular solution to the general solution of the related homogeneous equation we have $y_2 = c_2 e^{3x} + c_3 xe^{3x}$.

The first equation has the general solution $y_1 = c_1 e^{-4x}$.

Thus the general solution to the triangular system $Y' = BY$ is $Z = \begin{bmatrix} c_1 e^{-4x} \\ c_2 e^{3x} + c_3 xe^{3x} \\ c_3 e^{3x} \end{bmatrix}.$

Therefore the general solution to the original system $Y' = AY$ is

$$PZ = \begin{bmatrix} \frac{6}{49} & \frac{20}{7} & \frac{43}{49} \\ -\frac{8}{49} & -\frac{8}{7} & \frac{8}{49} \\ -\frac{3}{49} & \frac{4}{7} & \frac{3}{49} \end{bmatrix} \begin{bmatrix} c_1 e^{-4x} \\ c_2 e^{3x} + c_3 xe^{3x} \\ c_3 e^{3x} \end{bmatrix} = \begin{bmatrix} \frac{6}{49}c_1 e^{-4x} + \frac{20}{7}(c_2 e^{3x} + c_3 xe^{3x}) + \frac{43}{49}c_3 e^{3x} \\ -\frac{8}{49}c_1 e^{-4x} - \frac{8}{7}(c_2 e^{3x} + c_3 xe^{3x}) + \frac{8}{49}c_3 e^{3x} \\ -\frac{3}{49}c_1 e^{-4x} + \frac{4}{7}(c_2 e^{3x} + c_3 xe^{3x}) + \frac{3}{49}c_3 e^{3x} \end{bmatrix}.$$

Thus,

$$y_1 = \frac{6}{49}c_1e^{-4x} + \frac{20}{7}\left(c_2e^{3x} + c_3xe^{3x}\right) + \frac{43}{49}c_3e^{3x},$$

$$y_2 = -\frac{8}{49}c_1e^{-4x} - \frac{8}{7}\left(c_2e^{3x} + c_3xe^{3x}\right) + \frac{8}{49}c_3e^{3x},$$

$$y_3 = -\frac{3}{49}c_1e^{-4x} + \frac{4}{7}\left(c_2e^{3x} + c_3xe^{3x}\right) + \frac{3}{49}c_3e^{3x}.$$

Incorporating the fraction $\frac{1}{49}$ into each of the constants c_1, c_2 and c_3 this system is equivalent to

$$
\begin{aligned}
y_1 &= 6c_1e^{-4x} + 140\left(c_2e^{3x} + c_3xe^{3x}\right) + 43c_3e^{3x}, \\
y_2 &= -8c_1e^{-4x} - 56\left(c_2e^{3x} + c_3xe^{3x}\right) + 8c_3e^{3x}, \\
y_3 &= -3c_1e^{-4x} + 28\left(c_2e^{3x} + c_3xe^{3x}\right) + 3c_3e^{3x}.
\end{aligned}
$$

7. We have the system $Y' = \begin{bmatrix} 3 & 10 & 16 & -26 \\ 15 & 8 & 8 & -28 \\ 0 & 10 & 17 & -24 \\ 5 & 10 & 15 & -27 \end{bmatrix} Y$. We use Maple to find a Jordan canonical form, B, for matrix A and a

matrix P such that $P^{-1}AP = B$. To do this we use the commands:

```
A:=matrix(4,4,[3,10,16,-26,15,8,8,-28,0,10,17,-24,5,10,15,-27]);
B:=jordan(A,'P');
print(P);
```

Doing this we find

$$
B = \begin{bmatrix} -2 & 0 & 0 & 0 \\ 0 & 3 & 0 & 0 \\ 0 & 0 & i & 0 \\ 0 & 0 & 0 & -i \end{bmatrix} \text{ and } P = \begin{bmatrix} 2 & 2 & -\frac{3}{2}-\frac{1}{2}i & -\frac{3}{2}+\frac{1}{2}i \\ 1 & 2 & -\frac{3}{2}-\frac{11}{2}i & -\frac{3}{2}+\frac{11}{2}i \\ 2 & 2 & -2+i & -2-i \\ 2 & 2 & -2-\frac{3}{2}i & -2+\frac{3}{2}i \end{bmatrix}.
$$

Since B is a diagonal matrix we may use the technique of Section 6.2 to finish the problem. Here we have that matrix A is similar to the diagonal matrix B. The columns of P are the basis vectors for the eigenspaces. Two real valued solutions to $Y' = BY$ are $\begin{bmatrix} e^{-2x} \\ 0 \\ 0 \\ 0 \end{bmatrix}$ and $\begin{bmatrix} 0 \\ e^{3x} \\ 0 \\ 0 \end{bmatrix}$. Thus

$$
P\begin{bmatrix} e^{-2x} \\ 0 \\ 0 \\ 0 \end{bmatrix} = \begin{bmatrix} 2 & 2 & -\frac{3}{2}-\frac{1}{2}i & -\frac{3}{2}+\frac{1}{2}i \\ 1 & 2 & -\frac{3}{2}-\frac{11}{2}i & -\frac{3}{2}+\frac{11}{2}i \\ 2 & 2 & -2+i & -2-i \\ 2 & 2 & -2-\frac{3}{2}i & -2+\frac{3}{2}i \end{bmatrix} \begin{bmatrix} e^{-2x} \\ 0 \\ 0 \\ 0 \end{bmatrix} = \begin{bmatrix} 2e^{-2x} \\ e^{-2x} \\ 2e^{-2x} \\ 2e^{-2x} \end{bmatrix}
$$

and

$$
P\begin{bmatrix} 0 \\ e^{3x} \\ 0 \\ 0 \end{bmatrix} = \begin{bmatrix} 2 & 2 & -\frac{3}{2}-\frac{1}{2}i & -\frac{3}{2}+\frac{1}{2}i \\ 1 & 2 & -\frac{3}{2}-\frac{11}{2}i & -\frac{3}{2}+\frac{11}{2}i \\ 2 & 2 & -2+i & -2-i \\ 2 & 2 & -2-\frac{3}{2}i & -2+\frac{3}{2}i \end{bmatrix} \begin{bmatrix} 0 \\ e^{3x} \\ 0 \\ 0 \end{bmatrix} = \begin{bmatrix} 2e^{3x} \\ 2e^{3x} \\ 2e^{3x} \\ 2e^{3x} \end{bmatrix}
$$

will be two solutions to $Y' = AY$.

One complex valued solution to $Y' = DY$ is $Z = \begin{bmatrix} 0 \\ 0 \\ e^{ix} \\ 0 \end{bmatrix} = \begin{bmatrix} 0 \\ 0 \\ \cos x + i\sin x \\ 0 \end{bmatrix}$. By the first part of Theorem 6.7,

$$PZ = \begin{bmatrix} 2 & 2 & -\frac{3}{2}-\frac{1}{2}i & -\frac{3}{2}+\frac{1}{2}i \\ 1 & 2 & -\frac{3}{2}-\frac{11}{2}i & -\frac{3}{2}+\frac{11}{2}i \\ 2 & 2 & -2+i & -2-i \\ 2 & 2 & -2-\frac{3}{2}i & -2+\frac{3}{2}i \end{bmatrix} \begin{bmatrix} 0 \\ 0 \\ \cos x + i\sin x \\ 0 \end{bmatrix}$$

$$= \begin{bmatrix} -\frac{3}{2}\cos x + \frac{1}{2}\sin x + i\left(-\frac{3}{2}\sin x - \frac{1}{2}\cos x\right) \\ -\frac{3}{2}\cos x + \frac{11}{2}\sin x + i\left(-\frac{3}{2}\sin x - \frac{11}{2}\cos x\right) \\ -2\cos x - \sin x + i(-2\sin x + \cos x) \\ -2\cos x + \frac{3}{2}\sin x + i\left(-2\sin x - \frac{3}{2}\cos x\right) \end{bmatrix}$$

$$= \begin{bmatrix} -\frac{3}{2}\cos x + \frac{1}{2}\sin x \\ -\frac{3}{2}\cos x + \frac{11}{2}\sin x \\ -2\cos x - \sin x \\ -2\cos x + \frac{3}{2}\sin x \end{bmatrix} + i \begin{bmatrix} -\frac{3}{2}\sin x - \frac{1}{2}\cos x \\ -\frac{3}{2}\sin x - \frac{11}{2}\cos x \\ -2\sin x + \cos x \\ -2\sin x - \frac{3}{2}\cos x \end{bmatrix}$$

is a complex valued solution to $Y' = AY$. By Theorem 6.8 $\begin{bmatrix} -\frac{3}{2}\cos x + \frac{1}{2}\sin x \\ -\frac{3}{2}\cos x + \frac{11}{2}\sin x \\ -2\cos x - \sin x \\ -2\cos x + \frac{3}{2}\sin x \end{bmatrix}$ and $\begin{bmatrix} -\frac{3}{2}\sin x - \frac{1}{2}\cos x \\ -\frac{3}{2}\sin x - \frac{11}{2}\cos x \\ -2\sin x + \cos x \\ -2\sin x - \frac{3}{2}\cos x \end{bmatrix}$ are real

valued solutions to $Y' = AY$.

The general solution to $Y' = AY$ is then

$$c_1 \begin{bmatrix} 2e^{-2x} \\ e^{-2x} \\ 2e^{-2x} \\ 2e^{-2x} \end{bmatrix} + c_2 \begin{bmatrix} 2e^{3x} \\ 2e^{3x} \\ 2e^{3x} \\ 2e^{3x} \end{bmatrix} + c_3 \begin{bmatrix} -\frac{3}{2}\cos x + \frac{1}{2}\sin x \\ -\frac{3}{2}\cos x + \frac{11}{2}\sin x \\ -2\cos x - \sin x \\ -2\cos x + \frac{3}{2}\sin x \end{bmatrix} + c_4 \begin{bmatrix} -\frac{3}{2}\sin x - \frac{1}{2}\cos x \\ -\frac{3}{2}\sin x - \frac{11}{2}\cos x \\ -2\sin x + \cos x \\ -2\sin x - \frac{3}{2}\cos x \end{bmatrix}$$

$$= \begin{bmatrix} 2c_1 e^{-2x} + 2c_2 e^{3x} + c_3(-\frac{3}{2}\cos x + \frac{1}{2}\sin x) + c_4(-\frac{3}{2}\sin x - \frac{1}{2}\cos x) \\ c_1 e^{-2x} + 2c_2 e^{3x} + c_3(-\frac{3}{2}\cos x + \frac{11}{2}\sin x) + c_4(-\frac{3}{2}\sin x - \frac{11}{2}\cos x) \\ 2c_1 e^{-2x} + 2c_2 e^{3x} + c_3(-2\cos x - \sin x) + c_4(-2\sin x + \cos x) \\ 2c_1 e^{-2x} + 2c_2 e^{3x} + c_3(-2\cos x + \frac{3}{2}\sin x) + c_4(-2\sin x - \frac{3}{2}\cos x) \end{bmatrix}.$$

Thus,

$$y_1 = 2c_1 e^{-2x} + 2c_2 e^{3x} + c_3\left(-\frac{3}{2}\cos x + \frac{1}{2}\sin x\right) + c_4\left(-\frac{3}{2}\sin x - \frac{1}{2}\cos x\right),$$

$$y_2 = c_1 e^{-2x} + 2c_2 e^{3x} + c_3\left(-\frac{3}{2}\cos x + \frac{11}{2}\sin x\right) + c_4\left(-\frac{3}{2}\sin x - \frac{11}{2}\cos x\right),$$

$$y_3 = 2c_1 e^{-2x} + 2c_2 e^{3x} + c_3\left(-2\cos x - \sin x\right) + c_4\left(-2\sin x + \cos x\right),$$

$$y_4 = 2c_1 e^{-2x} + 2c_2 e^{3x} + c_3\left(-2\cos x + \frac{3}{2}\sin x\right) + c_4\left(-2\sin x - \frac{3}{2}\cos x\right).$$

Note: Here we give real solutions. In the text the solutions are complex.

9. From Exercise 1 we have $Y(x) = \begin{bmatrix} 6c_1 e^{4x} + c_2 e^{4x} + 6c_2 x e^{4x} \\ 4c_1 e^{4x} + 4c_2 x e^{4x} \end{bmatrix}$. We are now given the initial condition

$$Y(0) = \begin{bmatrix} 2 \\ 1 \end{bmatrix}.$$

Thus we have $Y(0) = \begin{bmatrix} 2 \\ 1 \end{bmatrix} = \begin{bmatrix} 6c_1 + c_2 \\ 4c_1 \end{bmatrix}$. The solution to this system is $c_1 = \frac{1}{4}$, $c_2 = \frac{1}{2}$. Thus the solution to the initial

value problem is $Y(x) = \begin{bmatrix} 6(\frac{1}{4})e^{4x} + \frac{1}{2}e^{4x} + 6(\frac{1}{2})xe^{4x} \\ 4(\frac{1}{4})e^{4x} + 4(\frac{1}{2})xe^{4x} \end{bmatrix} = \begin{bmatrix} e^{4x}(2+3x) \\ e^{4x}(1+2x) \end{bmatrix}$. Equivalently,

$$y_1 = e^{4x}(2+3x), \; y_2 = e^{4x}(1+2x).$$

11. From Exercise 5 we have $Y(x) = \begin{bmatrix} \frac{6}{49}c_1 e^{-4x} + \frac{20}{7}(c_2 e^{3x} + c_3 xe^{3x}) + \frac{43}{49}c_3 e^{3x} \\ -\frac{8}{49}c_1 e^{-4x} - \frac{8}{7}(c_2 e^{3x} + c_3 xe^{3x}) + \frac{8}{49}c_3 e^{3x} \\ -\frac{3}{49}c_1 e^{-4x} + \frac{4}{7}(c_2 e^{3x} + c_3 xe^{3x}) + \frac{3}{49}c_3 e^{3x} \end{bmatrix}$. We are now given the initial condition

$$Y(0) = \begin{bmatrix} 0 \\ 2 \\ -1 \end{bmatrix}.$$

Thus we have $Y(0) = \begin{bmatrix} 0 \\ 2 \\ -1 \end{bmatrix} = \begin{bmatrix} \frac{6}{49}c_1 + \frac{20}{7}c_2 + \frac{43}{49}c_3 \\ -\frac{8}{49}c_1 - \frac{8}{7}c_2 + \frac{8}{49}c_3 \\ -\frac{3}{49}c_1 + \frac{4}{7}c_2 + \frac{3}{49}c_3 \end{bmatrix}$. The solution to this system is

$$c_1 = 5, \; c_2 = -\frac{7}{4}, \; c_3 = 5.$$

Thus the solution to the initial value problem is

$$Y(x) = \begin{bmatrix} \frac{6}{49}(5)e^{-4x} + \frac{20}{7}(-\frac{7}{4}e^{3x} + 5xe^{3x}) + \frac{43}{49}(5)e^{3x} \\ -\frac{8}{49}(5)e^{-4x} - \frac{8}{7}(-\frac{7}{4}e^{3x} + 5xe^{3x}) + \frac{8}{49}(5)e^{3x} \\ -\frac{3}{49}(5)e^{-4x} + \frac{4}{7}(-\frac{7}{4}e^{3x} + 5xe^{3x}) + \frac{3}{49}(5)e^{3x} \end{bmatrix} = \begin{bmatrix} e^{3x}(-\frac{30}{49} + \frac{100}{7}x) + \frac{30}{49}e^{-4x} \\ e^{3x}(\frac{138}{49} - \frac{40}{7}x) - \frac{40}{49}e^{-4x} \\ e^{3x}(-\frac{34}{49} + \frac{20}{7}x) - \frac{15}{49}e^{-4x} \end{bmatrix}.$$

Equivalently,

$$y_1 = e^{3x}\left(-\frac{30}{49} + \frac{100}{7}x\right) + \frac{30}{49}e^{-4x}, \; y_2 = e^{3x}\left(\frac{138}{49} - \frac{40}{7}x\right) - \frac{40}{49}e^{-4x}, \; y_3 = e^{3x}\left(-\frac{34}{49} + \frac{20}{7}x\right) - \frac{15}{49}e^{-4x}.$$

17. We have the system $v = \begin{bmatrix} -5 & -1 \\ 1 & -7 \end{bmatrix}\begin{bmatrix} x \\ y \end{bmatrix}$. We are asked to find the position, s, of the object at time t if $s(0) = (1,0)$.

We let $A = \begin{bmatrix} -5 & -1 \\ 1 & -7 \end{bmatrix}$. We use Maple to find a Jordan canonical form, B, for matrix A and a matrix P such that $P^{-1}AP = B$.
To do this we use the commands:

```
A:=matrix(2,2,[-5,-1,1,-7]);
B:=jordan(A,'P');
print(P);
```

Doing this we find $B = \begin{bmatrix} -6 & 1 \\ 0 & -6 \end{bmatrix}$ and $P = \begin{bmatrix} 1 & 1 \\ 1 & 0 \end{bmatrix}$. We now solve the triangular system $v = B\begin{bmatrix} x \\ y \end{bmatrix}$, which is

$x' = -6x + y$, $y' = -6y$. The second equation has the general solution $y = c_2 e^{-6t}$. Substituting this into the first equation we have $x' = -6x + c_2 e^{-6t}$. The related homogeneous equation $x' + 6x = 0$ has the solution $x_H = c_1 e^{-6t}$. We may then use the method of undetermined coefficients to find the particular solution of the form $x_P = Ate^{-6t}$. Here, $x_P' = Ae^{-6t} - 6Ate^{-6t}$. Substituting these into the equation $x' = -6x + c_2 e^{-6t}$ we obtain

$$Ae^{-6t} - 6Ate^{-6t} = -6(Ate^{-6t}) + c_2 e^{-6t}.$$

Simplifying this we see that $Ae^{-6t} = c_2 e^{-6t}$. Thus $A = c_2$. Adding the particular solution to the general solution of the related homogeneous equation we have $x = c_1 e^{-6t} + c_2 te^{-6t}$. Thus the general solution to the triangular system $v = B\begin{bmatrix} x \\ y \end{bmatrix}$

is $Z = \begin{bmatrix} c_1 e^{-6t} + c_2 te^{-6t} \\ c_2 e^{-6t} \end{bmatrix}$.

Therefore the general solution to the original system is

$$PZ = \begin{bmatrix} 1 & 1 \\ 1 & 0 \end{bmatrix} \begin{bmatrix} c_1 e^{-6t} + c_2 t e^{-6t} \\ c_2 e^{-6t} \end{bmatrix} = \begin{bmatrix} c_1 e^{-6t} + c_2 e^{-6t} + c_2 t e^{-6t} \\ c_1 e^{-6t} + c_2 t e^{-6t} \end{bmatrix}.$$

We now use given the initial position $(1,0)$. Thus we have $\begin{bmatrix} 1 \\ 0 \end{bmatrix} = \begin{bmatrix} c_1 + c_2 \\ c_1 \end{bmatrix}$. The solution to this system is $c_1 = 0$, $c_2 = 1$.

Thus the solution to the initial value problem is $s = \begin{bmatrix} (1+t)e^{-6t} \\ te^{-6t} \end{bmatrix}$. Equivalently,

$$x = (1+t)e^{-6t},\ y = te^{-6t}.$$

6.4   Nonhomogeneous Linear Systems

Exercises 6.4, pp. 318

1. We have the system $Y' = \begin{bmatrix} 3 & 0 \\ 8 & -1 \end{bmatrix} Y + G(x)$, where $G(x) = \begin{bmatrix} 2 \\ x \end{bmatrix}$. From Exercise 1 of Section 6.2 we know the general

solution to the corresponding homogeneous system is $Y_H = \begin{bmatrix} c_1 e^{3x} \\ 2c_1 e^{3x} + c_2 e^{-x} \end{bmatrix}$.

To find a particular solution we will use the formula $Y_P = M \int M^{-1} G(x)\, dx$, where M is a matrix of fundamental solutions.

Here $M = \begin{bmatrix} e^{3x} & 0 \\ 2e^{3x} & e^{-x} \end{bmatrix}$. We have

$$\begin{bmatrix} e^{3x} & 0 & \vdots & 1 & 0 \\ 2e^{3x} & e^{-x} & \vdots & 0 & 1 \end{bmatrix} \to \begin{bmatrix} e^{3x} & 0 & \vdots & 1 & 0 \\ 0 & e^{-x} & \vdots & -2 & 1 \end{bmatrix}.$$

Multiplying the first row by e^{-3x} and the second by e^x we obtain $M^{-1} = \begin{bmatrix} e^{-3x} & 0 \\ -2e^x & e^x \end{bmatrix}$.

The product inside the integral is then

$$M^{-1}G(x) = \begin{bmatrix} e^{-3x} & 0 \\ -2e^x & e^x \end{bmatrix} \begin{bmatrix} 2 \\ x \end{bmatrix} = \begin{bmatrix} 2e^{-3x} \\ -4e^x + xe^x \end{bmatrix}.$$

Integrating each component of this column vector gives us

$$\int M^{-1}G(x)\, dx = \begin{bmatrix} \int 2e^{-3x}\, dx \\ \int (-4e^x + xe^x)\, dx \end{bmatrix} = \begin{bmatrix} -\frac{2}{3}e^{-3x} \\ xe^x - 5e^x \end{bmatrix}.$$

Thus we have

$$Y_P = M \int M^{-1}G(x)\, dx = \begin{bmatrix} e^{3x} & 0 \\ 2e^{3x} & e^{-x} \end{bmatrix} \begin{bmatrix} -\frac{2}{3}e^{-3x} \\ xe^x - 5e^x \end{bmatrix} = \begin{bmatrix} -\frac{2}{3} \\ x - \frac{19}{3} \end{bmatrix}.$$

The general solution is then

$$Y = Y_H + Y_P = \begin{bmatrix} c_1 e^{3x} \\ 2c_1 e^{3x} + c_2 e^{-x} \end{bmatrix} + \begin{bmatrix} -\frac{2}{3} \\ x - \frac{19}{3} \end{bmatrix} = \begin{bmatrix} c_1 e^{3x} - \frac{2}{3} \\ 2c_1 e^{3x} + c_2 e^{-x} + x - \frac{19}{3} \end{bmatrix}.$$

Equivalently, $y_1 = c_1 e^{3x} - \frac{2}{3}$, $y_2 = 2c_1 e^{3x} + c_2 e^{-x} + x - \frac{19}{3}$.

3. We have the system $Y' = \begin{bmatrix} 3 & -2 \\ 4 & -1 \end{bmatrix} Y + G(x)$, where $G(x) = \begin{bmatrix} 5 \\ 1-x \end{bmatrix}$. From Exercise 9 of Section 6.2 we know the general solution to the corresponding homogeneous system is

$$Y_H = \begin{bmatrix} c_1\left(e^x\cos 2x - e^x\sin 2x\right) + c_2\left(e^x\sin 2x + e^x\cos 2x\right) \\ 2c_1 e^x\cos 2x + 2c_2 e^x\sin 2x \end{bmatrix}.$$

To find a particular solution we will use the formula $Y_P = M\int M^{-1}G(x)\,dx$, where M is a matrix of fundamental solutions. Here $M = \begin{bmatrix} e^x\cos 2x - e^x\sin 2x & e^x\sin 2x + e^x\cos 2x \\ 2e^x\cos 2x & 2e^x\sin 2x \end{bmatrix}$. We compute M^{-1} using the adjoint as in Corollary 1.27 of the text

$$
\begin{aligned}
M^{-1} &= \frac{1}{\det M}\operatorname{adj}(M) = \frac{1}{(e^x\cos 2x - e^x\sin 2x)(2e^x\sin 2x) - (e^x\sin 2x + e^x\cos 2x)(2e^x\cos 2x)}\operatorname{adj}(M)\\
&= \frac{1}{-2e^{2x}}\begin{bmatrix} 2e^x\sin 2x & -(e^x\sin 2x + e^x\cos 2x) \\ -2e^x\cos 2x & e^x\cos 2x - e^x\sin 2x \end{bmatrix} = \begin{bmatrix} -e^{-x}\sin 2x & \frac{1}{2}(e^x\sin 2x + e^x\cos 2x) \\ e^{-x}\cos 2x & \frac{1}{2}(e^x\sin 2x - e^x\cos 2x) \end{bmatrix}.
\end{aligned}
$$

The product inside the integral is then

$$
\begin{aligned}
M^{-1}G(x) &= \begin{bmatrix} -e^{-x}\sin 2x & \frac{1}{2}(e^x\sin 2x + e^x\cos 2x) \\ e^{-x}\cos 2x & \frac{1}{2}(e^x\sin 2x - e^x\cos 2x) \end{bmatrix}\begin{bmatrix} 5 \\ 1-x \end{bmatrix}\\
&= \begin{bmatrix} e^{-x}\left(-5\sin 2x + \frac{1}{2}(1-x)(\cos 2x + \sin 2x)\right) \\ e^{-x}\left(5\cos 2x + \frac{1}{2}(1-x)(\sin 2x - \cos 2x)\right) \end{bmatrix}.
\end{aligned}
$$

Integrating each component of this column vector gives us

$$
\begin{aligned}
\int M^{-1}G(x)\,dx &= \begin{bmatrix} \int e^{-x}\left(-5\sin 2x + \frac{1}{2}(1-x)(\cos 2x + \sin 2x)\right)\,dx \\ \int e^{-x}\left(5\cos 2x + \frac{1}{2}(1-x)(\sin 2x - \cos 2x)\right)\,dx \end{bmatrix}\\
&= \begin{bmatrix} \frac{1}{50}e^{-x}(48\sin 2x + 86\cos 2x + 15x\cos 2x - 5x\sin 2x) \\ \frac{1}{50}e^{-x}(86\sin 2x - 48\cos 2x + 5x\cos 2x + 15x\sin 2x) \end{bmatrix}.
\end{aligned}
$$

Thus we have

$$
\begin{aligned}
Y_P &= M\int M^{-1}G(x)\,dx\\
&= \begin{bmatrix} e^x\cos 2x - e^x\sin 2x & e^x\sin 2x + e^x\cos 2x \\ 2e^x\cos 2x & 2e^x\sin 2x \end{bmatrix}\begin{bmatrix} \frac{1}{50}e^{-x}(48\sin 2x + 86\cos 2x + 15x\cos 2x - 5x\sin 2x) \\ \frac{1}{50}e^{-x}(86\sin 2x - 48\cos 2x + 5x\cos 2x + 15x\sin 2x) \end{bmatrix}\\
&= \begin{bmatrix} \frac{19}{25} + \frac{2}{5}x \\ \frac{86}{25} + \frac{3}{5}x \end{bmatrix}.
\end{aligned}
$$

The general solution is then

$$
\begin{aligned}
Y &= Y_H + Y_P = \begin{bmatrix} c_1(e^x\cos 2x - e^x\sin 2x) + c_2(e^x\sin 2x + e^x\cos 2x) \\ 2c_1 e^x\cos 2x + 2c_2 e^x\sin 2x \end{bmatrix} + \begin{bmatrix} \frac{19}{25} + \frac{2}{5}x \\ \frac{86}{25} + \frac{3}{5}x \end{bmatrix}\\
&= \begin{bmatrix} c_1(e^x\cos 2x - e^x\sin 2x) + c_2(e^x\sin 2x + e^x\cos 2x) + \frac{19}{25} + \frac{2}{5}x \\ 2c_1 e^x\cos 2x + 2c_2 e^x\sin 2x + \frac{86}{25} + \frac{3}{5}x \end{bmatrix}.
\end{aligned}
$$

Equivalently,

$$y_1 = c_1(e^x\cos 2x - e^x\sin 2x) + c_2(e^x\sin 2x + e^x\cos 2x) + \frac{19}{25} + \frac{2}{5}x, \quad y_2 = 2c_1 e^x\cos 2x + 2c_2 e^x\sin 2x + \frac{86}{25} + \frac{3}{5}x.$$

Please note the answer to this problem in the back of the text is incorrect.

5. We have the system $Y' = \begin{bmatrix} 4 & 0 & 1 \\ -2 & 1 & 0 \\ -2 & 0 & 1 \end{bmatrix} Y + G(x)$, where $G(x) = \begin{bmatrix} 1-2x \\ xe^{-x} \\ 0 \end{bmatrix}$. From Exercise 5 of Section 6.2 we know

the general solution to the corresponding homogeneous system is $Y_H = \begin{bmatrix} c_2e^{2x} - c_3e^{3x} \\ c_1e^x - 2c_2e^{2x} + c_3e^{3x} \\ -2c_2e^{2x} + c_3e^{3x} \end{bmatrix}$.

To find a particular solution we will use the formula $Y_P = M \int M^{-1}G(x)\,dx$, where M is a matrix of fundamental solutions.

Here $M = \begin{bmatrix} 0 & e^{2x} & -e^{3x} \\ e^x & -2e^{2x} & e^{3x} \\ 0 & -2e^{2x} & e^{3x} \end{bmatrix}$. We have

$$\left[\begin{array}{ccc:ccc} 0 & e^{2x} & -e^{3x} & 1 & 0 & 0 \\ e^x & -2e^{2x} & e^{3x} & 0 & 1 & 0 \\ 0 & -2e^{2x} & e^{3x} & 0 & 0 & 1 \end{array}\right] \rightarrow \left[\begin{array}{ccc:ccc} e^x & -2e^{2x} & e^{3x} & 0 & 1 & 0 \\ 0 & e^{2x} & -e^{3x} & 1 & 0 & 0 \\ 0 & 0 & -e^{3x} & 2 & 0 & 1 \end{array}\right]$$

$$\rightarrow \left[\begin{array}{ccc:ccc} e^x & -2e^{2x} & 0 & 2 & 1 & 1 \\ 0 & e^{2x} & 0 & -1 & 0 & -1 \\ 0 & 0 & e^{3x} & -2 & 0 & -1 \end{array}\right] \rightarrow \left[\begin{array}{ccc:ccc} e^x & 0 & 0 & 0 & 1 & -1 \\ 0 & e^{2x} & 0 & -1 & 0 & -1 \\ 0 & 0 & e^{3x} & -2 & 0 & -1 \end{array}\right].$$

Multiplying the first row by e^{-x}, the second by e^{-2x} and the third by e^{-3x} we obtain

$$M^{-1} = \begin{bmatrix} 0 & e^{-x} & -e^{-x} \\ -e^{-2x} & 0 & -e^{-2x} \\ -2e^{-3x} & 0 & -e^{-3x} \end{bmatrix}.$$

The product inside the integral is then

$$M^{-1}G(x) = \begin{bmatrix} 0 & e^{-x} & -e^{-x} \\ -e^{-2x} & 0 & -e^{-2x} \\ -2e^{-3x} & 0 & -e^{-3x} \end{bmatrix} \begin{bmatrix} 1-2x \\ xe^{-x} \\ 0 \end{bmatrix} = \begin{bmatrix} xe^{-2x} \\ (2x-1)e^{-2x} \\ (4x-2)e^{-3x} \end{bmatrix}.$$

Integrating each component of this column vector gives us

$$\int M^{-1}G(x)\,dx = \begin{bmatrix} \int xe^{-2x}\,dx \\ \int(2x-1)e^{-2x}\,dx \\ \int(4x-2)e^{-3x}\,dx \end{bmatrix} = \begin{bmatrix} -\left(\frac{1}{2}x+\frac{1}{4}\right)e^{-2x} \\ -xe^{-2x} \\ \left(-\frac{4}{3}x+\frac{2}{9}\right)e^{-3x} \end{bmatrix}.$$

Thus we have

$$Y_P = M\int M^{-1}G(x)\,dx = \begin{bmatrix} 0 & e^{2x} & -e^{3x} \\ e^x & -2e^{2x} & e^{3x} \\ 0 & -2e^{2x} & e^{3x} \end{bmatrix} \begin{bmatrix} -\left(\frac{1}{2}x+\frac{1}{4}\right)e^{-2x} \\ -xe^{-2x} \\ \left(-\frac{4}{3}x+\frac{2}{9}\right)e^{-3x} \end{bmatrix} = \begin{bmatrix} \frac{1}{3}x-\frac{2}{9} \\ -\left(\frac{1}{2}x+\frac{1}{4}\right)e^{-x}+\frac{2}{3}x+\frac{2}{9} \\ \frac{2}{3}x+\frac{2}{9} \end{bmatrix}.$$

The general solution is then

$$\begin{aligned} Y &= Y_H + Y_P = \begin{bmatrix} c_2e^{2x} - c_3e^{3x} \\ c_1e^x - 2c_2e^{2x} + c_3e^{3x} \\ -2c_2e^{2x} + c_3e^{3x} \end{bmatrix} + \begin{bmatrix} \frac{1}{3}x-\frac{2}{9} \\ -\left(\frac{1}{2}x+\frac{1}{4}\right)e^{-x}+\frac{2}{3}x+\frac{2}{9} \\ \frac{2}{3}x+\frac{2}{9} \end{bmatrix} \\[2mm] &= \begin{bmatrix} c_2e^{2x} - c_3e^{3x} + \frac{1}{3}x-\frac{2}{9} \\ c_1e^x - 2c_2e^{2x} + c_3e^{3x} - \left(\frac{1}{2}x+\frac{1}{4}\right)e^{-x}+\frac{2}{3}x+\frac{2}{9} \\ -2c_2e^{2x} + c_3e^{3x} + \frac{2}{3}x+\frac{2}{9} \end{bmatrix}. \end{aligned}$$

Equivalently,

$$
\begin{aligned}
y_1 &= c_2 e^{2x} - c_3 e^{3x} + \frac{1}{3}x - \frac{2}{9}, \\
y_2 &= c_1 e^x - 2c_2 e^{2x} + c_3 e^{3x} - \left(\frac{1}{2}x + \frac{1}{4}\right)e^{-x} + \frac{2}{3}x + \frac{2}{9}, \\
y_3 &= -2c_2 e^{2x} + c_3 e^{3x} + \frac{2}{3}x + \frac{2}{9}.
\end{aligned}
$$

7. We have the system $Y' = \begin{bmatrix} 10 & -9 \\ 4 & -2 \end{bmatrix} Y + G(x)$, where $G(x) = \begin{bmatrix} 2 \\ x \end{bmatrix}$. From Exercise 1 of Section 6.3 we know the general

solution to the corresponding homogeneous system is $Y_H = \begin{bmatrix} 6c_1 e^{4x} + c_2 e^{4x} + 6c_2 x e^{4x} \\ 4c_1 e^{4x} + 4c_2 x e^{4x} \end{bmatrix}$.

To find a particular solution we will use the formula $Y_P = M \int M^{-1} G(x)\, dx$, where M is a matrix of fundamental solutions.

Here $M = \begin{bmatrix} 6e^{4x} & (1+6x)e^{4x} \\ 4e^{4x} & 4xe^{4x} \end{bmatrix}$. We compute M^{-1} using the adjoint as in Corollary 1.27 of the text

$$
\begin{aligned}
M^{-1} &= \frac{1}{\det M}\operatorname{adj}(M) = \frac{1}{(6e^{4x})(4xe^{4x}) - (1+6x)e^{4x}(4e^{4x})}\operatorname{adj}(M) \\
&= \frac{1}{-4e^{8x}}\begin{bmatrix} 4xe^{4x} & -(1+6x)e^{4x} \\ -4e^{4x} & 6e^{4x} \end{bmatrix} = \begin{bmatrix} -xe^{-4x} & \frac{1}{4}(1+6x)e^{-4x} \\ e^{-4x} & -\frac{3}{2}e^{-4x} \end{bmatrix}.
\end{aligned}
$$

The product inside the integral is then

$$
M^{-1}G(x) = \begin{bmatrix} -xe^{-4x} & \frac{1}{4}(1+6x)e^{-4x} \\ e^{-4x} & -\frac{3}{2}e^{-4x} \end{bmatrix}\begin{bmatrix} 2 \\ x \end{bmatrix} = \begin{bmatrix} -2xe^{-4x} + \frac{1}{4}x(1+6x)e^{-4x} \\ 2e^{-4x} - \frac{3}{2}xe^{-4x} \end{bmatrix}.
$$

Integrating each component of this column vector gives us

$$
\int M^{-1}G(x)\, dx = \begin{bmatrix} \int \left(-2xe^{-4x} + \frac{1}{4}x(1+6x)e^{-4x}\right) dx \\ \int \left(2e^{-4x} - \frac{3}{2}xe^{-4x}\right) dx \end{bmatrix} = \begin{bmatrix} e^{-4x}\left(-\frac{3}{8}x^2 + \frac{1}{4}x + \frac{1}{16}\right) \\ e^{-4x}\left(\frac{3}{8}x - \frac{13}{32}\right) \end{bmatrix}.
$$

Thus we have

$$
Y_P = M \int M^{-1}G(x)\, dx = \begin{bmatrix} 6e^{4x} & (1+6x)e^{4x} \\ 4e^{4x} & 4xe^{4x} \end{bmatrix}\begin{bmatrix} e^{-4x}\left(-\frac{3}{8}x^2 + \frac{1}{4}x + \frac{1}{16}\right) \\ e^{-4x}\left(\frac{3}{8}x - \frac{13}{32}\right) \end{bmatrix} = \begin{bmatrix} -\frac{9}{16} - \frac{1}{32} \\ -\frac{5}{8}x + \frac{1}{4} \end{bmatrix}.
$$

The general solution is then

$$
Y = Y_H + Y_P = \begin{bmatrix} 6c_1 e^{4x} + c_2 e^{4x} + 6c_2 x e^{4x} \\ 4c_1 e^{4x} + 4c_2 x e^{4x} \end{bmatrix} + \begin{bmatrix} -\frac{9}{16} - \frac{1}{32} \\ -\frac{5}{8}x + \frac{1}{4} \end{bmatrix} = \begin{bmatrix} 6c_1 e^{4x} + c_2 e^{4x} + 6c_2 x e^{4x} - \frac{9}{16} - \frac{1}{32} \\ 4c_1 e^{4x} + 4c_2 x e^{4x} - \frac{5}{8}x + \frac{1}{4} \end{bmatrix}.
$$

Equivalently, $y_1 = 6c_1 e^{4x} + c_2 e^{4x} + 6c_2 x e^{4x} - \frac{9}{16} - \frac{1}{32}$, $y_2 = 4c_1 e^{4x} + 4c_2 x e^{4x} - \frac{5}{8}x + \frac{1}{4}$.

9. We have the system $Y' = \begin{bmatrix} 5 & 6 & 2 \\ 0 & -1 & -8 \\ 1 & 0 & -2 \end{bmatrix} Y + G(x)$, where $G(x) = \begin{bmatrix} 0 \\ e^{-2x} \\ 0 \end{bmatrix}$. From Exercise 5 of Section 6.3 we know

the general solution to the corresponding homogeneous system is

$$
Y_H = \begin{bmatrix} 6c_1 e^{-4x} + 140(c_2 e^{3x} + c_3 x e^{3x}) + 43c_3 e^{3x} \\ -8c_1 e^{-4x} - 56(c_2 e^{3x} + c_3 x e^{3x}) + 8c_3 e^{3x} \\ -3c_1 e^{-4x} + 28(c_2 e^{3x} + c_3 x e^{3x}) + 3c_3 e^{3x} \end{bmatrix}.
$$

To find a particular solution we will use the formula $Y_P = M \int M^{-1} G(x)\, dx$, where M is a matrix of fundamental solutions.

Here $M = \begin{bmatrix} 6e^{-4x} & 140e^{3x} & (140x+43)e^{3x} \\ -8e^{-4x} & -56e^{3x} & (-56x+8)e^{3x} \\ -3e^{-4x} & 28e^{3x} & (28x+3)e^{3x} \end{bmatrix}$. To find M^{-1} using Maple we use the commands:

```
with(linalg);
M:=matrix(3,3,[6*exp(-4*x),140*exp(3*x),(140*x+43)*exp(3*x),-8*exp(-4*x),
    -56*exp(3*x),(-56*x+8)*exp(3*x),-3*exp(-4*x),28*exp(3*x),(28*x+3)*exp(3*x)]);
inverse(M);
```

Here we obtain $M^{-1} = \begin{bmatrix} \frac{1}{49}e^{4x} & -\frac{2}{49}e^{4x} & -\frac{9}{49}e^{4x} \\ \frac{1}{49}xe^{-3x} & \left(-\frac{3}{98}x - \frac{3}{392}\right)e^{-3x} & \left(\frac{2}{49}x + \frac{1}{49}\right)e^{-3x}e^{4x} \\ \frac{1}{49}e^{-3x} & \frac{3}{98}e^{-3x} & -\frac{2}{49}e^{-3x} \end{bmatrix}$.

The product inside the integral is then

$$M^{-1}G(x) = \begin{bmatrix} \frac{1}{49}e^{4x} & -\frac{2}{49}e^{4x} & -\frac{9}{49}e^{4x} \\ \frac{1}{49}xe^{-3x} & \left(-\frac{3}{98}x - \frac{3}{392}\right)e^{-3x} & \left(\frac{2}{49}x + \frac{1}{49}\right)e^{-3x}e^{4x} \\ \frac{1}{49}e^{-3x} & \frac{3}{98}e^{-3x} & -\frac{2}{49}e^{-3x} \end{bmatrix} \begin{bmatrix} 0 \\ e^{-2x} \\ 0 \end{bmatrix} = \begin{bmatrix} -\frac{2}{49}e^{2x} \\ \left(\frac{3}{98}x - \frac{3}{392}\right)e^{-5x} \\ \frac{3}{98}e^{-5x} \end{bmatrix} .$$

Integrating each component of this column vector gives us

$$\int M^{-1}G(x)\, dx = \begin{bmatrix} \int -\frac{2}{49}e^{2x}\, dx \\ \int \left(\frac{3}{98}x - \frac{3}{392}\right)e^{-5x}\, dx \\ \int \frac{3}{98}e^{-5x}\, dx \end{bmatrix} = \begin{bmatrix} -\frac{1}{49}e^{2x} \\ \frac{3}{490}xe^{-5x} + \frac{27}{9800}e^{-5x} \\ -\frac{3}{490}e^{-5x} \end{bmatrix} .$$

Thus we have

$$Y_P = M \int M^{-1}G(x)\, dx = \begin{bmatrix} 6e^{-4x} & 140e^{3x} & (140x+43)e^{3x} \\ -8e^{-4x} & -56e^{3x} & (-56x+8)e^{3x} \\ -3e^{-4x} & 28e^{3x} & (28x+3)e^{3x} \end{bmatrix} \begin{bmatrix} -\frac{1}{49}e^{2x} \\ \frac{3}{490}xe^{-5x} + \frac{27}{9800}e^{-5x} \\ -\frac{3}{490}e^{-5x} \end{bmatrix} = \begin{bmatrix} 0 \\ -\frac{1}{25}e^{-2x} \\ \frac{3}{25}e^{-2x} \end{bmatrix} .$$

The general solution is then

$$Y = Y_H + Y_P = \begin{bmatrix} 6c_1e^{-4x} + 140(c_2e^{3x} + c_3xe^{3x}) + 43c_3e^{3x} \\ -8c_1e^{-4x} - 56(c_2e^{3x} + c_3xe^{3x}) + 8c_3e^{3x} \\ -3c_1e^{-4x} + 28(c_2e^{3x} + c_3xe^{3x}) + 3c_3e^{3x} \end{bmatrix} + \begin{bmatrix} 0 \\ -\frac{1}{25}e^{-2x} \\ \frac{3}{25}e^{-2x} \end{bmatrix}$$

$$= \begin{bmatrix} 6c_1e^{-4x} + 140(c_2e^{3x} + c_3xe^{3x}) + 43c_3e^{3x} \\ -8c_1e^{-4x} - 56(c_2e^{3x} + c_3xe^{3x}) + 8c_3e^{3x} - \frac{1}{25}e^{-2x} \\ -3c_1e^{-4x} + 28(c_2e^{3x} + c_3xe^{3x}) + 3c_3e^{3x} + \frac{3}{25}e^{-2x} \end{bmatrix} .$$

Equivalently,

$$y_1 = 6c_1e^{-4x} + 140(c_2e^{3x} + c_3xe^{3x}) + 43c_3e^{3x},$$

$$y_2 = -8c_1e^{-4x} - 56(c_2e^{3x} + c_3xe^{3x}) + 8c_3e^{3x} - \frac{1}{25}e^{-2x},$$

$$y_3 = -3c_1e^{-4x} + 28(c_2e^{3x} + c_3xe^{3x}) + 3c_3e^{3x} + \frac{3}{25}e^{-2x}.$$

11. From Exercise 1 we have $Y(x) = \begin{bmatrix} c_1e^{3x} - \frac{2}{3} \\ 2c_1e^{3x} + c_2e^{-x} + x - \frac{19}{3} \end{bmatrix}$. We are now given the initial condition

$$Y(0) = \begin{bmatrix} 2 \\ 1 \end{bmatrix} .$$

Thus we have $Y(0) = \begin{bmatrix} 2 \\ 1 \end{bmatrix} = \begin{bmatrix} c_1 - \frac{2}{3} \\ 2c_1 + c_2 - \frac{19}{3} \end{bmatrix}$. The solution to this system is $c_1 = \frac{8}{3}$, $c_2 = 2$. Thus the solution to the

initial value problem is $Y(x) = \begin{bmatrix} \frac{8}{3}e^{3x} - \frac{2}{3} \\ \frac{16}{3}e^{3x} + 2e^{-x} + x - \frac{19}{3} \end{bmatrix}$. Equivalently,

$$y_1 = \frac{8}{3}e^{3x} - \frac{2}{3}, \; y_2 = \frac{16}{3}e^{3x} + 2e^{-x} + x - \frac{19}{3}.$$

13. From Exercise 5 we have $Y(x) = \begin{bmatrix} c_2 e^{2x} - c_3 e^{3x} + \frac{1}{3}x - \frac{2}{9} \\ c_1 e^x - 2c_2 e^{2x} + c_3 e^{3x} - \left(\frac{1}{2}x + \frac{1}{4}\right)e^{-x} + \frac{2}{3}x + \frac{2}{9} \\ -2c_2 e^{2x} + c_3 e^{3x} + \frac{2}{3}x + \frac{2}{9} \end{bmatrix}$. We are now given the initial condition

$$Y(0) = \begin{bmatrix} 0 \\ 1 \\ -1 \end{bmatrix}.$$

Thus we have $Y(0) = \begin{bmatrix} 0 \\ 1 \\ -1 \end{bmatrix} = \begin{bmatrix} c_2 - c_3 - \frac{2}{9} \\ c_1 - 2c_2 + c_3 + \frac{2}{9} - \frac{1}{4} \\ -2c_2 + c_3 + \frac{2}{9} \end{bmatrix}$. The solution to this system is

$$c_1 = \frac{9}{4}, \; c_2 = 1, \; c_3 = \frac{7}{9}.$$

Thus the solution to the initial value problem is $Y(x) = \begin{bmatrix} e^{2x} - \frac{7}{9}e^{3x} + \frac{1}{3}x - \frac{2}{9} \\ \frac{9}{4}e^x - 2e^{2x} + \frac{7}{9}e^{3x} - \left(\frac{1}{2}x + \frac{1}{4}\right)e^{-x} + \frac{2}{3}x + \frac{2}{9} \\ -2e^{2x} + \frac{7}{9}e^{3x} + \frac{2}{3}x + \frac{2}{9} \end{bmatrix}$.

Equivalently,

$$y_1 = e^{2x} - \frac{7}{9}e^{3x} + \frac{1}{3}x - \frac{2}{9},$$

$$y_2 = \frac{9}{4}e^x - 2e^{2x} + \frac{7}{9}e^{3x} - \left(\frac{1}{2}x + \frac{1}{4}\right)e^{-x} + \frac{2}{3}x + \frac{2}{9},$$

$$y_3 = -2e^{2x} + \frac{7}{9}e^{3x} + \frac{2}{3}x + \frac{2}{9}.$$

15. We have the velocity vector $v = \begin{bmatrix} 2x+y-t \\ 2x+3y+t \end{bmatrix}$. This may be written as $v = A\begin{bmatrix} x \\ y \end{bmatrix} + G(t)$ where $A = \begin{bmatrix} 2 & 1 \\ 2 & 3 \end{bmatrix}$ and

$G(t) = \begin{bmatrix} -t \\ t \end{bmatrix}$. We are asked to find the position, s, of the object at time t if $s(0) = (1,1)$.

The characteristic equation of A is

$$\det(\lambda I - A) = \begin{vmatrix} \lambda - 2 & -1 \\ -2 & \lambda - 3 \end{vmatrix}$$

$$= (\lambda - 2)(\lambda - 3) - 2 = \lambda^2 - 5\lambda + 4 = (\lambda - 1)(\lambda - 4) = 0.$$

Therefore, the eigenvalues of A are $\lambda = 1$ and $\lambda = 4$. We now wish to find bases for the corresponding eigenspaces.

For $\lambda = 1$ we need to find a basis for the null space of $I - A = \begin{bmatrix} -1 & -1 \\ -2 & -2 \end{bmatrix}$. Reducing the matrix for the associated homogeneous system,

$$\begin{bmatrix} -1 & -1 & \vdots & 0 \\ -2 & -2 & \vdots & 0 \end{bmatrix} \rightarrow \begin{bmatrix} 1 & 1 & \vdots & 0 \\ 0 & 0 & \vdots & 0 \end{bmatrix}$$

we see our solutions are

$$\begin{bmatrix} x \\ y \end{bmatrix} = \begin{bmatrix} x \\ -x \end{bmatrix} = x \begin{bmatrix} 1 \\ -1 \end{bmatrix}$$

and hence the column vector $\begin{bmatrix} 1 \\ -1 \end{bmatrix}$ forms a basis for E_1.

For $\lambda = 4$ we need to find a basis for the null space of $4I - A = \begin{bmatrix} 2 & -1 \\ -2 & 1 \end{bmatrix}$. Reducing the matrix for the associated homogeneous system,

$$\begin{bmatrix} 2 & -1 & \vdots & 0 \\ -2 & 1 & \vdots & 0 \end{bmatrix} \rightarrow \begin{bmatrix} 2 & -1 & \vdots & 0 \\ 0 & 0 & \vdots & 0 \end{bmatrix}$$

we see our solutions are

$$\begin{bmatrix} x \\ y \end{bmatrix} = \begin{bmatrix} x \\ 2x \end{bmatrix} = x \begin{bmatrix} 1 \\ 2 \end{bmatrix}$$

and hence the column vector $\begin{bmatrix} 1 \\ 2 \end{bmatrix}$ forms a basis for E_4.

Matrix A is similar to the diagonal matrix

$$D = P^{-1}AP = \begin{bmatrix} 1 & 0 \\ 0 & 4 \end{bmatrix} \text{ where } P = \begin{bmatrix} 1 & 1 \\ -1 & 2 \end{bmatrix}.$$

The columns of P are the basis vectors for the eigenspaces. We also know that $\begin{bmatrix} c_1 e^t \\ c_2 e^{4t} \end{bmatrix}$ is the general solution to

$$v = D \begin{bmatrix} x \\ y \end{bmatrix}.$$

Therefore the general solution to the associated homogeneous equation $v = A \begin{bmatrix} x \\ y \end{bmatrix}$ is

$$s_H = P \begin{bmatrix} c_1 e^t \\ c_2 e^{4t} \end{bmatrix} = \begin{bmatrix} 1 & 1 \\ -1 & 2 \end{bmatrix} \begin{bmatrix} c_1 e^t \\ c_2 e^{4t} \end{bmatrix} = \begin{bmatrix} c_1 e^t + c_2 e^{4t} \\ -c_1 e^t + 2c_2 e^{4t} \end{bmatrix}.$$

To find a particular solution we will use the formula $s_P = M \int M^{-1} G(t) \, dt$, where M is a matrix of fundamental solutions. Here $M = \begin{bmatrix} e^t & e^{4t} \\ -e^t & 2e^{4t} \end{bmatrix}$. We have

$$\begin{bmatrix} e^t & e^{4t} & \vdots & 1 & 0 \\ -e^t & 2e^{4t} & \vdots & 0 & 1 \end{bmatrix} \rightarrow \begin{bmatrix} e^t & e^{4t} & \vdots & 1 & 0 \\ 0 & 3e^{4t} & \vdots & 1 & 1 \end{bmatrix} \rightarrow \begin{bmatrix} e^t & e^{4t} & \vdots & 1 & 0 \\ 0 & e^{4t} & \vdots & \frac{1}{3} & \frac{1}{3} \end{bmatrix} \rightarrow \begin{bmatrix} e^t & 0 & \vdots & \frac{2}{3} & -\frac{1}{3} \\ 0 & e^{4t} & \vdots & \frac{1}{3} & \frac{1}{3} \end{bmatrix}.$$

Multiplying the first row by e^{-t} and the second by e^{-4t} we obtain $M^{-1} = \begin{bmatrix} \frac{2}{3}e^{-t} & -\frac{1}{3}e^{-t} \\ \frac{1}{3}e^{-4t} & \frac{1}{3}e^{-4t} \end{bmatrix}$.

The product inside the integral is then

$$M^{-1}G(t) = \begin{bmatrix} \frac{2}{3}e^{-t} & -\frac{1}{3}e^{-t} \\ \frac{1}{3}e^{-4t} & \frac{1}{3}e^{-4t} \end{bmatrix} \begin{bmatrix} -t \\ t \end{bmatrix} = \begin{bmatrix} -te^{-t} \\ 0 \end{bmatrix}.$$

Integrating each component of this column vector gives us

$$\int M^{-1}G(t) \, dt = \begin{bmatrix} -\int te^{-t} \, dt \\ 0 \end{bmatrix} = \begin{bmatrix} te^{-t} + e^{-t} \\ 0 \end{bmatrix}.$$

Thus we have

$$s_P = M \int M^{-1} G(t)\, dt = \begin{bmatrix} e^t & e^{4t} \\ -e^t & 2e^{4t} \end{bmatrix} \begin{bmatrix} te^{-t} + e^{-t} \\ 0 \end{bmatrix} = \begin{bmatrix} t+1 \\ -t-1 \end{bmatrix}.$$

The general solution is then

$$s = s_H + s_P = \begin{bmatrix} c_1 e^t + c_2 e^{4t} \\ -c_1 e^t + 2c_2 e^{4t} \end{bmatrix} + \begin{bmatrix} t+1 \\ -t-1 \end{bmatrix} = \begin{bmatrix} c_1 e^t + c_2 e^{4t} + t + 1 \\ -c_1 e^t + 2c_2 e^{4t} - t - 1 \end{bmatrix}.$$

We now use given the initial position $s(0) = (1,1)$. Thus we have $\begin{bmatrix} 1 \\ 1 \end{bmatrix} = \begin{bmatrix} c_1 + c_2 + 1 \\ -c_1 + 2c_2 - 1 \end{bmatrix}$. The solution to this system

is $c_1 = -\frac{2}{3}$, $c_2 = \frac{2}{3}$. Thus the solution to the initial value problem is $s = \begin{bmatrix} -\frac{2}{3}e^t + \frac{2}{3}e^{4t} + t + 1 \\ \frac{2}{3}e^t + \frac{4}{3}e^{4t} - t - 1 \end{bmatrix}$. Equivalently,

$$x = -\frac{2}{3}e^t + \frac{2}{3}e^{4t} + t + 1, \; y = \frac{2}{3}e^t + \frac{4}{3}e^{4t} - t - 1.$$

6.5   Converting Differential Equations to First Order Systems

Exercises 6.5, pp. 322

1. In Example 1 of Section 6.5 the text explains how the linear differential equation $y'' + 3y' + 2y = 0$ can be converted to the system of first order differential equation 's

$$\begin{bmatrix} v_1' \\ v_2' \end{bmatrix} = \begin{bmatrix} 0 & 1 \\ -2 & -3 \end{bmatrix} \begin{bmatrix} v_1 \\ v_2 \end{bmatrix}.$$

Furthermore the characteristic equation $\lambda^2 + 3\lambda + 2 = 0$ is also the characteristic equation of the matrix

$$A = \begin{bmatrix} 0 & 1 \\ -2 & -3 \end{bmatrix}.$$

The eigenvalues of A are $\lambda = -1$ and $\lambda = -2$. We now wish to find bases for the corresponding eigenspaces.

For $\lambda = -1$ we need to find a basis for the null space of $-I - A = \begin{bmatrix} -1 & -1 \\ 2 & 2 \end{bmatrix}$. Reducing the matrix for the associated

homogeneous system, $\begin{bmatrix} -1 & -1 & \vdots & 0 \\ 2 & 2 & \vdots & 0 \end{bmatrix} \rightarrow \begin{bmatrix} 1 & 1 & \vdots & 0 \\ 0 & 0 & \vdots & 0 \end{bmatrix}$, we see our solutions are

$$\begin{bmatrix} x \\ y \end{bmatrix} = \begin{bmatrix} x \\ -x \end{bmatrix} = x \begin{bmatrix} 1 \\ -1 \end{bmatrix}$$

and hence the column vector $\begin{bmatrix} 1 \\ -1 \end{bmatrix}$ forms a basis for E_{-1}.

For $\lambda = -2$ we need to find a basis for the null space of $-2I - A = \begin{bmatrix} -2 & -1 \\ 2 & 1 \end{bmatrix}$. Reducing the matrix for the associated

homogeneous system, $\begin{bmatrix} -2 & -1 & \vdots & 0 \\ 2 & 1 & \vdots & 0 \end{bmatrix} \rightarrow \begin{bmatrix} 2 & 1 & \vdots & 0 \\ 0 & 0 & \vdots & 0 \end{bmatrix}$, we see our solutions are

$$\begin{bmatrix} x \\ y \end{bmatrix} = \begin{bmatrix} x \\ -2x \end{bmatrix} = x \begin{bmatrix} 1 \\ -2 \end{bmatrix}$$

and hence the column vector $\begin{bmatrix} 1 \\ -2 \end{bmatrix}$ forms a basis for E_{-2}.

Matrix A is similar to the diagonal matrix $D = P^{-1}AP = \begin{bmatrix} -1 & 0 \\ 0 & -2 \end{bmatrix}$ where $P = \begin{bmatrix} 1 & 1 \\ -1 & -2 \end{bmatrix}$. The columns of P are the

basis vectors for the eigenspaces. We also know that $\begin{bmatrix} c_1 e^{-x} \\ c_2 e^{-2x} \end{bmatrix}$ is the general solution to

$$\begin{bmatrix} v_1' \\ v_2' \end{bmatrix} = D \begin{bmatrix} v_1 \\ v_2 \end{bmatrix}.$$

Therefore the general solution to $\begin{bmatrix} v_1' \\ v_2' \end{bmatrix} = A \begin{bmatrix} v_1 \\ v_2 \end{bmatrix}$ is

$$P \begin{bmatrix} c_1 e^{-x} \\ c_2 e^{-2x} \end{bmatrix} = \begin{bmatrix} 1 & 1 \\ -1 & -2 \end{bmatrix} \begin{bmatrix} c_1 e^{-x} \\ c_2 e^{-2x} \end{bmatrix} = \begin{bmatrix} c_1 e^{-x} + c_2 e^{-2x} \\ -c_1 e^{-x} - 2c_2 e^{-2x} \end{bmatrix}.$$

Thus, $v_1 = c_1 e^{-x} + c_2 e^{-2x}$, $v_2 = -c_1 e^{-x} - 2c_2 e^{-2x}$. Since $v_1 = y$ the solution to the original differential equation is

$$y = c_1 e^{-x} + c_2 e^{-2x}.$$

3. In Example 3 of Section 6.5 the text explains how the linear differential equation $y'' - y' - 2y = \sin x$ can be converted to the system of first order differential equation's

$$\begin{bmatrix} v_1' \\ v_2' \end{bmatrix} = \begin{bmatrix} 0 & 1 \\ 2 & 1 \end{bmatrix} \begin{bmatrix} v_1 \\ v_2 \end{bmatrix} + \begin{bmatrix} 0 \\ \sin x \end{bmatrix}.$$

We let $A = \begin{bmatrix} 0 & 1 \\ -2 & -3 \end{bmatrix}$. The characteristic equation of A is equal to the characteristic equation of the original differential

equation, $\lambda^2 - \lambda - 2 = (\lambda + 1)(\lambda - 2) = 0$. Therefore, the eigenvalues of A are $\lambda = -1$ and $\lambda = 2$. We now wish to find bases for the corresponding eigenspaces.

For $\lambda = -1$ we need to find a basis for the null space of $-I - A = \begin{bmatrix} -1 & -1 \\ -2 & -2 \end{bmatrix}$. Reducing the matrix for the associated

homogeneous system, $\begin{bmatrix} -1 & -1 & \vdots & 0 \\ -2 & -2 & \vdots & 0 \end{bmatrix} \rightarrow \begin{bmatrix} 1 & 1 & \vdots & 0 \\ 0 & 0 & \vdots & 0 \end{bmatrix}$, we see our solutions are

$$\begin{bmatrix} x \\ y \end{bmatrix} = \begin{bmatrix} x \\ -x \end{bmatrix} = x \begin{bmatrix} 1 \\ -1 \end{bmatrix}$$

and hence the column vector $\begin{bmatrix} 1 \\ -1 \end{bmatrix}$ forms a basis for E_{-1}.

For $\lambda = 2$ we need to find a basis for the null space of $2I - A = \begin{bmatrix} 2 & -1 \\ -2 & 1 \end{bmatrix}$. Reducing the matrix for the associated

homogeneous system, $\begin{bmatrix} 2 & -1 & \vdots & 0 \\ -2 & 1 & \vdots & 0 \end{bmatrix} \rightarrow \begin{bmatrix} 2 & -1 & \vdots & 0 \\ 0 & 0 & \vdots & 0 \end{bmatrix}$, we see our solutions are

$$\begin{bmatrix} x \\ y \end{bmatrix} = \begin{bmatrix} x \\ 2x \end{bmatrix} = x \begin{bmatrix} 1 \\ 2 \end{bmatrix}$$

and hence the column vector $\begin{bmatrix} 1 \\ 2 \end{bmatrix}$ forms a basis for E_2.

Matrix A is similar to the diagonal matrix $D = P^{-1}AP = \begin{bmatrix} -1 & 0 \\ 0 & 2 \end{bmatrix}$ where $P = \begin{bmatrix} 1 & 1 \\ -1 & 2 \end{bmatrix}$. The columns of P are the basis

vectors for the eigenspaces. We also know that $\begin{bmatrix} c_1 e^{-x} \\ c_2 e^{2x} \end{bmatrix}$ is the general solution to the corresponding homogeneous system

$$\begin{bmatrix} v_1' \\ v_2' \end{bmatrix} = D \begin{bmatrix} v_1 \\ v_2 \end{bmatrix}.$$

Therefore the general solution to $\begin{bmatrix} v_1' \\ v_2' \end{bmatrix} = A \begin{bmatrix} v_1 \\ v_2 \end{bmatrix}$ is

$$V_H = P \begin{bmatrix} c_1 e^{-x} \\ c_2 e^{2x} \end{bmatrix} = \begin{bmatrix} 1 & 1 \\ -1 & 2 \end{bmatrix} \begin{bmatrix} c_1 e^{-x} \\ c_2 e^{2x} \end{bmatrix} = \begin{bmatrix} c_1 e^{-x} + c_2 e^{2x} \\ -c_1 e^{-x} + 2c_2 e^{2x} \end{bmatrix}.$$

To find a particular solution we will use the formula $V_P = M \int M^{-1} G(x)\, dx$, where M is a matrix of fundamental solutions and $G(x) = \begin{bmatrix} 0 \\ \sin x \end{bmatrix}$.

Here $M = \begin{bmatrix} e^{-x} & e^{2x} \\ -e^{-x} & 2e^{2x} \end{bmatrix}$. We have

$$\begin{bmatrix} e^{-x} & e^{2x} & : & 1 & 0 \\ -e^{-x} & 2e^{2x} & : & 0 & 1 \end{bmatrix} \rightarrow \begin{bmatrix} e^{-x} & e^{2x} & : & 1 & 0 \\ 0 & 3e^{2x} & : & 1 & 1 \end{bmatrix} \rightarrow \begin{bmatrix} e^{-x} & 0 & : & \frac{2}{3} & -\frac{1}{3} \\ 0 & e^{2x} & : & \frac{1}{3} & \frac{1}{3} \end{bmatrix}.$$

Multiplying the first row by e^x and the second by e^{-2x} we obtain $M^{-1} = \begin{bmatrix} \frac{2}{3} e^x & -\frac{1}{3} e^x \\ \frac{1}{3} e^{-2x} & \frac{1}{3} e^{-2x} \end{bmatrix}$.

The product inside the integral is then

$$M^{-1} G(x) = \begin{bmatrix} \frac{2}{3} e^x & -\frac{1}{3} e^x \\ \frac{1}{3} e^{-2x} & \frac{1}{3} e^{-2x} \end{bmatrix} \begin{bmatrix} 0 \\ \sin x \end{bmatrix} = \begin{bmatrix} -\frac{1}{3} e^x \sin x \\ \frac{1}{3} e^{-2x} \sin x \end{bmatrix}.$$

Integrating each component of this column vector gives us

$$\int M^{-1} G(x)\, dx = \begin{bmatrix} -\int \frac{1}{3} e^x \sin x\, dx \\ \int \frac{1}{3} e^{-2x} \sin x\, dx \end{bmatrix} = \begin{bmatrix} \frac{1}{6} e^x \cos x - \frac{1}{6} e^x \sin x \\ -\frac{1}{15} e^{-2x} \cos x - \frac{2}{15} e^{-2x} \sin x \end{bmatrix}.$$

Thus we have

$$V_P = M \int M^{-1} G(x)\, dx = \begin{bmatrix} e^{-x} & e^{2x} \\ -e^{-x} & 2e^{2x} \end{bmatrix} \begin{bmatrix} \frac{1}{6} e^x \cos x - \frac{1}{6} e^x \sin x \\ -\frac{1}{15} e^{-2x} \cos x - \frac{2}{15} e^{-2x} \sin x \end{bmatrix} = \begin{bmatrix} \frac{1}{10} \cos x - \frac{3}{10} \sin x \\ -\frac{3}{10} \cos x - \frac{1}{10} \sin x \end{bmatrix}.$$

The general solution is then

$$V = V_H + V_P = \begin{bmatrix} c_1 e^{-x} + c_2 e^{2x} \\ -c_1 e^{-x} + 2c_2 e^{2x} \end{bmatrix} + \begin{bmatrix} \frac{1}{10} \cos x - \frac{3}{10} \sin x \\ -\frac{3}{10} \cos x - \frac{1}{10} \sin x \end{bmatrix} = \begin{bmatrix} c_1 e^{-x} + c_2 e^{2x} + \frac{1}{10} \cos x - \frac{3}{10} \sin x \\ -c_1 e^{-x} + 2c_2 e^{2x} - \frac{3}{10} \cos x - \frac{1}{10} \sin x \end{bmatrix}.$$

Equivalently, $v_1 = c_1 e^{-x} + c_2 e^{2x} + \frac{1}{10} \cos x - \frac{3}{10} \sin x$, $v_2 = -c_1 e^{-x} + 2c_2 e^{2x} - \frac{3}{10} \cos x - \frac{1}{10} \sin x$.

Since $v_1 = y$ the solution to the original differential equation is

$$y = c_1 e^{-x} + c_2 e^{2x} + \frac{1}{10} \cos x - \frac{3}{10} \sin x.$$

5. In Exercise 5 of Section 4.2 we are given the differential equation $y''' + 4y'' - y' - 4y = 0$. Letting $v_1 = y$, $v_2 = y'$ and $v_3 = y''$ leads to

$$\begin{aligned} v_1' &= y' = v_2 \\ v_2' &= y'' = v_3 \\ v_3' &= y''' = 4y + y' - 4y'' = 4v_1 + v_2 - 4v_3 \end{aligned}$$

which can be rewritten as the system

$$\begin{bmatrix} v_1' \\ v_2' \\ v_3' \end{bmatrix} = \begin{bmatrix} 0 & 1 & 0 \\ 0 & 0 & 1 \\ 4 & 1 & -4 \end{bmatrix} \begin{bmatrix} v_1 \\ v_2 \\ v_3 \end{bmatrix}.$$

We let $A = \begin{bmatrix} 0 & 1 & 0 \\ 0 & 0 & 1 \\ 4 & 1 & -4 \end{bmatrix}$. The characteristic equation of A is equal to the characteristic equation of the original differential

equation, $\lambda^3 + 4\lambda^2 + \lambda - 4 = (\lambda - 1)(\lambda + 1)(\lambda + 4) = 0$. Therefore, the eigenvalues of A are $\lambda = \pm 1$ and $\lambda = -4$. We now wish to find bases for the corresponding eigenspaces.

For $\lambda = 1$ we need to find a basis for the null space of $I - A = \begin{bmatrix} 1 & -1 & 0 \\ 0 & 1 & -1 \\ -4 & -1 & 5 \end{bmatrix}$. Reducing the matrix for the associated

homogeneous system,

$$\begin{bmatrix} 1 & -1 & 0 & \vdots & 0 \\ 0 & 1 & -1 & \vdots & 0 \\ -4 & -1 & 5 & \vdots & 0 \end{bmatrix} \rightarrow \begin{bmatrix} 1 & -1 & 0 & \vdots & 0 \\ 0 & 1 & -1 & \vdots & 0 \\ 0 & -5 & 5 & \vdots & 0 \end{bmatrix} \rightarrow \begin{bmatrix} 1 & -1 & 0 & \vdots & 0 \\ 0 & 1 & -1 & \vdots & 0 \\ 0 & 0 & 0 & \vdots & 0 \end{bmatrix} \rightarrow \begin{bmatrix} 1 & 0 & -1 & \vdots & 0 \\ 0 & 1 & -1 & \vdots & 0 \\ 0 & 0 & 0 & \vdots & 0 \end{bmatrix},$$

we see our solutions are

$$\begin{bmatrix} x \\ y \\ z \end{bmatrix} = \begin{bmatrix} z \\ z \\ z \end{bmatrix} = z \begin{bmatrix} 1 \\ 1 \\ 1 \end{bmatrix}$$

and hence the column vector $\begin{bmatrix} 1 \\ 1 \\ 1 \end{bmatrix}$ forms a basis for E_1.

For $\lambda = -1$ we need to find a basis for the null space of $-I - A = \begin{bmatrix} -1 & -1 & 0 \\ 0 & -1 & -1 \\ -4 & -1 & 3 \end{bmatrix}$. Reducing the matrix for the associated

homogeneous system,

$$\begin{bmatrix} -1 & -1 & 0 & \vdots & 0 \\ 0 & -1 & -1 & \vdots & 0 \\ -4 & -1 & 3 & \vdots & 0 \end{bmatrix} \rightarrow \begin{bmatrix} 1 & 1 & 0 & \vdots & 0 \\ 0 & 1 & 1 & \vdots & 0 \\ 0 & 3 & 3 & \vdots & 0 \end{bmatrix} \rightarrow \begin{bmatrix} 1 & 1 & 0 & \vdots & 0 \\ 0 & 1 & 1 & \vdots & 0 \\ 0 & 0 & 0 & \vdots & 0 \end{bmatrix} \rightarrow \begin{bmatrix} 1 & 0 & -1 & \vdots & 0 \\ 0 & 1 & 1 & \vdots & 0 \\ 0 & 0 & 0 & \vdots & 0 \end{bmatrix},$$

we see our solutions are

$$\begin{bmatrix} x \\ y \\ z \end{bmatrix} = \begin{bmatrix} z \\ -z \\ z \end{bmatrix} = z \begin{bmatrix} 1 \\ -1 \\ 1 \end{bmatrix}$$

and hence the column vector $\begin{bmatrix} 1 \\ -1 \\ 1 \end{bmatrix}$ forms a basis for E_{-1}.

For $\lambda = -4$ we need to find a basis for the null space of $-4I - A = \begin{bmatrix} -4 & -1 & 0 \\ 0 & -4 & -1 \\ -4 & -1 & 0 \end{bmatrix}$. Reducing the matrix for the

associated homogeneous system,

$$\begin{bmatrix} -4 & -1 & 0 & \vdots & 0 \\ 0 & -4 & -1 & \vdots & 0 \\ -4 & -1 & 0 & \vdots & 0 \end{bmatrix} \rightarrow \begin{bmatrix} 4 & 1 & 0 & \vdots & 0 \\ 0 & 4 & 1 & \vdots & 0 \\ 0 & 0 & 0 & \vdots & 0 \end{bmatrix} \rightarrow \begin{bmatrix} 4 & 1 & 0 & \vdots & 0 \\ -16 & 0 & 1 & \vdots & 0 \\ 0 & 0 & 0 & \vdots & 0 \end{bmatrix},$$

we see our solutions are

$$\begin{bmatrix} x \\ y \\ z \end{bmatrix} = \begin{bmatrix} x \\ -4x \\ 16x \end{bmatrix} = x \begin{bmatrix} 1 \\ -4 \\ 16 \end{bmatrix}$$

and hence the column vector $\begin{bmatrix} 1 \\ -4 \\ 16 \end{bmatrix}$ forms a basis for E_{-4}.

Matrix A is similar to the diagonal matrix $D = P^{-1}AP = \begin{bmatrix} 1 & 0 & 0 \\ 0 & -1 & 0 \\ 0 & 0 & -4 \end{bmatrix}$ where $P = \begin{bmatrix} 1 & 1 & 1 \\ 1 & -1 & -4 \\ 1 & 1 & 16 \end{bmatrix}$. The columns of

P are the basis vectors for the eigenspaces. We also know that $\begin{bmatrix} c_1 e^x \\ c_2 e^{-x} \\ c_3 e^{-4x} \end{bmatrix}$ is the solution to the system

$$\begin{bmatrix} v_1' \\ v_2' \\ v_3' \end{bmatrix} = D \begin{bmatrix} v_1 \\ v_2 \\ v_3 \end{bmatrix}.$$

Therefore the general solution to $\begin{bmatrix} v_1' \\ v_2' \\ v_3' \end{bmatrix} = A \begin{bmatrix} v_1 \\ v_2 \\ v_3 \end{bmatrix}$ is

$$P \begin{bmatrix} c_1 e^x \\ c_2 e^{-x} \\ c_3 e^{-4x} \end{bmatrix} = \begin{bmatrix} 1 & 1 & 1 \\ 1 & -1 & -4 \\ 1 & 1 & 16 \end{bmatrix} \begin{bmatrix} c_1 e^x \\ c_2 e^{-x} \\ c_3 e^{-4x} \end{bmatrix} = \begin{bmatrix} c_1 e^x + c_2 e^{-x} + c_3 e^{-4x} \\ c_1 e^x - c_2 e^{-x} - 4c_3 e^{-4x} \\ c_1 e^x + c_2 e^{-x} + 16c_3 e^{-4x} \end{bmatrix}.$$

Equivalently, $v_1 = c_1 e^x + c_2 e^{-x} + c_3 e^{-4x}$, $v_2 = c_1 e^x - c_2 e^{-x} - 4c_3 e^{-4x}$, $v_3 = c_1 e^x + c_2 e^{-x} + 16c_3 e^{-4x}$.

Since $v_1 = y$ the solution to the original differential equation is

$$y = c_1 e^x + c_2 e^{-x} + c_3 e^{-4x}.$$

7. In Exercise 13 of Section 4.2 we are given the differential equation $y''' - 2y' + 4y = 0$. Letting $v_1 = y$, $v_2 = y'$ and $v_3 = y''$ leads to

$$\begin{aligned} v_1' &= y' = v_2 \\ v_2' &= y'' = v_3 \\ v_3' &= y''' = -4y + 2y' = -4v_1 + 2v_2 \end{aligned}$$

which can be rewritten as the system

$$\begin{bmatrix} v_1' \\ v_2' \\ v_3' \end{bmatrix} = \begin{bmatrix} 0 & 1 & 0 \\ 0 & 0 & 1 \\ -4 & 2 & 0 \end{bmatrix} \begin{bmatrix} v_1 \\ v_2 \\ v_3 \end{bmatrix}.$$

We let $A = \begin{bmatrix} 0 & 1 & 0 \\ 0 & 0 & 1 \\ -4 & 2 & 0 \end{bmatrix}$. The characteristic equation of A is equal to the characteristic equation of the original differential

equation, $\lambda^3 + 2\lambda + 4 = (\lambda + 2)(\lambda^2 - 2\lambda + 2) = 0$. The roots of $\lambda^2 - 2\lambda + 2 = 0$ are $\lambda = \frac{2 \pm \sqrt{4-8}}{2} = 1 \pm i$. Therefore, the eigenvalues of A are $\lambda = -2$ and $\lambda = 1 \pm i$. We now wish to find bases for the corresponding eigenspaces.

For $\lambda = -2$ we need to find a basis for the null space of $-2I - A = \begin{bmatrix} -2 & -1 & 0 \\ 0 & -2 & -1 \\ 4 & -2 & -2 \end{bmatrix}$. Reducing the matrix for the

associated homogeneous system,

$$\begin{bmatrix} -2 & -1 & 0 & : & 0 \\ 0 & -2 & -1 & : & 0 \\ 4 & -2 & -2 & : & 0 \end{bmatrix} \rightarrow \begin{bmatrix} 2 & 1 & 0 & : & 0 \\ 0 & 2 & 1 & : & 0 \\ 0 & -4 & -2 & : & 0 \end{bmatrix} \rightarrow \begin{bmatrix} 2 & 1 & 0 & : & 0 \\ -4 & 0 & 1 & : & 0 \\ 0 & 0 & 0 & : & 0 \end{bmatrix},$$

we see our solutions are

$$\begin{bmatrix} x \\ y \\ z \end{bmatrix} = \begin{bmatrix} x \\ -2x \\ 4x \end{bmatrix} = x \begin{bmatrix} 1 \\ -2 \\ 4 \end{bmatrix}$$

and hence the column vector $\begin{bmatrix} 1 \\ -2 \\ 4 \end{bmatrix}$ forms a basis for E_{-2}.

For $\lambda = 1 + i$ we need to find a basis for the null space of $(1+i)I - A = \begin{bmatrix} 1+i & -1 & 0 \\ 0 & 1+i & -1 \\ 4 & -2 & 1+i \end{bmatrix}$. Reducing the matrix for

the associated homogeneous system,

$$\begin{bmatrix} 1+i & -1 & 0 & \vdots & 0 \\ 0 & 1+i & -1 & \vdots & 0 \\ 4 & -2 & 1+i & \vdots & 0 \end{bmatrix} \rightarrow \begin{bmatrix} 1+i & -1 & 0 & \vdots & 0 \\ (1+i)^2 & 0 & -1 & \vdots & 0 \\ 0 & 0 & 0 & \vdots & 0 \end{bmatrix},$$

we see our solutions are

$$\begin{bmatrix} x \\ y \\ z \end{bmatrix} = \begin{bmatrix} x \\ (1+i)x \\ (1+i)^2 x \end{bmatrix} = x \begin{bmatrix} 1 \\ 1+i \\ (1+i)^2 \end{bmatrix}$$

and hence the column vector $\begin{bmatrix} 1 \\ 1+i \\ (1+i)^2 \end{bmatrix}$ forms a basis for E_{1+i}.

For $\lambda = 1 - i$ we need to find a basis for the null space of $(1-i)I - A = \begin{bmatrix} 1-i & -1 & 0 \\ 0 & 1-i & -1 \\ 4 & -2 & 1-i \end{bmatrix}$. Reducing the matrix for

the associated homogeneous system,

$$\begin{bmatrix} 1-i & -1 & 0 & \vdots & 0 \\ 0 & 1-i & -1 & \vdots & 0 \\ 4 & -2 & 1-i & \vdots & 0 \end{bmatrix} \rightarrow \begin{bmatrix} 1-i & -1 & 0 & \vdots & 0 \\ (1-i)^2 & 0 & -1 & \vdots & 0 \\ 0 & 0 & 0 & \vdots & 0 \end{bmatrix},$$

we see our solutions are

$$\begin{bmatrix} x \\ y \\ z \end{bmatrix} = \begin{bmatrix} x \\ (1-i)x \\ (1-i)^2 x \end{bmatrix} = x \begin{bmatrix} 1 \\ 1-i \\ (1-i)^2 \end{bmatrix}$$

and hence the column vector $\begin{bmatrix} 1 \\ 1-i \\ (1-i)^2 \end{bmatrix}$ forms a basis for E_{1-i}.

Matrix A is similar to the diagonal matrix $D = P^{-1}AP = \begin{bmatrix} -2 & 0 & 0 \\ 0 & 1+i & 0 \\ 0 & 0 & 1-i \end{bmatrix}$ where

$$P = \begin{bmatrix} 1 & 1 & 1 \\ -2 & 1+1 & 1-i \\ 4 & (1+i)^2 & (1-i)^2 \end{bmatrix}.$$

The columns of P are the basis vectors for the eigenspaces.

From the diagonal entry -2 of matrix D, we obtain one solution to $\begin{bmatrix} v_1' \\ v_2' \\ v_3' \end{bmatrix} = D \begin{bmatrix} v_1 \\ v_2 \\ v_3 \end{bmatrix} : \begin{bmatrix} e^{-2x} \\ 0 \\ 0 \end{bmatrix}$. This gives us the

solution $P \begin{bmatrix} e^{-2x} \\ 0 \\ 0 \end{bmatrix} = \begin{bmatrix} 1 & 1 & 1 \\ -2 & 1+1 & 1-i \\ 4 & (1+i)^2 & (1-i)^2 \end{bmatrix} \begin{bmatrix} e^{-2x} \\ 0 \\ 0 \end{bmatrix} = \begin{bmatrix} e^{-2x} \\ -2e^{-2x} \\ 4e^{-2x} \end{bmatrix}$ to $\begin{bmatrix} v_1' \\ v_2' \\ v_3' \end{bmatrix} = A \begin{bmatrix} v_1 \\ v_2 \\ v_3 \end{bmatrix}$.

From the diagonal entry $1+i$ of D we obtain the complex valued solution to $\begin{bmatrix} v_1' \\ v_2' \\ v_3' \end{bmatrix} = D \begin{bmatrix} v_1 \\ v_2 \\ v_3 \end{bmatrix}$,

$$Z = \begin{bmatrix} 0 \\ e^{(1+i)x} \\ 0 \end{bmatrix} = \begin{bmatrix} 0 \\ e^x(\cos x + i\sin x) \\ 0 \end{bmatrix}.$$

By the first part of Theorem 6.7,

$$PZ = \begin{bmatrix} 1 & 1 & 1 \\ -2 & 1+1 & 1-i \\ 4 & 2i & -2i \end{bmatrix} \begin{bmatrix} 0 \\ e^x(\cos x + i\sin x) \\ 0 \end{bmatrix}$$

$$= \begin{bmatrix} e^x\cos x + ie^x\sin x \\ e^x\cos x - e^x\sin x + i(e^x\cos x + e^x\sin x) \\ -2e^x\sin x + 2ie^x\cos x \end{bmatrix}$$

$$= \begin{bmatrix} e^x\cos x \\ e^x\cos x - e^x\sin x \\ -2e^x\sin x \end{bmatrix} + i \begin{bmatrix} e^x\sin x \\ e^x\cos x + e^x\sin x \\ 2e^x\cos x \end{bmatrix}$$

is a complex valued solution to $\begin{bmatrix} v_1' \\ v_2' \\ v_3' \end{bmatrix} = A \begin{bmatrix} v_1 \\ v_2 \\ v_3 \end{bmatrix}$. By Theorem 6.8

$$\begin{bmatrix} e^x\cos x \\ e^x\cos x - e^x\sin x \\ -2e^x\sin x \end{bmatrix} \quad \text{and} \quad \begin{bmatrix} e^x\sin x \\ e^x\cos x + e^x\sin x \\ 2e^x\cos x \end{bmatrix}$$

are real valued solutions to $\begin{bmatrix} v_1' \\ v_2' \\ v_3' \end{bmatrix} = A \begin{bmatrix} v_1 \\ v_2 \\ v_3 \end{bmatrix}$.

The general solution to $\begin{bmatrix} v_1' \\ v_2' \\ v_3' \end{bmatrix} = A \begin{bmatrix} v_1 \\ v_2 \\ v_3 \end{bmatrix}$ is then

$$c_1 \begin{bmatrix} e^{-2x} \\ -2e^{-2x} \\ 4e^{-2x} \end{bmatrix} + c_2 \begin{bmatrix} e^x\cos x \\ e^x\cos x - e^x\sin x \\ -2e^x\sin x \end{bmatrix} + c_3 \begin{bmatrix} e^x\sin x \\ e^x\cos x + e^x\sin x \\ 2e^x\cos x \end{bmatrix} =$$

$$\begin{bmatrix} c_1e^{-2x} + c_2e^x\cos x + c_3e^x\sin x \\ -2c_1e^{-2x} + c_2(e^x\cos x - e^x\sin x) + c_3(e^x\cos x + e^x\sin x) \\ +4c_1e^{-2x} - 2c_2e^x\sin x + 2c_3e^x\cos x \end{bmatrix}.$$

Thus, $v_1 = c_1e^{-2x} + c_2e^x\cos x + c_3e^x\sin x$, $v_2 = -2c_1e^{-2x} + c_2(e^x\cos x - e^x\sin x) + c_3(e^x\cos x + e^x\sin x)$, $v_3 = 4c_1e^{-2x} - 2c_2e^x\sin x + 2c_3e^x\cos x$.

Since $v_1 = y$ the solution to the original differential equation is

$$y = c_1e^{-2x} + c_2e^x\cos x + c_3e^x\sin x.$$

Note: Although the form is slightly different, this answer is equivalent to the answer in the text.

9. In Exercise 11 of Section 4.3 we are given the differential equation $y'' + 8y' = 2x^2 - 7x + 3$. Letting $v_1 = y$ and $v_2 = y'$ leads to

$$\begin{aligned} v_1' &= y' = v_2 \\ v_2' &= y'' = -8y' + 2x^2 - 7x + 3 = -8v_2 + 2x^2 - 7x + 3 \end{aligned}$$

which can be rewritten as the system

$$\begin{bmatrix} v_1' \\ v_2' \end{bmatrix} = \begin{bmatrix} 0 & 1 \\ 0 & -8 \end{bmatrix} \begin{bmatrix} v_1 \\ v_2 \end{bmatrix} + \begin{bmatrix} 0 \\ 2x^2 - 7x + 3 \end{bmatrix}.$$

We let

$$V = \begin{bmatrix} v_1 \\ v_2 \end{bmatrix}, \quad A = \begin{bmatrix} 0 & 1 \\ 0 & -8 \end{bmatrix}, \quad \text{and } G(x) = \begin{bmatrix} 0 \\ 2x^2 - 7x + 3 \end{bmatrix}.$$

The characteristic equation of A is equal to the characteristic equation of the original homogeneous differential equation,

$$\lambda^2 + 8\lambda = \lambda(\lambda + 8) = 0.$$

Therefore, the eigenvalues of A are $\lambda = 0$ and $\lambda = -8$. We now wish to find bases for the corresponding eigenspaces.

For $\lambda = 0$ we need to find a basis for the null space of $A = \begin{bmatrix} 0 & 1 \\ 0 & -8 \end{bmatrix}$. Reducing the matrix for the associated homogeneous system,

$$\begin{bmatrix} 0 & 1 & \vdots & 0 \\ 0 & -8 & \vdots & 0 \end{bmatrix} \rightarrow \begin{bmatrix} 0 & 1 & \vdots & 0 \\ 0 & 0 & \vdots & 0 \end{bmatrix},$$

we see our solutions are

$$\begin{bmatrix} x \\ 0 \end{bmatrix} = x \begin{bmatrix} 1 \\ 0 \end{bmatrix}$$

and hence the column vector $\begin{bmatrix} 1 \\ 0 \end{bmatrix}$ forms a basis for E_0.

For $\lambda = -8$ we need to find a basis for the null space of $-8I - A = \begin{bmatrix} -8 & -1 \\ 0 & 0 \end{bmatrix}$. Reducing the matrix for the associated homogeneous system,

$$\begin{bmatrix} -8 & -1 & \vdots & 0 \\ 0 & 0 & \vdots & 0 \end{bmatrix} \rightarrow \begin{bmatrix} 8 & 1 & \vdots & 0 \\ 0 & 0 & \vdots & 0 \end{bmatrix},$$

we see our solutions are

$$\begin{bmatrix} x \\ y \end{bmatrix} = \begin{bmatrix} x \\ -8x \end{bmatrix} = x \begin{bmatrix} 1 \\ -8 \end{bmatrix}$$

and hence the column vector $\begin{bmatrix} 1 \\ -8 \end{bmatrix}$ forms a basis for E_{-8}.

Matrix A is similar to the diagonal matrix $D = P^{-1}AP = \begin{bmatrix} 0 & 0 \\ 0 & -8 \end{bmatrix}$ where $P = \begin{bmatrix} 1 & 1 \\ 0 & 8 \end{bmatrix}$. The columns of P are the basis vectors for the eigenspaces. We also know that $\begin{bmatrix} c_1 \\ c_2 e^{-8x} \end{bmatrix}$ is the solution to the system $V' = DV$. Therefore the general solution to the related homogeneous system $V' = AV$ is

$$V_H = P \begin{bmatrix} c_1 \\ c_2 e^{-8x} \end{bmatrix} = \begin{bmatrix} 1 & 1 \\ 0 & 8 \end{bmatrix} \begin{bmatrix} c_1 \\ c_2 e^{-8x} \end{bmatrix} = \begin{bmatrix} c_1 + c_2 e^{-8x} \\ -8c_2 e^{-8x} \end{bmatrix}.$$

To find a particular solution we will use the formula $V_P = M \int M^{-1} G(x)\, dx$, where M is a matrix of fundamental solutions. Here $M = \begin{bmatrix} 1 & e^{-8x} \\ 0 & -8e^{-8x} \end{bmatrix}$. We have

$$\begin{bmatrix} 1 & e^{-8x} & \vdots & 1 & 0 \\ 0 & -8e^{-8x} & \vdots & 0 & 1 \end{bmatrix} \rightarrow \begin{bmatrix} 1 & e^{-8x} & \vdots & 1 & 0 \\ 0 & e^{-8x} & \vdots & 0 & -\frac{1}{8} \end{bmatrix} \rightarrow \begin{bmatrix} 1 & 0 & \vdots & 1 & \frac{1}{8} \\ 0 & e^{-8x} & \vdots & 0 & -\frac{1}{8} \end{bmatrix}.$$

Multiplying the second row by e^{8x} we obtain $M^{-1} = \begin{bmatrix} 1 & \frac{1}{8} \\ 0 & -\frac{1}{8}e^{8x} \end{bmatrix}$.

The product inside the integral is then

$$M^{-1}G(x) = \begin{bmatrix} 1 & \frac{1}{8} \\ 0 & -\frac{1}{8}e^{8x} \end{bmatrix}\begin{bmatrix} 0 \\ 2x^2 - 7x + 3 \end{bmatrix} = \begin{bmatrix} \frac{1}{4}x^2 - \frac{7}{8}x + \frac{3}{8} \\ \left(-\frac{1}{4}x^2 + \frac{7}{8}x - \frac{3}{8}\right)e^{8x} \end{bmatrix}.$$

Integrating each component of this column vector gives us

$$\int M^{-1}G(x)\,dx = \begin{bmatrix} \int\left(\frac{1}{4}x^2 - \frac{7}{8}x + \frac{3}{8}\right)dx \\ \int\left(-\frac{1}{4}x^2 + \frac{7}{8}x - \frac{3}{8}\right)e^{8x}\,dx \end{bmatrix} = \begin{bmatrix} \frac{1}{12}x^3 - \frac{7}{16}x^2 + \frac{3}{8}x \\ \left(-\frac{1}{32}x^3 + \frac{15}{128}x - \frac{63}{1024}\right)e^{8x} \end{bmatrix}.$$

Thus we have

$$V_P = M\int M^{-1}G(x)\,dx = \begin{bmatrix} 1 & e^{-8x} \\ 0 & -8e^{-8x} \end{bmatrix}\begin{bmatrix} \frac{1}{12}x^3 - \frac{7}{16}x^2 + \frac{3}{8}x \\ \left(-\frac{1}{32}x^3 + \frac{15}{128}x - \frac{63}{1024}\right)e^{8x} \end{bmatrix} = \begin{bmatrix} \frac{1}{12}x^3 - \frac{15}{32}x^2 + \frac{63}{128}x - \frac{63}{1024} \\ \frac{1}{4}x^2 - \frac{15}{16}x + \frac{63}{128} \end{bmatrix}.$$

The general solution is then

$$\begin{aligned} V &= V_H + V_P = \begin{bmatrix} c_1 + c_2 e^{-8x} \\ -8c_2 e^{-8x} \end{bmatrix} + \begin{bmatrix} \frac{1}{12}x^3 - \frac{15}{32}x^2 + \frac{63}{128}x - \frac{63}{1024} \\ \frac{1}{4}x^2 - \frac{15}{16}x + \frac{63}{128} \end{bmatrix} \\ &= \begin{bmatrix} c_1 + c_2 e^{-8x} + \frac{1}{12}x^3 - \frac{15}{32}x^2 + \frac{63}{128}x - \frac{63}{1024} \\ -8c_2 e^{-8x} + \frac{1}{4}x^2 - \frac{15}{16}x + \frac{63}{128} \end{bmatrix}. \end{aligned}$$

Equivalently, $v_1 = c_1 + c_2 e^{-8x} + \frac{1}{12}x^3 - \frac{15}{32}x^2 + \frac{63}{128}x - \frac{63}{1024}$, $v_2 = -8c_2 e^{-8x} + \frac{1}{4}x^2 - \frac{15}{16}x + \frac{63}{128}$.

Since $v_1 = y$ the solution to the original differential equation is

$$y = c_1 + c_2 e^{-8x} + \frac{1}{12}x^3 - \frac{15}{32}x^2 + \frac{63}{128}x.$$

Please note we have "absorbed" the constant $-\frac{63}{1024}$ into the constant c_1.

13. We are given the system

$$\begin{aligned} y_1'' &= 2y_1' + y_2, \\ y_2'' &= 6y_1' + 3y_2 \end{aligned}$$

Letting $v_1 = y_1$, $v_2 = y_1'$, $v_3 = y_2$ and $v_4 = y_2'$ leads to

$$\begin{aligned} v_1' &= y_1' = v_2 \\ v_2' &= y_1'' = 2y_1' + y_2 = 2v_2 + v_3 \\ v_3' &= y_2' = v_4 \\ v_4' &= y_2'' = 6y_1' + 3y_2 = 6v_2 + 3v_3 \end{aligned}$$

which can be rewritten as the system

$$\begin{bmatrix} v_1' \\ v_2' \\ v_3' \\ v_4' \end{bmatrix} = \begin{bmatrix} 0 & 1 & 0 & 0 \\ 0 & 2 & 1 & 0 \\ 0 & 0 & 0 & 1 \\ 0 & 6 & 3 & 0 \end{bmatrix}\begin{bmatrix} v_1 \\ v_2 \\ v_3 \\ v_4 \end{bmatrix}.$$

We let

$$V = \begin{bmatrix} v_1 \\ v_2 \\ v_3 \\ v_4 \end{bmatrix} \quad\text{and}\quad A = \begin{bmatrix} 0 & 1 & 0 & 0 \\ 0 & 2 & 1 & 0 \\ 0 & 0 & 0 & 1 \\ 0 & 6 & 3 & 0 \end{bmatrix}.$$

The characteristic equation of A is

$$\det(\lambda I - A) = \begin{vmatrix} \lambda & -1 & 0 & 0 \\ 0 & \lambda - 2 & -1 & 0 \\ 0 & 0 & \lambda & -1 \\ 0 & -6 & -3 & \lambda \end{vmatrix}$$

$$= \lambda^2(\lambda + 1)(\lambda - 3) = 0.$$

Therefore, the eigenvalues of A are $\lambda = 0, -1$ and 3.

We use Maple to find a Jordan canonical form, B, for matrix A and a matrix P such that $P^{-1}AP = B$. To do this we use the commands:

```
A:=matrix(4,4,[0,1,0,0,0,2,1,0,0,0,0,1,0,6,3,0]);
B:=jordan(A,'P');
print(P);
```

Doing this we find $B = \begin{bmatrix} 0 & 1 & 0 & 0 \\ 0 & 0 & 0 & 0 \\ 0 & 0 & -1 & 0 \\ 0 & 0 & 0 & 3 \end{bmatrix}$ and $P = \begin{bmatrix} 1 & -\frac{2}{3} & \frac{1}{2} & \frac{1}{6} \\ 0 & 1 & -\frac{1}{2} & \frac{1}{2} \\ 0 & -2 & \frac{3}{2} & \frac{1}{2} \\ 0 & 0 & -\frac{3}{2} & \frac{3}{2} \end{bmatrix}.$

We now solve the triangular system $V' = BV$, which is $v_1' = v_2$, $v_2' = 0$, $v_3' = -v_3$, $v_4 = 3v_4$. We quickly see that

$$v_2 = c_2, \ v_3 = c_3 e^{-x} \text{ and } v_4 = c_4 e^{3x}.$$

Substituting $v_2 = c_2$ into the first equation we have $v_1' = c_2$ and so, $v_1 = c_1 + c_2 x$.

Thus the general solution to the triangular system $V' = BV$ is $Z = \begin{bmatrix} c_1 + c_2 x \\ c_2 \\ c_3 e^{-x} \\ c_4 e^{3x} \end{bmatrix}.$

Therefore the general solution to the original system $V' = AV$ is

$$PZ = \begin{bmatrix} 1 & -\frac{2}{3} & \frac{1}{2} & \frac{1}{6} \\ 0 & 1 & -\frac{1}{2} & \frac{1}{2} \\ 0 & -2 & \frac{3}{2} & \frac{1}{2} \\ 0 & 0 & -\frac{3}{2} & \frac{3}{2} \end{bmatrix} \begin{bmatrix} c_1 + c_2 x \\ c_2 \\ c_3 e^{-x} \\ c_4 e^{3x} \end{bmatrix} = \begin{bmatrix} c_1 + c_2 x - \frac{2}{3}c_2 + \frac{1}{2}c_3 e^{-x} + \frac{1}{6}c_4 e^{3x} \\ c_2 - \frac{1}{2}c_3 e^{-x} + \frac{1}{2}c_4 e^{3x} \\ -2c_2 + \frac{3}{2}c_3 e^{-x} + \frac{1}{2}c_4 e^{3x} \\ -\frac{3}{2}c_3 e^{-x} + \frac{3}{2}c_4 e^{3x} \end{bmatrix}.$$

Thus,

$$v_1 = c_1 + c_2 x - \frac{2}{3}c_2 + \frac{1}{2}c_3 e^{-x} + \frac{1}{6}c_4 e^{3x},$$

$$v_2 = c_2 - \frac{1}{2}c_3 e^{-x} + \frac{1}{2}c_4 e^{3x},$$

$$v_3 = -2c_2 + \frac{3}{2}c_3 e^{-x} + \frac{1}{2}c_4 e^{3x},$$

$$v_4 = -\frac{3}{2}c_3 e^{-x} + \frac{3}{2}c_4 e^{3x}.$$

Since $v_1 = y_1$ and $v_3 = y_2$ the solution to the original system is

$$y_1 = c_1 + c_2 x - \frac{2}{3}c_2 + \frac{1}{2}c_3 e^{-x} + \frac{1}{6}c_4 e^{3x}.$$

$$y_2 = -2c_2 + \frac{3}{2}c_3 e^{-x} + \frac{1}{2}c_4 e^{3x}.$$

6.6   Applications Involving Systems of Linear Differential Equations

1. First note that the volume of solution in each tank remains constant over time. Let q_1 and q_2 be the amount of salt in the first tank and the second tank, respectively. We have

$$\frac{dq_1}{dt} = \text{the rate of salt going into tank 1 - the rate of salt going out of tank 1,}$$

$$\frac{dq_2}{dt} = \text{the rate of salt going into tank 2 - the rate of salt going out of tank 2.}$$

Using the flow rates given, it follows that

$$\frac{dq_1}{dt} = \frac{q_2 \text{ lb}}{500 \text{ gal}} \cdot 25 \frac{\text{gal}}{\text{min}} - \frac{q_1 \text{ lb}}{500 \text{ gal}} \cdot 25 \frac{\text{gal}}{\text{min}} \quad \text{and}$$

$$\frac{dq_2}{dt} = \frac{q_1 \text{ lb}}{500 \text{ gal}} \cdot 25 \frac{\text{gal}}{\text{min}} - \frac{q_2 \text{ lb}}{500 \text{ gal}} \cdot 25 \frac{\text{gal}}{\text{min}}.$$

Letting $Q = \begin{bmatrix} q_1 \\ q_2 \end{bmatrix}$ we have $Q' = AQ$ where $A = \begin{bmatrix} -\frac{1}{20} & \frac{1}{20} \\ \frac{1}{20} & -\frac{1}{20} \end{bmatrix}$.

The characteristic equation of A is

$$\det(\lambda I - A) = \begin{vmatrix} \lambda + \frac{1}{20} & -\frac{1}{20} \\ -\frac{1}{20} & \lambda + \frac{1}{20} \end{vmatrix}$$

$$= \left(\lambda + \frac{1}{20}\right)\left(\lambda + \frac{1}{20}\right) - \frac{1}{400} = \lambda^2 + \frac{1}{10}\lambda + \frac{1}{400} - \frac{1}{400} = \lambda^2 + \frac{1}{10}\lambda = \lambda\left(\lambda + \frac{1}{10}\right) = 0.$$

Therefore, the eigenvalues of A are $\lambda = 0$ and $\lambda = -\frac{1}{10}$. We now wish to find bases for the corresponding eigenspaces.

For $\lambda = 0$ we need to find a basis for the null space of $A = \begin{bmatrix} -\frac{1}{20} & \frac{1}{20} \\ \frac{1}{20} & -\frac{1}{20} \end{bmatrix}$. Reducing the matrix for the associated homo-

geneous system, $\begin{bmatrix} -\frac{1}{20} & \frac{1}{20} & \vdots & 0 \\ \frac{1}{20} & -\frac{1}{20} & \vdots & 0 \end{bmatrix} \rightarrow \begin{bmatrix} 1 & -1 & \vdots & 0 \\ 0 & 0 & \vdots & 0 \end{bmatrix}$, we see our solutions are

$$\begin{bmatrix} x \\ y \end{bmatrix} = \begin{bmatrix} x \\ x \end{bmatrix} = x\begin{bmatrix} 1 \\ 1 \end{bmatrix}$$

and hence the column vector $\begin{bmatrix} 1 \\ 1 \end{bmatrix}$ forms a basis for E_0.

For $\lambda = -\frac{1}{10}$ we need to find a basis for the null space of $-\frac{1}{10}I - A = \begin{bmatrix} -\frac{1}{20} & -\frac{1}{20} \\ -\frac{1}{20} & -\frac{1}{20} \end{bmatrix}$. Reducing the matrix for the associated

homogeneous system, $\begin{bmatrix} -\frac{1}{20} & -\frac{1}{20} & \vdots & 0 \\ -\frac{1}{20} & -\frac{1}{20} & \vdots & 0 \end{bmatrix} \rightarrow \begin{bmatrix} 1 & 1 & \vdots & 0 \\ 0 & 0 & \vdots & 0 \end{bmatrix}$, we see our solutions are

$$\begin{bmatrix} x \\ y \end{bmatrix} = \begin{bmatrix} x \\ -x \end{bmatrix} = x\begin{bmatrix} 1 \\ -1 \end{bmatrix}$$

and hence the column vector $\begin{bmatrix} 1 \\ -1 \end{bmatrix}$ forms a basis for $E_{-\frac{1}{10}}$.

Matrix A is similar to the diagonal matrix $D = P^{-1}AP = \begin{bmatrix} 0 & 0 \\ 0 & -\frac{1}{10} \end{bmatrix}$ where $P = \begin{bmatrix} 1 & 1 \\ 1 & -1 \end{bmatrix}$. The columns of P are the

basis vectors for the eigenspaces. We also know that $\begin{bmatrix} c_1 \\ c_2 e^{-\frac{1}{10}t} \end{bmatrix}$ is the general solution to $Q' = DQ$. Therefore the general solution to $Q' = AQ$ is

$$Q = P \begin{bmatrix} c_1 \\ c_2 e^{-\frac{1}{10}t} \end{bmatrix} = \begin{bmatrix} 1 & 1 \\ 1 & -1 \end{bmatrix} \begin{bmatrix} c_1 \\ c_2 e^{-\frac{1}{10}t} \end{bmatrix} = \begin{bmatrix} c_1 + c_2 e^{-\frac{1}{10}t} \\ c_1 - c_2 e^{-\frac{1}{10}t} \end{bmatrix}.$$

Since the solution in tank 1 starts with 10 lb ammonia and the solution in tank 2 contain 1 lb ammonia we have the initial conditions $q_1(0) = 10$, $q_2(0) = 1$. We have the system of equations

$$q_1(0) = 10 = c_1 + c_2, \quad q_2(0) = 1 = c_1 - c_2.$$

This system has the solution $c_1 = \frac{11}{2}$, $c_2 = \frac{9}{2}$. Thus the solution to the initial value problem is

$$q_1 = \frac{11}{2} + \frac{9}{2} e^{-\frac{1}{10}t}, \quad q_2 = \frac{11}{2} - \frac{9}{2} e^{-\frac{1}{10}t}.$$

3. First note that even though 2 gallons of fluid per minute is now entering tank 1 from an addition inlet pipe and 2 gallons of fluid per minute is draining from tank 2, the volume of solution in each tank remains constant over time due to the adjusted flow rates between the tanks. Let q_1 and q_2 be the amount of salt in the first tank and the second tank, respectively. We have

$$\frac{dq_1}{dt} = \text{the rate of salt going into tank 1 - the rate of salt going out of tank 1,}$$

$$\frac{dq_2}{dt} = \text{the rate of salt going into tank 2 - the rate of salt going out of tank 2.}$$

Using the flow rates given, it follows that

$$\frac{dq_1}{dt} = \frac{1 \text{ lb}}{\text{gal}} \cdot 2\frac{\text{gal}}{\text{min}} + \frac{q_2 \text{ lb}}{100 \text{ gal}} \cdot 4\frac{\text{gal}}{\text{min}} - \frac{q_1 \text{ lb}}{100 \text{ gal}} \cdot 6\frac{\text{gal}}{\text{min}} \quad \text{and}$$

$$\frac{dq_2}{dt} = \frac{q_1 \text{ lb}}{100 \text{ gal}} \cdot 6\frac{\text{gal}}{\text{min}} - \frac{q_2 \text{ lb}}{100 \text{ gal}} \cdot 4\frac{\text{gal}}{\text{min}} - \frac{q_2 \text{ lb}}{100 \text{ gal}} \cdot 2\frac{\text{gal}}{\text{min}}.$$

Letting $Q = \begin{bmatrix} q_1 \\ q_2 \end{bmatrix}$, $A = \begin{bmatrix} -\frac{3}{50} & \frac{1}{25} \\ \frac{3}{50} & -\frac{3}{50} \end{bmatrix}$ and $G(x) = \begin{bmatrix} 2 \\ 0 \end{bmatrix}$ we have $Q' = AQ + G(x)$. The characteristic equation of A is

$$\det(\lambda I - A) = \begin{vmatrix} \lambda + \frac{3}{50} & -\frac{1}{25} \\ -\frac{3}{50} & \lambda + \frac{3}{50} \end{vmatrix}$$

$$= \left(\lambda + \frac{3}{50}\right)\left(\lambda + \frac{3}{50}\right) - \frac{3}{1250} = \lambda^2 + \frac{3}{25}\lambda + \frac{3}{2500} = 0.$$

Therefore, the eigenvalues of A are $\lambda = \frac{-\frac{3}{25} \pm \sqrt{\frac{9}{625} - \frac{12}{2500}}}{2} = \frac{-3 \pm \sqrt{6}}{50}$. We now wish to find bases for the corresponding eigenspaces.

For $\lambda = \frac{-3+\sqrt{6}}{50}$ we need to find a basis for the null space of $\left(\frac{-3+\sqrt{6}}{50}\right)I - A = \begin{bmatrix} \frac{\sqrt{6}}{50} & -\frac{1}{25} \\ -\frac{3}{50} & \frac{\sqrt{6}}{50} \end{bmatrix}$. Reducing the matrix for the

associated homogeneous system, $\begin{bmatrix} \frac{\sqrt{6}}{50} & -\frac{1}{25} & : & 0 \\ -\frac{3}{50} & \frac{\sqrt{6}}{50} & : & 0 \end{bmatrix} \rightarrow \begin{bmatrix} \sqrt{6} & -2 & : & 0 \\ 0 & 0 & : & 0 \end{bmatrix}$, we see our solutions are

$$\begin{bmatrix} x \\ y \end{bmatrix} = \begin{bmatrix} x \\ \frac{\sqrt{6}}{2}x \end{bmatrix} = \frac{1}{2}x \begin{bmatrix} 2 \\ \sqrt{6} \end{bmatrix}$$

and hence the column vector $\begin{bmatrix} 2 \\ \sqrt{6} \end{bmatrix}$ forms a basis for $E_{\frac{-3+\sqrt{6}}{50}}$.

For $\lambda = \frac{-3-\sqrt{6}}{50}$ we need to find a basis for the null space of $\left(\frac{-3-\sqrt{6}}{50}\right)I - A = \begin{bmatrix} -\frac{\sqrt{6}}{50} & -\frac{1}{25} \\ -\frac{3}{50} & -\frac{\sqrt{6}}{50} \end{bmatrix}$. Reducing the matrix for

the associated homogeneous system, $\begin{bmatrix} -\frac{\sqrt{6}}{50} & -\frac{1}{25} & \vdots & 0 \\ -\frac{3}{50} & -\frac{\sqrt{6}}{50} & \vdots & 0 \end{bmatrix} \rightarrow \begin{bmatrix} \sqrt{6} & 2 & \vdots & 0 \\ 0 & 0 & \vdots & 0 \end{bmatrix}$, we see our solutions are

$$\begin{bmatrix} x \\ y \end{bmatrix} = \begin{bmatrix} x \\ -\frac{\sqrt{6}}{2}x \end{bmatrix} = \frac{1}{2}x\begin{bmatrix} 2 \\ -\sqrt{6} \end{bmatrix}$$

and hence the column vector $\begin{bmatrix} 2 \\ -\sqrt{6} \end{bmatrix}$ forms a basis for $E_{\frac{-3-\sqrt{6}}{50}}$.

Matrix A is similar to the diagonal matrix $D = P^{-1}AP = \begin{bmatrix} \frac{-3+\sqrt{6}}{50} & 0 \\ 0 & \frac{-3+\sqrt{6}}{50} \end{bmatrix}$ where $P = \begin{bmatrix} 2 & 2 \\ \sqrt{6} & -\sqrt{6} \end{bmatrix}$. The columns of

P are the basis vectors for the eigenspaces. We also know that $\begin{bmatrix} c_1 e^{\frac{-3+\sqrt{6}}{50}x} \\ c_2 e^{\frac{-3-\sqrt{6}}{50}x} \end{bmatrix}$ is the general solution to $Q' = DQ$. Therefore

the general solution to $Q' = AQ$ is

$$Q_H = P\begin{bmatrix} c_1 e^{\frac{-3+\sqrt{6}}{50}x} \\ c_2 e^{\frac{-3-\sqrt{6}}{50}x} \end{bmatrix} = \begin{bmatrix} 2 & 2 \\ \sqrt{6} & -\sqrt{6} \end{bmatrix}\begin{bmatrix} c_1 e^{\frac{-3+\sqrt{6}}{50}x} \\ c_2 e^{\frac{-3-\sqrt{6}}{50}x} \end{bmatrix} = \begin{bmatrix} 2c_1 e^{\frac{-3+\sqrt{6}}{50}x} + 2c_2 e^{\frac{-3-\sqrt{6}}{50}x} \\ \sqrt{6}c_1 e^{\frac{-3+\sqrt{6}}{50}x} - \sqrt{6}c_2 e^{\frac{-3-\sqrt{6}}{50}x} \end{bmatrix}.$$

To find a particular solution we will use the formula $Q_P = M \int M^{-1} G(x)\, dx$, where M is a matrix of fundamental solutions.
Here $M = \begin{bmatrix} 2e^{\frac{-3+\sqrt{6}}{50}x} & 2e^{\frac{-3-\sqrt{6}}{50}x} \\ \sqrt{6}e^{\frac{-3+\sqrt{6}}{50}x} & -\sqrt{6}e^{\frac{-3-\sqrt{6}}{50}x} \end{bmatrix}$. We have

$$\begin{bmatrix} 2e^{\frac{-3+\sqrt{6}}{50}x} & 2e^{\frac{-3-\sqrt{6}}{50}x} & \vdots & 1 & 0 \\ \sqrt{6}e^{\frac{-3+\sqrt{6}}{50}x} & -\sqrt{6}e^{\frac{-3-\sqrt{6}}{50}x} & \vdots & 0 & 1 \end{bmatrix} \rightarrow \begin{bmatrix} e^{\frac{-3+\sqrt{6}}{50}x} & e^{\frac{-3-\sqrt{6}}{50}x} & \vdots & \frac{1}{2} & 0 \\ e^{\frac{-3+\sqrt{6}}{50}x} & -e^{\frac{-3-\sqrt{6}}{50}x} & \vdots & 0 & \frac{\sqrt{6}}{6} \end{bmatrix}$$

$$\rightarrow \begin{bmatrix} e^{\frac{-3+\sqrt{6}}{50}x} & e^{\frac{-3-\sqrt{6}}{50}x} & \vdots & \frac{1}{2} & 0 \\ 0 & -2e^{\frac{-3-\sqrt{6}}{50}x} & \vdots & -\frac{1}{2} & \frac{\sqrt{6}}{6} \end{bmatrix} \rightarrow \begin{bmatrix} e^{\frac{-3+\sqrt{6}}{50}x} & e^{\frac{-3-\sqrt{6}}{50}x} & \vdots & \frac{1}{2} & 0 \\ 0 & e^{\frac{-3-\sqrt{6}}{50}x} & \vdots & \frac{1}{4} & -\frac{\sqrt{6}}{12} \end{bmatrix}$$

$$\rightarrow \begin{bmatrix} e^{\frac{-3+\sqrt{6}}{50}x} & 0 & \vdots & \frac{1}{4} & \frac{\sqrt{6}}{12} \\ 0 & e^{\frac{-3-\sqrt{6}}{50}x} & \vdots & \frac{1}{4} & -\frac{\sqrt{6}}{12} \end{bmatrix}.$$

Multiplying the first row by $e^{\frac{3-\sqrt{6}}{50}x}$ and the second by $e^{\frac{3+\sqrt{6}}{50}x}$ we obtain $M^{-1} = \begin{bmatrix} \frac{1}{4}e^{\frac{3-\sqrt{6}}{50}x} & \frac{\sqrt{6}}{12}e^{\frac{3-\sqrt{6}}{50}x} \\ \frac{1}{4}e^{\frac{3+\sqrt{6}}{50}x} & \frac{\sqrt{6}}{12}e^{\frac{3+\sqrt{6}}{50}x} \end{bmatrix}.$

The product inside the integral is then

$$M^{-1}G(x) = \begin{bmatrix} \frac{1}{4}e^{\frac{3-\sqrt{6}}{50}x} & \frac{\sqrt{6}}{12}e^{\frac{3-\sqrt{6}}{50}x} \\ \frac{1}{4}e^{\frac{3+\sqrt{6}}{50}x} & \frac{\sqrt{6}}{12}e^{\frac{3+\sqrt{6}}{50}x} \end{bmatrix}\begin{bmatrix} 2 \\ 0 \end{bmatrix} = \begin{bmatrix} \frac{1}{2}e^{\frac{3-\sqrt{6}}{50}x} \\ \frac{1}{2}e^{\frac{3+\sqrt{6}}{50}x} \end{bmatrix}.$$

Integrating each component of this column vector gives us

$$\int M^{-1}G(x)\, dx = \begin{bmatrix} \int \frac{1}{2}e^{\frac{3-\sqrt{6}}{50}x}\, dx \\ \int \frac{1}{2}e^{\frac{3+\sqrt{6}}{50}x}\, dx \end{bmatrix} = \begin{bmatrix} \frac{25}{3-\sqrt{6}}e^{\frac{3-\sqrt{6}}{50}x} \\ \frac{25}{3+\sqrt{6}}e^{\frac{3+\sqrt{6}}{50}x} \end{bmatrix}.$$

Thus we have

$$Q_P = M \int M^{-1} G(x)\, dx = \left[\begin{array}{cc} 2e^{\frac{-3+\sqrt6}{50}x} & 2e^{\frac{-3-\sqrt6}{50}x} \\ \sqrt6 e^{\frac{-3+\sqrt6}{50}x} & -\sqrt6 e^{\frac{-3-\sqrt6}{50}x} \end{array} \right] \left[\begin{array}{c} \frac{25}{3-\sqrt6} e^{\frac{3-\sqrt6}{50}x} \\ \frac{25}{3+\sqrt6} e^{\frac{3+\sqrt6}{50}x} \end{array} \right] = \left[\begin{array}{c} \frac{50}{3-\sqrt6} + \frac{50}{3+\sqrt6} \\ \frac{25\sqrt6}{3-\sqrt6} - \frac{25\sqrt6}{3+\sqrt6} \end{array} \right] = \left[\begin{array}{c} 100 \\ 100 \end{array} \right].$$

The general solution is then

$$Q = Q_H + Q_P = \left[\begin{array}{c} 2c_1 e^{\frac{-3+\sqrt6}{50}x} + 2c_2 e^{\frac{-3-\sqrt6}{50}x} \\ \sqrt6 c_1 e^{\frac{-3+\sqrt6}{50}x} - \sqrt6 c_2 e^{\frac{-3-\sqrt6}{50}x} \end{array} \right] + \left[\begin{array}{c} 100 \\ 100 \end{array} \right] = \left[\begin{array}{c} 2c_1 e^{\frac{-3+\sqrt6}{50}x} + 2c_2 e^{\frac{-3-\sqrt6}{50}x} + 100 \\ \sqrt6 c_1 e^{\frac{-3+\sqrt6}{50}x} - \sqrt6 c_2 e^{\frac{-3-\sqrt6}{50}x} + 100 \end{array} \right].$$

Since the solution in tank 1 starts with 5 lb salt and the solution in tank 2 initially contains pure water we have the initial conditions $q_1(0) = 5$, $q_2(0) = 0$. We have the system of equations

$$q_1(0) = 0 = 2c_1 + 2c_2 + 100, \quad q_2(0) = 0 = \sqrt6 c_1 - \sqrt6 c_2.$$

This system has the solution $c_1 = -\frac{100\sqrt6 + 285}{12}$, $c_2 = \frac{100\sqrt6 - 285}{12}$. Thus the solution to the initial value problem is

$$q_1 = -\frac{100\sqrt6 + 285}{6} e^{\frac{-3+\sqrt6}{50}x} + \frac{100\sqrt6 - 285}{6} e^{\frac{-3-\sqrt6}{50}x} + 100,$$

$$q_2 = -\frac{200 + 95\sqrt6}{4} e^{\frac{-3+\sqrt6}{50}x} + \frac{-200 + 95\sqrt6}{4} e^{\frac{-3+\sqrt6}{50}x} + 100.$$

5. Given the double loop closed circuit shown in Figure 6.8 we obtain the following system of differential equations for the currents i_1 and i_2 where the derivatives are with respect to t:

$$R(i'_1 - i'_2) = -\frac{1}{C} i_1, \quad Li_2 = R(i_1 - i_2).$$

In matrix form, this system becomes

$$\left[\begin{array}{cc} R & -R \\ 0 & L \end{array} \right] \left[\begin{array}{c} i'_1 \\ i'_2 \end{array} \right] = \left[\begin{array}{cc} -\frac{1}{C} & 0 \\ R & -R \end{array} \right] \left[\begin{array}{c} i_1 \\ i_2 \end{array} \right].$$

Solving this system for i'_1 and i'_2 by multiplying each side of this equation on the left by

$$\left[\begin{array}{cc} R & -R \\ 0 & L \end{array} \right]^{-1} = \left[\begin{array}{cc} \frac{1}{R} & \frac{1}{L} \\ 0 & \frac{1}{L} \end{array} \right]$$

give us the homogeneous system

$$\left[\begin{array}{c} i'_1 \\ i'_2 \end{array} \right] = \left[\begin{array}{cc} \frac{R}{L} - \frac{1}{RC} & -\frac{R}{L} \\ \frac{R}{L} & -\frac{R}{L} \end{array} \right] \left[\begin{array}{c} i_1 \\ i_2 \end{array} \right]$$

modeling the circuit. With $L = 1$, $R = 2$ and $C = \frac{1}{16}$ we have

$$\left[\begin{array}{c} i'_1 \\ i'_2 \end{array} \right] = \left[\begin{array}{cc} -6 & -2 \\ 2 & -2 \end{array} \right] \left[\begin{array}{c} i_1 \\ i_2 \end{array} \right].$$

We let $A = \left[\begin{array}{cc} -6 & -2 \\ 2 & -2 \end{array} \right]$ and $I = \left[\begin{array}{c} i_1 \\ i_2 \end{array} \right]$.

The characteristic equation of A is

$$\begin{aligned} \det(\lambda I - A) &= \left| \begin{array}{cc} \lambda + 6 & 2 \\ -2 & \lambda + 2 \end{array} \right| \\ &= (\lambda + 6)(\lambda + 2) + 4 = \lambda^2 + 8\lambda + 16 = (\lambda + 4)^2 = 0. \end{aligned}$$

Therefore, the sole eigenvalue of A is $\lambda = -4$. We now wish to find basis for the corresponding eigenspace.

We use Maple to find a Jordan canonical form, B, for matrix A and a matrix P such that $P^{-1}AP = B$. To do this we use the commands:

```
A:=matrix(2,2,[-6,-2,2,-2]);
B:=jordan(A,'P');
print(P);
```

Doing this we find $B = \begin{bmatrix} -4 & 1 \\ 0 & -4 \end{bmatrix}$ and $P = \begin{bmatrix} -2 & 1 \\ 2 & 0 \end{bmatrix}$. We now solve the triangular system $I' = BI$, which is

$$i_1' = -4i_1 + i_2, \ i_2' = -4i_2.$$

The second equation has the general solution $i_2 = c_2 e^{-4t}$. Substituting this into the first equation we have $i_1' = -4i_1 + c_2 e^{-4t}$. The related homogeneous equation $i_1' + 4i_1 = 0$ has the solution $i_{1H} = c_1 e^{-4t}$. We may then use the method of undetermined coefficients to find the particular solution of the form $i_{1P} = Ate^{-4t}$. Here, $i_{1P}' = Ae^{-4t} - 4Ate^{-4t}$. Substituting these into the equation $i_1' = -4i_1 + c_2 e^{-4t}$ we obtain

$$Ae^{-4t} - 4Ate^{-4t} = -4(Ate^{-4t}) + c_2 e^{-4t}.$$

Simplifying this we see that $Ae^{-4t} = c_2 e^{-4t}$. Thus $A = c_2$. Adding the particular solution to the general solution of the related homogeneous equation we have $i_1 = c_1 e^{-4t} + c_2 te^{-4t}$. Thus the general solution to the triangular system $I' = BI$ is
$Z = \begin{bmatrix} c_1 e^{-4t} + c_2 te^{-4t} \\ c_2 e^{-4t} \end{bmatrix}$.

Therefore the general solution to the original system $I' = AI$ is

$$PZ = \begin{bmatrix} -2 & 1 \\ 2 & 0 \end{bmatrix} \begin{bmatrix} c_1 e^{-4t} + c_2 te^{-4t} \\ c_2 e^{-4t} \end{bmatrix} = \begin{bmatrix} -2c_1 e^{-4t} + c_2 e^{-4t} - 2c_2 te^{-4t} \\ 2c_1 e^{-4t} + 2c_2 te^{-4t} \end{bmatrix}.$$

We now use the given initial conditions $i_1(0) = 1$, $i_2(0) = 0$. We have the system of equations

$$i_1(0) = 1 = -2c_1 + c_2, \ i_2(0) = 0 = 2c_1.$$

This system has the solution $c_1 = 0$, $c_2 = 1$. Thus the solution to the initial value problem is

$$i_1 = (1 - 2t)e^{-4t}, \ i_2 = 2te^{-4t}.$$

7. Since Figure 6.10 is identical to Figure 6.4 the same nonhomogeneous system that precedes Example 2 models the circuit:

$$\begin{bmatrix} i_1' \\ i_2' \end{bmatrix} = \begin{bmatrix} -\frac{R}{L} & \frac{R}{L} \\ -\frac{R}{L} & \frac{R}{L} - \frac{1}{RC} \end{bmatrix} \begin{bmatrix} i_1 \\ i_2 \end{bmatrix} + \begin{bmatrix} \frac{E(t)}{L} \\ \frac{E(t)}{L} \end{bmatrix}.$$

With $L = 2$, $R = 2$, $C = \frac{1}{4}$ and $E(t) = 10\sin t$ we have

$$\begin{bmatrix} i_1' \\ i_2' \end{bmatrix} = \begin{bmatrix} -1 & 1 \\ -1 & -1 \end{bmatrix} \begin{bmatrix} i_1 \\ i_2 \end{bmatrix} + \begin{bmatrix} 5\sin t \\ 5\sin t \end{bmatrix}.$$

We let $I = \begin{bmatrix} i_1 \\ i_2 \end{bmatrix}$, $A = \begin{bmatrix} -1 & 1 \\ -1 & -1 \end{bmatrix}$ and $G(t) = \begin{bmatrix} 5\sin t \\ 5\sin t \end{bmatrix}$.

The characteristic equation of A is

$$\det(\lambda I - A) = \begin{vmatrix} \lambda + 1 & -1 \\ 1 & \lambda + 1 \end{vmatrix} = \lambda^2 + 2\lambda + 2 = 0.$$

Therefore, the eigenvalues of A are $\lambda = \frac{-2 \pm \sqrt{4-8}}{2} = -1 \pm i$. We now wish to find bases for the corresponding eigenspaces.

For $\lambda = -1 + i$ we need to find a basis for the null space of $(-1+i)I - A = \begin{bmatrix} i & -1 \\ 1 & i \end{bmatrix}$. Reducing the matrix for the

associated homogeneous system, $\begin{bmatrix} i & -1 & \vdots & 0 \\ 1 & i & \vdots & 0 \end{bmatrix} \rightarrow \begin{bmatrix} i & -1 & \vdots & 0 \\ 0 & 0 & \vdots & 0 \end{bmatrix}$, we see our solutions are

$$\begin{bmatrix} x \\ y \end{bmatrix} = \begin{bmatrix} x \\ ix \end{bmatrix} = x \begin{bmatrix} 1 \\ i \end{bmatrix}$$

and hence the column vector $\begin{bmatrix} 1 \\ i \end{bmatrix}$ forms a basis for E_{-1+i}.

For $\lambda = -1 - i$ we need to find a basis for the null space of $(-1-i)I - A = \begin{bmatrix} -i & -1 \\ 1 & -i \end{bmatrix}$. Reducing the matrix for the

associated homogeneous system, $\begin{bmatrix} -i & -1 & \vdots & 0 \\ 1 & -i & \vdots & 0 \end{bmatrix} \rightarrow \begin{bmatrix} i & 1 & \vdots & 0 \\ 0 & 0 & \vdots & 0 \end{bmatrix}$, we see our solutions are

$$\begin{bmatrix} x \\ y \end{bmatrix} = \begin{bmatrix} x \\ -ix \end{bmatrix} = x \begin{bmatrix} 1 \\ -i \end{bmatrix}$$

and hence the column vector $\begin{bmatrix} 1 \\ -i \end{bmatrix}$ forms a basis for E_{-1-i}.

Matrix A is similar to the diagonal matrix $D = P^{-1}AP = \begin{bmatrix} -1+i & 0 \\ 0 & -1-i \end{bmatrix}$ where $P = \begin{bmatrix} 1 & 1 \\ i & -i \end{bmatrix}$. The columns of P are the basis vectors for the eigenspaces.

One complex valued solution to $I' = DI$ is $Z = \begin{bmatrix} e^{(-1+i)t} \\ 0 \end{bmatrix} = \begin{bmatrix} e^{-t}\cos t + ie^{-t}\sin t \\ 0 \end{bmatrix}$. By the first part of Theorem 6.7,

$$
\begin{aligned}
PZ &= \begin{bmatrix} 1 & 1 \\ i & -i \end{bmatrix} \begin{bmatrix} e^{-t}\cos t + ie^{-t}\sin t \\ 0 \end{bmatrix} \\
&= \begin{bmatrix} e^{-t}\cos t + ie^{-t}\sin t \\ -e^{-t}\sin t + ie^{-t}\cos t \end{bmatrix} \\
&= \begin{bmatrix} e^{-t}\cos t \\ -e^{-t}\sin t \end{bmatrix} + i\begin{bmatrix} e^{-t}\sin t \\ e^{-t}\cos t \end{bmatrix}
\end{aligned}
$$

is a complex valued solution to $I' = AI$. By Theorem 6.8

$$\begin{bmatrix} e^{-t}\cos t \\ -e^{-t}\sin t \end{bmatrix} \text{ and } \begin{bmatrix} e^{-t}\sin t \\ e^{-t}\cos t \end{bmatrix}$$

are real valued solutions to $I' = AI$.

Therefore the general solution to $I' = AI$ is

$$I_H = c_1 \begin{bmatrix} e^{-t}\cos t \\ -e^{-t}\sin t \end{bmatrix} + c_2 \begin{bmatrix} e^{-t}\sin t \\ e^{-t}\cos t \end{bmatrix} = \begin{bmatrix} c_1 e^{-t}\cos t + c_2 e^{-t}\sin t \\ -c_1 e^{-t}\sin t + c_2 e^{-t}\cos t \end{bmatrix}.$$

To find a particular solution we will use the formula $I_P = M \int M^{-1}G(t)\,dt$, where M is a matrix of fundamental solutions.

Here $M = \begin{bmatrix} e^{-t}\cos t & e^{-t}\sin t \\ -e^{-t}\sin t & e^{-t}\cos t \end{bmatrix}$. We compute M^{-1} using the adjoint as in Corollary 1.27 of the text

$$
\begin{aligned}
M^{-1} &= \frac{1}{\det M}\text{adj}(M) = \frac{1}{\left(e^{-2t}(\cos^2 t + \sin^2 t)\right)} \begin{bmatrix} e^{-t}\cos t & -e^{-t}\sin t \\ e^{-t}\sin t & e^{-t}\cos t \end{bmatrix} = e^{2t}\begin{bmatrix} e^{-t}\cos t & -e^{-t}\sin t \\ e^{-t}\sin t & e^{-t}\cos t \end{bmatrix} \\
&= \begin{bmatrix} e^t\cos t & -e^t\sin t \\ e^t\sin t & e^t\cos t \end{bmatrix}.
\end{aligned}
$$

The product inside the integral is then

$$M^{-1}G(t) = \begin{bmatrix} e^t \cos t & -e^t \sin t \\ e^t \sin t & e^t \cos t \end{bmatrix} \begin{bmatrix} 5\sin t \\ 5\sin t \end{bmatrix} = \begin{bmatrix} 5e^t \sin t \cos t - 5e^t \sin^2 t \\ 5e^t \sin t \cos t + 5e^t \sin^2 t \end{bmatrix}.$$

Integrating each component of this column vector gives us

$$\int M^{-1}G(t)\,dt = \begin{bmatrix} \int (5e^t \sin t \cos t - 5e^t \sin^2 t)\,dt \\ \int (5e^t \sin t \cos t + 5e^t \sin^2 t)\,dt \end{bmatrix} = \begin{bmatrix} e^t(3\sin t \cos t - \cos^2 t - 2) \\ e^t(2 - \sin t \cos t - \cos^2 t + 2\sin^2 t) \end{bmatrix}.$$

Thus we have

$$I_P = M \int M^{-1}G(t)\,dt = \begin{bmatrix} e^{-t}\cos t & e^{-t}\sin t \\ -e^{-t}\sin t & e^{-t}\cos t \end{bmatrix} \begin{bmatrix} e^t(3\sin t \cos t - \cos^2 t - 2) \\ e^t(2 - \sin t \cos t - \cos^2 t + 2\sin^2 t) \end{bmatrix} = \begin{bmatrix} 4\sin t - 3\cos t \\ 2\sin t + \cos t \end{bmatrix}.$$

The general solution is then

$$I = I_H + I_P = \begin{bmatrix} c_1 e^{-t}\cos t + c_2 e^{-t}\sin t \\ -c_1 e^{-t}\sin t + c_2 e^{-t}\cos t \end{bmatrix} + \begin{bmatrix} 4\sin t - 3\cos t \\ 2\sin t + \cos t \end{bmatrix} = \begin{bmatrix} c_1 e^{-t}\cos t + c_2 e^{-t}\sin t + 4\sin t - 3\cos t \\ -c_1 e^{-t}\sin t + c_2 e^{-t}\cos t + 2\sin t + \cos t \end{bmatrix}.$$

We are now given the initial condition

$$I(0) = \begin{bmatrix} 0 \\ 0 \end{bmatrix}.$$

Thus we have $I(0) = \begin{bmatrix} 0 \\ 0 \end{bmatrix} = \begin{bmatrix} c_1 - 3 \\ c_2 + 1 \end{bmatrix}$. The solution to this system is $c_1 = 3$, $c_2 = -1$. We now have the solution to the

initial value problem, $I = \begin{bmatrix} 3e^{-t}\cos t - e^{-t}\sin t + 4\sin t - 3\cos t \\ -3e^{-t}\sin t - e^{-t}\cos t + 2\sin t + \cos t \end{bmatrix}.$

Equivalently,

$$i_1 = 3e^{-t}\cos t - e^{-t}\sin t + 4\sin t - 3\cos t, \quad i_2 = -3e^{-t}\sin t - e^{-t}\cos t + 2\sin t + \cos t.$$

9. Since $F = ma$ or $a = \frac{F}{m}$ where a is the acceleration vector and m is the mass of the object, and since $a = \begin{bmatrix} \frac{d^2x}{dt^2} \\ \frac{d^2y}{dt^2} \end{bmatrix}$, we have

the second order system

$$\frac{d^2x}{dt^2} = 100x - 50y.$$
$$\frac{d^2y}{dt^2} = 25y.$$

Letting $v_1 = x$, $v_2 = \frac{dx}{dt}$, $v_3 = y$ and $v_4 = \frac{dy}{dt}$ we have the first order system $V = AV'$ where

$$V = \begin{bmatrix} v_1 \\ v_2 \\ v_3 \\ v_4 \end{bmatrix} \quad \text{and } A = \begin{bmatrix} 0 & 1 & 0 & 0 \\ 100 & 0 & -50 & 0 \\ 0 & 0 & 0 & 1 \\ 0 & 0 & 25 & 0 \end{bmatrix}.$$

The characteristic equation of A is

$$\det(\lambda I - A) = \begin{vmatrix} \lambda & -1 & 0 & 0 \\ -100 & \lambda & 50 & 0 \\ 0 & 0 & \lambda & -1 \\ 0 & 0 & -25 & \lambda \end{vmatrix}$$
$$= (\lambda + 5)(\lambda - 5)(\lambda + 10)(\lambda - 10) = 0.$$

Therefore, the eigenvalues of A are $\lambda = \pm 5$ and $\lambda = \pm 10$. We now wish to find bases for the corresponding eigenspaces.

For $\lambda = -5$ we need to find a basis for the null space of $-5I - A = \begin{bmatrix} -5 & -1 & 0 & 0 \\ -100 & -5 & 50 & 0 \\ 0 & 0 & -5 & -1 \\ 0 & 0 & -25 & -5 \end{bmatrix}$. Reducing the matrix for

the associated homogeneous system,

$$\left[\begin{array}{cccc:c} -5 & -1 & 0 & 0 & 0 \\ -100 & -5 & 50 & 0 & 0 \\ 0 & 0 & -5 & -1 & 0 \\ 0 & 0 & -25 & -5 & 0 \end{array}\right] \rightarrow \left[\begin{array}{cccc:c} 5 & 1 & 0 & 0 & 0 \\ -75 & 0 & 50 & 0 & 0 \\ 0 & 0 & 5 & 1 & 0 \\ 0 & 0 & 0 & 0 & 0 \end{array}\right] \rightarrow \left[\begin{array}{cccc:c} 5 & 1 & 0 & 0 & 0 \\ -\frac{3}{2} & 0 & 1 & 0 & 0 \\ \frac{15}{2} & 0 & 0 & 1 & 0 \\ 0 & 0 & 0 & 0 & 0 \end{array}\right],$$

we see our solutions are

$$\begin{bmatrix} v_1 \\ v_2 \\ v_3 \\ v_4 \end{bmatrix} = \begin{bmatrix} v_1 \\ -5v_1 \\ \frac{3}{2}v_1 \\ -\frac{15}{2}v_1 \end{bmatrix} = v_1 \begin{bmatrix} 1 \\ -5 \\ \frac{3}{2} \\ -\frac{15}{2} \end{bmatrix}$$

and hence the column vector $\begin{bmatrix} 1 \\ -5 \\ \frac{3}{2} \\ -\frac{15}{2} \end{bmatrix}$ forms a basis for E_{-5}.

For $\lambda = 5$ we need to find a basis for the null space of $5I - A = \begin{bmatrix} 5 & -1 & 0 & 0 \\ -100 & 5 & 50 & 0 \\ 0 & 0 & 5 & -1 \\ 0 & 0 & -25 & 5 \end{bmatrix}$. Reducing the matrix for the

associated homogeneous system,

$$\left[\begin{array}{cccc:c} 5 & -1 & 0 & 0 & 0 \\ -100 & 5 & 50 & 0 & 0 \\ 0 & 0 & 5 & -1 & 0 \\ 0 & 0 & -25 & 5 & 0 \end{array}\right] \rightarrow \left[\begin{array}{cccc:c} 5 & -1 & 0 & 0 & 0 \\ -75 & 0 & 50 & 0 & 0 \\ 0 & 0 & 5 & -1 & 0 \\ 0 & 0 & 0 & 0 & 0 \end{array}\right] \rightarrow \left[\begin{array}{cccc:c} 5 & 1 & 0 & 0 & 0 \\ -\frac{3}{2} & 0 & 1 & 0 & 0 \\ -\frac{15}{2} & 0 & 0 & 1 & 0 \\ 0 & 0 & 0 & 0 & 0 \end{array}\right],$$

we see our solutions are

$$\begin{bmatrix} v_1 \\ v_2 \\ v_3 \\ v_4 \end{bmatrix} = \begin{bmatrix} v_1 \\ 5v_1 \\ \frac{3}{2}v_1 \\ \frac{15}{2}v_1 \end{bmatrix} = v_1 \begin{bmatrix} 1 \\ 5 \\ \frac{3}{2} \\ \frac{15}{2} \end{bmatrix}$$

and hence the column vector $\begin{bmatrix} 1 \\ 5 \\ \frac{3}{2} \\ \frac{15}{2} \end{bmatrix}$ forms a basis for E_5.

For $\lambda = -10$ we need to find a basis for the null space of $-10I - A = \begin{bmatrix} -10 & -1 & 0 & 0 \\ -100 & -10 & 50 & 0 \\ 0 & 0 & -10 & -1 \\ 0 & 0 & -25 & -10 \end{bmatrix}$. Reducing the matrix for the associated homogeneous system,

$$
\begin{bmatrix}
-10 & -1 & 0 & 0 & \vdots & 0 \\
-100 & -10 & 50 & 0 & \vdots & 0 \\
0 & 0 & -10 & -1 & \vdots & 0 \\
0 & 0 & -25 & -10 & \vdots & 0
\end{bmatrix}
\rightarrow
\begin{bmatrix}
10 & 1 & 0 & 0 & \vdots & 0 \\
0 & 0 & 1 & 0 & \vdots & 0 \\
0 & 0 & 0 & 1 & \vdots & 0 \\
0 & 0 & 0 & 0 & \vdots & 0
\end{bmatrix},
$$

we see our solutions are

$$
\begin{bmatrix} v_1 \\ v_2 \\ v_3 \\ v_4 \end{bmatrix}
=
\begin{bmatrix} v_1 \\ -10v_1 \\ 0 \\ 0 \end{bmatrix}
= v_1
\begin{bmatrix} 1 \\ -10 \\ 0 \\ 0 \end{bmatrix}
$$

and hence the column vector $\begin{bmatrix} 1 \\ -10 \\ 0 \\ 0 \end{bmatrix}$ forms a basis for E_{-10}.

For $\lambda = 10$ we need to find a basis for the null space of $10I - A = \begin{bmatrix} 10 & -1 & 0 & 0 \\ -100 & 10 & 50 & 0 \\ 0 & 0 & 10 & -1 \\ 0 & 0 & -25 & 10 \end{bmatrix}$. Reducing the matrix for

the associated homogeneous system,

$$
\begin{bmatrix}
10 & -1 & 0 & 0 & \vdots & 0 \\
-100 & 10 & 50 & 0 & \vdots & 0 \\
0 & 0 & 10 & -1 & \vdots & 0 \\
0 & 0 & -25 & 10 & \vdots & 0
\end{bmatrix}
\rightarrow
\begin{bmatrix}
10 & -1 & 0 & 0 & \vdots & 0 \\
0 & 0 & 1 & 0 & \vdots & 0 \\
0 & 0 & 0 & 1 & \vdots & 0 \\
0 & 0 & 0 & 0 & \vdots & 0
\end{bmatrix},
$$

we see our solutions are

$$
\begin{bmatrix} v_1 \\ v_2 \\ v_3 \\ v_4 \end{bmatrix}
=
\begin{bmatrix} v_1 \\ 10v_1 \\ 0 \\ 0 \end{bmatrix}
= v_1
\begin{bmatrix} 1 \\ 10 \\ 0 \\ 0 \end{bmatrix}
$$

and hence the column vector $\begin{bmatrix} 1 \\ 10 \\ 0 \\ 0 \end{bmatrix}$ forms a basis for E_{10}.

Matrix A is similar to the diagonal matrix $D = P^{-1}AP = \begin{bmatrix} -5 & 0 & 0 & 0 \\ 0 & 5 & 0 & 0 \\ 0 & 0 & -10 & 0 \\ 0 & 0 & 0 & 10 \end{bmatrix}$ where

$$
P = \begin{bmatrix}
1 & 1 & 1 & 1 \\
-5 & 5 & -10 & 10 \\
\frac{3}{2} & \frac{3}{2} & 0 & 0 \\
-\frac{15}{2} & \frac{15}{2} & 0 & 0
\end{bmatrix}.
$$

The columns of P are the basis vectors for the eigenspaces.

We also know that $\begin{bmatrix} c_1 e^{-5t} \\ c_2 e^{5t} \\ c_3 e^{-10t} \\ c_4 e^{10t} \end{bmatrix}$ is the general solution to $V' = DV$. Therefore the general solution to $V' = AV$ is

$$P \begin{bmatrix} c_1 e^{-5t} \\ c_2 e^{5t} \\ c_3 e^{-10t} \\ c_4 e^{10t} \end{bmatrix} = \begin{bmatrix} 1 & 1 & 1 & 1 \\ -5 & 5 & -10 & 10 \\ \frac{3}{2} & \frac{3}{2} & 0 & 0 \\ -\frac{15}{2} & \frac{15}{2} & 0 & 0 \end{bmatrix} \begin{bmatrix} c_1 e^{-5t} \\ c_2 e^{5t} \\ c_3 e^{-10t} \\ c_4 e^{10t} \end{bmatrix} = \begin{bmatrix} c_1 e^{-5t} + c_2 e^{5t} + c_3 e^{-10t} + c_4 e^{10t} \\ -5c_1 e^{-5t} + 5c_2 e^{5t} - 10c_3 e^{-10t} + 10c_4 e^{10t} \\ \frac{3}{2} c_1 e^{-5t} + \frac{3}{2} c_2 e^{5t} \\ -\frac{15}{2} c_1 e^{-5t} + \frac{15}{2} c_2 e^{5t} \end{bmatrix}.$$

Finally we use the initial position and initial velocity of the mass. The initial postion is the point $(2,6)$. Thus $v_1(0) = x(0) = 2$ and $v_3(0) = y(0) = 6$. The initial velocity vector is the zero vector, thus $v_2(0) = x'(0) = 0$ and $v_4(0) = y'(0) = 0$. We have the system of equations

$$\begin{aligned} 2 &= c_1 + c_2 + c_3 + c_4 \\ 0 &= -5c_1 + 5c_2 - 10c_3 + 10c_4 \\ 6 &= \frac{3}{2} c_1 + \frac{3}{2} c_2 \\ 0 &= -\frac{15}{2} c_1 + \frac{15}{2} c_2 \end{aligned}$$

which has the solution $c_1 = 2$, $c_2 = 2$, $c_3 = -1$, $c_4 = -1$. We have the solution

$$v_1 = x = 2e^{-5t} + 2e^{5t} - e^{-10t} - e^{10t}$$

$$v_3 = y = 3e^{-5t} + 3e^{5t}.$$

11. We have the system

$$V' = \begin{bmatrix} 0 & 1 & 0 & 0 \\ -\frac{k_1 + k_2}{m_1} & -\frac{f_1}{m_1} & \frac{k_2}{m_1} & 0 \\ 0 & 0 & 0 & 1 \\ \frac{k_2}{m_2} & 0 & -\frac{k_2}{m_2} & -\frac{f_2}{m_2} \end{bmatrix} V + \begin{bmatrix} 0 \\ \frac{1}{m_1} h_1(t) \\ 0 \\ \frac{1}{m_2} h_2(t) \end{bmatrix},$$

where $v_1(t) = x_1(t)$, $v_2(t) = x_1'(t)$, $v_3(t) = x_2(t)$ and $v_4(t) = x_2'(t)$. Since the two masses are at the equilibrium position and are given initial velocities of 1 m/sec at $t = 0$ we have $v_1(0) = v_3(0) = 0$ and $v_2 = v_4(0) = 1$.

We are given that $f_1 = f_2 = 0$, $m_1 = m_2 = 1$, $k_1 = 3$, $k_2 = 2$, $h_1(t) = 0$ and $h_2(t) = 0$. Thus, we have $V' = AV$, where $A = \begin{bmatrix} 0 & 1 & 0 & 0 \\ -5 & 0 & 2 & 0 \\ 0 & 0 & 0 & 1 \\ 2 & 0 & -2 & 0 \end{bmatrix}$. The characteristic equation of A is

$$\det(\lambda I - A) = \begin{vmatrix} \lambda & -1 & 0 & 0 \\ 5 & \lambda & -2 & 0 \\ 0 & 0 & \lambda & -1 \\ -2 & 0 & 2 & \lambda \end{vmatrix} = (\lambda^2 + 1)(\lambda^2 + 6) = 0.$$

Therefore, the eigenvalues of A are $\lambda = \pm i$ and $\lambda = \pm\sqrt{6}i$. We now wish to find bases for the corresponding eigenspaces.

For $\lambda = i$ we need to find a basis for the null space of $iI - A = \begin{bmatrix} i & -1 & 0 & 0 \\ 5 & i & -2 & 0 \\ 0 & 0 & i & -1 \\ -2 & 0 & 2 & i \end{bmatrix}$. Reducing the matrix for the

associated homogeneous system,

$$\begin{bmatrix} i & -1 & 0 & 0 & \vdots & 0 \\ 5 & i & -2 & 0 & \vdots & 0 \\ 0 & 0 & i & -1 & \vdots & 0 \\ -2 & 0 & 2 & i & \vdots & 0 \end{bmatrix} \rightarrow \begin{bmatrix} i & -1 & 0 & 0 & \vdots & 0 \\ 2 & 0 & -1 & 0 & \vdots & 0 \\ 2i & 0 & 0 & -1 & \vdots & 0 \\ 0 & 0 & 0 & 0 & \vdots & 0 \end{bmatrix},$$

we see our solutions are

$$\begin{bmatrix} v_1 \\ v_2 \\ v_3 \\ v_4 \end{bmatrix} = \begin{bmatrix} v_1 \\ iv_1 \\ 2v_1 \\ 2iv_1 \end{bmatrix} = v_1 \begin{bmatrix} 1 \\ i \\ 2 \\ 2i \end{bmatrix}$$

and hence the column vector $\begin{bmatrix} 1 \\ i \\ 2 \\ 2i \end{bmatrix}$ forms a basis for E_i.

For $\lambda = -i$ we need to find a basis for the null space of $-iI - A = \begin{bmatrix} -i & -1 & 0 & 0 \\ 5 & -i & -2 & 0 \\ 0 & 0 & -i & -1 \\ -2 & 0 & 2 & -i \end{bmatrix}$. Reducing the matrix for the

associated homogeneous system,

$$\begin{bmatrix} -i & -1 & 0 & 0 & : & 0 \\ 5 & -i & -2 & 0 & : & 0 \\ 0 & 0 & -i & -1 & : & 0 \\ -2 & 0 & 2 & -i & : & 0 \end{bmatrix} \rightarrow \begin{bmatrix} i & 1 & 0 & 0 & : & 0 \\ 2 & 0 & -1 & 0 & : & 0 \\ 2i & 0 & 0 & 1 & : & 0 \\ 0 & 0 & 0 & 0 & : & 0 \end{bmatrix},$$

we see our solutions are

$$\begin{bmatrix} v_1 \\ v_2 \\ v_3 \\ v_4 \end{bmatrix} = \begin{bmatrix} v_1 \\ -iv_1 \\ 2v_1 \\ -2iv_1 \end{bmatrix} = v_1 \begin{bmatrix} 1 \\ -i \\ 2 \\ -2i \end{bmatrix}$$

and hence the column vector $\begin{bmatrix} 1 \\ -i \\ 2 \\ -2i \end{bmatrix}$ forms a basis for E_{-i}.

For $\lambda = \sqrt{6}i$ we need to find a basis for the null space of $\sqrt{6}iI - A = \begin{bmatrix} \sqrt{6}i & -1 & 0 & 0 \\ 5 & \sqrt{6}i & -2 & 0 \\ 0 & 0 & \sqrt{6}i & -1 \\ -2 & 0 & 2 & \sqrt{6}i \end{bmatrix}$. Reducing the matrix for

the associated homogeneous system,

$$\begin{bmatrix} \sqrt{6}i & -1 & 0 & 0 & : & 0 \\ 5 & \sqrt{6}i & -2 & 0 & : & 0 \\ 0 & 0 & \sqrt{6}i & -1 & : & 0 \\ -2 & 0 & 2 & \sqrt{6}i & : & 0 \end{bmatrix} \rightarrow \begin{bmatrix} \sqrt{6}i & -1 & 0 & 0 & : & 0 \\ \frac{1}{2} & 0 & 1 & 0 & : & 0 \\ \frac{\sqrt{6}}{2}i & 0 & 0 & 1 & : & 0 \\ 0 & 0 & 0 & 0 & : & 0 \end{bmatrix},$$

we see our solutions are

$$\begin{bmatrix} v_1 \\ v_2 \\ v_3 \\ v_4 \end{bmatrix} = \begin{bmatrix} v_1 \\ \sqrt{6}iv_1 \\ -\frac{1}{2}v_1 \\ -\frac{\sqrt{6}}{2}iv_1 \end{bmatrix} = v_1 \begin{bmatrix} 1 \\ \sqrt{6}i \\ -\frac{1}{2} \\ -\frac{\sqrt{6}}{2}i \end{bmatrix}$$

and hence the column vector $\begin{bmatrix} 1 \\ \sqrt{6}i \\ -\frac{1}{2} \\ -\frac{\sqrt{6}}{2}i \end{bmatrix}$ forms a basis for $E_{\sqrt{6}i}$.

For $\lambda = -\sqrt{6}i$ we need to find a basis for the null space of $-\sqrt{6}iI - A = \begin{bmatrix} -\sqrt{6}i & -1 & 0 & 0 \\ 5 & -\sqrt{6}i & -2 & 0 \\ 0 & 0 & -\sqrt{6}i & -1 \\ -2 & 0 & 2 & -\sqrt{6}i \end{bmatrix}$. Reducing

the matrix for the associated homogeneous system,

$$\begin{bmatrix} -\sqrt{6}i & -1 & 0 & 0 & \vdots & 0 \\ 5 & -\sqrt{6}i & -2 & 0 & \vdots & 0 \\ 0 & 0 & -\sqrt{6}i & -1 & \vdots & 0 \\ -2 & 0 & 2 & -\sqrt{6}i & \vdots & 0 \end{bmatrix} \rightarrow \begin{bmatrix} \sqrt{6}i & 1 & 0 & 0 & \vdots & 0 \\ \frac{1}{2} & 0 & 1 & 0 & \vdots & 0 \\ -\frac{\sqrt{6}}{2}i & 0 & 0 & 1 & \vdots & 0 \\ 0 & 0 & 0 & 0 & \vdots & 0 \end{bmatrix},$$

we see our solutions are

$$\begin{bmatrix} v_1 \\ v_2 \\ v_3 \\ v_4 \end{bmatrix} = \begin{bmatrix} v_1 \\ -\sqrt{6}iv_1 \\ -\frac{1}{2}v_1 \\ \frac{\sqrt{6}}{2}iv_1 \end{bmatrix} = v_1 \begin{bmatrix} 1 \\ -\sqrt{6}i \\ -\frac{1}{2} \\ \frac{\sqrt{6}}{2}i \end{bmatrix}$$

and hence the column vector $\begin{bmatrix} 1 \\ -\sqrt{6}i \\ -\frac{1}{2} \\ \frac{\sqrt{6}}{2}i \end{bmatrix}$ forms a basis for $E_{-\sqrt{6}i}$.

Matrix A is similar to the diagonal matrix $D = P^{-1}AP = \begin{bmatrix} i & 0 & 0 & 0 \\ 0 & -i & 0 & 0 \\ 0 & 0 & \sqrt{6}i & 0 \\ 0 & 0 & 0 & -\sqrt{6}i \end{bmatrix}$ where

$$P = \begin{bmatrix} 1 & 1 & 1 & 1 \\ i & -i & \sqrt{6}i & -\sqrt{6}i \\ 2 & 2 & -\frac{1}{2} & -\frac{1}{2} \\ 2i & -2i & -\frac{\sqrt{6}}{2}i & \frac{\sqrt{6}}{2}i \end{bmatrix}.$$

The columns of P are the basis vectors for the eigenspaces. Two complex solutions to $V' = DV$ are

$$Z_1 = \begin{bmatrix} e^{it} \\ 0 \\ 0 \\ 0 \end{bmatrix} = \begin{bmatrix} \cos t + i\sin t \\ 0 \\ 0 \\ 0 \end{bmatrix}$$

and

$$Z_2 = \begin{bmatrix} 0 \\ 0 \\ e^{\sqrt{6}it} \\ 0 \end{bmatrix} = \begin{bmatrix} 0 \\ 0 \\ \cos\sqrt{6}t + i\sin\sqrt{6}t \\ 0 \end{bmatrix}.$$

By the first part of Theorem 6.7,

$$PZ_1 = \begin{bmatrix} 1 & 1 & 1 & 1 \\ i & -i & \sqrt{6}i & -\sqrt{6}i \\ 2 & 2 & -\frac{1}{2} & -\frac{1}{2} \\ 2i & -2i & -\frac{\sqrt{6}}{2}i & \frac{\sqrt{6}}{2}i \end{bmatrix} \begin{bmatrix} \cos t + i\sin t \\ 0 \\ 0 \\ 0 \end{bmatrix}$$

$$
= \begin{bmatrix} \cos t + i\sin t \\ -\sin t + i\cos t \\ 2\cos t + 2i\sin t \\ -2\sin t + 2i\cos t \end{bmatrix}
$$

$$
= \begin{bmatrix} \cos t \\ -\sin t \\ 2\cos t \\ -2\sin t \end{bmatrix} + i \begin{bmatrix} \sin t \\ \cos t \\ 2\sin t \\ 2\cos t \end{bmatrix}
$$

is a complex valued solution to $V' = AV$. By Theorem 6.8

$$
\begin{bmatrix} \cos t \\ -\sin t \\ 2\cos t \\ -2\sin t \end{bmatrix} \quad \text{and} \quad \begin{bmatrix} \sin t \\ \cos t \\ 2\sin t \\ 2\cos t \end{bmatrix}
$$

are two real valued solutions to $V' = AV$.

We obtain two more real valued solutions by considering

$$
PZ_2 = \begin{bmatrix} 1 & 1 & 1 & 1 \\ i & -i & \sqrt{6}i & -\sqrt{6}i \\ 2 & 2 & -\frac{1}{2} & -\frac{1}{2} \\ 2i & -2i & -\frac{\sqrt{6}}{2}i & \frac{\sqrt{6}}{2}i \end{bmatrix} \begin{bmatrix} 0 \\ 0 \\ \cos\sqrt{6}t + i\sin\sqrt{6}t \\ 0 \end{bmatrix}
$$

$$
= \begin{bmatrix} \cos\sqrt{6}t + i\sin\sqrt{6}t \\ -\sqrt{6}\sin\sqrt{6}t + i\sqrt{6}\cos\sqrt{6}t \\ -\frac{1}{2}\cos\sqrt{6}t - \frac{1}{2}i\sin\sqrt{6}t \\ \frac{\sqrt{6}}{2}\sin\sqrt{6}t - \frac{\sqrt{6}}{2}i\cos\sqrt{6}t \end{bmatrix}
$$

$$
= \begin{bmatrix} \cos\sqrt{6}t \\ -\sqrt{6}\sin\sqrt{6}t \\ -\frac{1}{2}\cos\sqrt{6}t \\ \frac{\sqrt{6}}{2}\sin\sqrt{6}t \end{bmatrix} + i \begin{bmatrix} \sin\sqrt{6}t \\ \sqrt{6}\cos\sqrt{6}t \\ -\frac{1}{2}\sin\sqrt{6}t \\ -\frac{\sqrt{6}}{2}\cos\sqrt{6}t \end{bmatrix} .
$$

The two new real valued solutions are

$$
\begin{bmatrix} \cos\sqrt{6}t \\ -\sqrt{6}\sin\sqrt{6}t \\ -\frac{1}{2}\cos\sqrt{6}t \\ \frac{\sqrt{6}}{2}\sin\sqrt{6}t \end{bmatrix} \quad \text{and} \quad \begin{bmatrix} \sin\sqrt{6}t \\ \sqrt{6}\cos\sqrt{6}t \\ -\frac{1}{2}\sin\sqrt{6}t \\ -\frac{\sqrt{6}}{2}\cos\sqrt{6}t \end{bmatrix} .
$$

Therefore the general solution to $V' = AV$ is

$$
V = c_1 \begin{bmatrix} \cos t \\ -\sin t \\ 2\cos t \\ -2\sin t \end{bmatrix} + c_2 \begin{bmatrix} \sin t \\ \cos t \\ 2\sin t \\ 2\cos t \end{bmatrix} + c_3 \begin{bmatrix} \cos\sqrt{6}t \\ -\sqrt{6}\sin\sqrt{6}t \\ -\frac{1}{2}\cos\sqrt{6}t \\ \frac{\sqrt{6}}{2}\sin\sqrt{6}t \end{bmatrix} + c_4 \begin{bmatrix} \sin\sqrt{6}t \\ \sqrt{6}\cos\sqrt{6}t \\ -\frac{1}{2}\sin\sqrt{6}t \\ -\frac{\sqrt{6}}{2}\cos\sqrt{6}t \end{bmatrix}
$$

$$
= \begin{bmatrix} c_1\cos t + c_2\sin t + c_3\cos\sqrt{6}t + c_4\sin\sqrt{6}t \\ -c_1\sin t + c_2\cos t - \sqrt{6}c_3\sin\sqrt{6}t + \sqrt{6}c_4\sqrt{6}\cos\sqrt{6}t \\ 2c_1\cos t + 2c_2\sin t - \frac{1}{2}c_3\cos\sqrt{6}t - \frac{1}{2}c_4\sin\sqrt{6}t \\ -2c_1\sin t + 2c_2\cos t + \frac{\sqrt{6}}{2}c_3\sin\sqrt{6}t - \frac{\sqrt{6}}{2}c_4\cos\sqrt{6}t \end{bmatrix} .
$$

Finally we use the initial positions and velocities of the masses. We have $v_1(0) = v_3(0) = 0$ and $v_2 = v_4(0) = 1$. We have the system of equations

$$
\begin{aligned}
0 &= c_1 + c_3 \\
1 &= c_2 + \sqrt{6}c_4 \\
0 &= 2c_1 - \frac{1}{2}c_3 \\
1 &= 2c_2 - \frac{\sqrt{6}}{2}c_4
\end{aligned}
$$

which has the solution $c_1 = 0$, $c_2 = \frac{3}{5}$, $c_3 = 0$, $c_4 = \frac{\sqrt{6}}{15}$. We have the solution

$$v_1 = x_1 = \frac{3}{5}\sin t + \frac{\sqrt{6}}{15}\sin\sqrt{6}t$$

$$v_3 = x_2 = \frac{6}{5}\sin t - \frac{\sqrt{6}}{20}\sin\sqrt{6}t.$$

13. We have the system

$$
V' = \begin{bmatrix}
0 & 1 & 0 & 0 \\
-\frac{k_1+k_2}{m_1} & -\frac{f_1}{m_1} & \frac{k_2}{m_1} & 0 \\
0 & 0 & 0 & 1 \\
\frac{k_2}{m_2} & 0 & -\frac{k_2}{m_2} & -\frac{f_2}{m_2}
\end{bmatrix} V + \begin{bmatrix}
0 \\
\frac{1}{m_1}h_1(t) \\
0 \\
\frac{1}{m_2}h_2(t)
\end{bmatrix},
$$

where $v_1(t) = x_1(t)$, $v_2(t) = x_1'(t)$, $v_3(t) = x_2(t)$ and $v_4(t) = x_2'(t)$. Since the two masses are pulled down $\frac{1}{10}$ m from equilibrium and released we have the initial conditions $v_1(0) = v_3(0) = 0.1$ and $v_2 = v_4(0) = 0$.

We are also given that $f_1 = f_2 = 0$, $m_1 = m_2 = 1$, $h_1(t) = 0$ and $h_2(t) = 0$. We need to find the spring constants k_1 and k_2.

Since $k = \frac{mg}{l}$ we have $k_1 = \frac{1\cdot 10}{1.25} = 8$ and $k_2 = \frac{1\cdot 10}{\frac{10}{3}} = 3$. Thus, we have $V' = AV$, where $A = \begin{bmatrix} 0 & 1 & 0 & 0 \\ -11 & 0 & 3 & 0 \\ 0 & 0 & 0 & 1 \\ 3 & 0 & -3 & 0 \end{bmatrix}$. The characteristic equation of A is

$$
\det(\lambda I - A) = \begin{vmatrix}
\lambda & -1 & 0 & 0 \\
11 & \lambda & -3 & 0 \\
0 & 0 & \lambda & -1 \\
-3 & 0 & 3 & \lambda
\end{vmatrix} = (\lambda^2 + 2)(\lambda^2 + 12) = 0.
$$

Therefore, the eigenvalues of A are $\lambda = \pm\sqrt{2}i$ and $\lambda = \pm 2\sqrt{3}i$. We now wish to find bases for the corresponding eigenspaces.

For $\lambda = \sqrt{2}i$ we need to find a basis for the null space of $\sqrt{2}iI - A = \begin{bmatrix} \sqrt{2}i & -1 & 0 & 0 \\ 11 & \sqrt{2}i & -3 & 0 \\ 0 & 0 & \sqrt{2}i & -1 \\ -3 & 0 & 3 & \sqrt{2}i \end{bmatrix}$. Reducing the matrix for

the associated homogeneous system,

$$
\begin{bmatrix}
\sqrt{2}i & -1 & 0 & 0 & \vdots & 0 \\
11 & \sqrt{2}i & -3 & 0 & \vdots & 0 \\
0 & 0 & \sqrt{2}i & -1 & \vdots & 0 \\
-3 & 0 & 3 & \sqrt{2}i & \vdots & 0
\end{bmatrix}
\rightarrow
\begin{bmatrix}
\sqrt{2}i & -1 & 0 & 0 & \vdots & 0 \\
3 & 0 & -1 & 0 & \vdots & 0 \\
3\sqrt{2}i & 0 & 0 & -1 & \vdots & 0 \\
0 & 0 & 0 & 0 & \vdots & 0
\end{bmatrix},
$$

we see our solutions are

$$
\begin{bmatrix} v_1 \\ v_2 \\ v_3 \\ v_4 \end{bmatrix} = \begin{bmatrix} v_1 \\ \sqrt{2}iv_1 \\ 3v_1 \\ 3\sqrt{2}iv_1 \end{bmatrix} = v_1 \begin{bmatrix} 1 \\ \sqrt{2}i \\ 3 \\ 3\sqrt{2}i \end{bmatrix}
$$

and hence the column vector $\begin{bmatrix} 1 \\ \sqrt{2}i \\ 3 \\ 3\sqrt{2}i \end{bmatrix}$ forms a basis for $E_{\sqrt{2}i}$.

For $\lambda = -\sqrt{2}i$ we need to find a basis for the null space of $-\sqrt{2}iI - A = \begin{bmatrix} -\sqrt{2}i & -1 & 0 & 0 \\ 11 & -\sqrt{2}i & -3 & 0 \\ 0 & 0 & -\sqrt{2}i & -1 \\ -3 & 0 & 3 & -\sqrt{2}i \end{bmatrix}$. Reducing

the matrix for the associated homogeneous system,

$$\begin{bmatrix} -\sqrt{2}i & -1 & 0 & 0 & \vdots & 0 \\ 11 & -\sqrt{2}i & -3 & 0 & \vdots & 0 \\ 0 & 0 & -\sqrt{2}i & -1 & \vdots & 0 \\ -3 & 0 & 3 & -\sqrt{2}i & \vdots & 0 \end{bmatrix} \rightarrow \begin{bmatrix} \sqrt{2}i & 1 & 0 & 0 & \vdots & 0 \\ 3 & 0 & -1 & 0 & \vdots & 0 \\ 3\sqrt{2}i & 0 & 0 & 1 & \vdots & 0 \\ 0 & 0 & 0 & 0 & \vdots & 0 \end{bmatrix},$$

we see our solutions are

$$\begin{bmatrix} v_1 \\ v_2 \\ v_3 \\ v_4 \end{bmatrix} = \begin{bmatrix} v_1 \\ -\sqrt{2}iv_1 \\ 3v_1 \\ -3\sqrt{2}iv_1 \end{bmatrix} = v_1 \begin{bmatrix} 1 \\ -\sqrt{2}i \\ 3 \\ -3\sqrt{2}i \end{bmatrix}$$

and hence the column vector $\begin{bmatrix} 1 \\ -\sqrt{2}i \\ 3 \\ -3\sqrt{2}i \end{bmatrix}$ forms a basis for $E_{-\sqrt{2}i}$.

For $\lambda = 2\sqrt{3}i$ we need to find a basis for the null space of $2\sqrt{3}iI - A = \begin{bmatrix} 2\sqrt{3}i & -1 & 0 & 0 \\ 11 & 2\sqrt{3}i & -3 & 0 \\ 0 & 0 & 2\sqrt{3}i & -1 \\ -3 & 0 & 3 & 2\sqrt{3}i \end{bmatrix}$. Reducing the

matrix for the associated homogeneous system,

$$\begin{bmatrix} 2\sqrt{3}i & -1 & 0 & 0 & \vdots & 0 \\ 11 & 2\sqrt{3}i & -3 & 0 & \vdots & 0 \\ 0 & 0 & 2\sqrt{3}i & -1 & \vdots & 0 \\ -3 & 0 & 3 & 2\sqrt{3}i & \vdots & 0 \end{bmatrix} \rightarrow \begin{bmatrix} 2\sqrt{3}i & -1 & 0 & 0 & \vdots & 0 \\ \frac{1}{3} & 0 & 1 & 0 & \vdots & 0 \\ \frac{2\sqrt{3}}{3}i & 0 & 0 & 1 & \vdots & 0 \\ 0 & 0 & 0 & 0 & \vdots & 0 \end{bmatrix},$$

we see our solutions are

$$\begin{bmatrix} v_1 \\ v_2 \\ v_3 \\ v_4 \end{bmatrix} = \begin{bmatrix} v_1 \\ 2\sqrt{3}iv_1 \\ -\frac{1}{3}v_1 \\ -\frac{2\sqrt{3}}{3}iv_1 \end{bmatrix} = v_1 \begin{bmatrix} 1 \\ 2\sqrt{3}i \\ -\frac{1}{3} \\ -\frac{2\sqrt{3}}{3}i \end{bmatrix}$$

and hence the column vector $\begin{bmatrix} 1 \\ 2\sqrt{3}i \\ -\frac{1}{3} \\ -\frac{2\sqrt{3}}{3}i \end{bmatrix}$ forms a basis for $E_{2\sqrt{3}i}$.

For $\lambda = -2\sqrt{3}i$ we need to find a basis for the null space of $-2\sqrt{3}iI - A = \begin{bmatrix} -2\sqrt{3}i & -1 & 0 & 0 \\ 11 & -2\sqrt{3}i & -3 & 0 \\ 0 & 0 & -2\sqrt{3}i & -1 \\ -3 & 0 & 3 & -2\sqrt{3}i \end{bmatrix}$.

Reducing the matrix for the associated homogeneous system,

$$
\begin{bmatrix}
-2\sqrt{3}i & -1 & 0 & 0 & \vdots & 0 \\
11 & -2\sqrt{3}i & -3 & 0 & \vdots & 0 \\
0 & 0 & -2\sqrt{3}i & -1 & \vdots & 0 \\
-3 & 0 & 3 & -2\sqrt{3}i & \vdots & 0
\end{bmatrix}
\rightarrow
\begin{bmatrix}
2\sqrt{3}i & 1 & 0 & 0 & \vdots & 0 \\
\frac{1}{3} & 0 & 1 & 0 & \vdots & 0 \\
\frac{2\sqrt{3}}{3}i & 0 & 0 & -1 & \vdots & 0 \\
0 & 0 & 0 & 0 & \vdots & 0
\end{bmatrix},
$$

we see our solutions are

$$
\begin{bmatrix} v_1 \\ v_2 \\ v_3 \\ v_4 \end{bmatrix}
=
\begin{bmatrix} v_1 \\ -2\sqrt{3}iv_1 \\ -\frac{1}{3}v_1 \\ \frac{2\sqrt{3}}{3}iv_1 \end{bmatrix}
= v_1
\begin{bmatrix} 1 \\ -2\sqrt{3}i \\ -\frac{1}{3} \\ \frac{2\sqrt{3}}{3}i \end{bmatrix}
$$

and hence the column vector $\begin{bmatrix} 1 \\ -2\sqrt{3}i \\ -\frac{1}{3} \\ \frac{2\sqrt{3}}{3}i \end{bmatrix}$ forms a basis for $E_{-2\sqrt{3}i}$.

Matrix A is similar to the diagonal matrix $D = P^{-1}AP = \begin{bmatrix} \sqrt{2}i & 0 & 0 & 0 \\ 0 & -\sqrt{2}i & 0 & 0 \\ 0 & 0 & 2\sqrt{3}i & 0 \\ 0 & 0 & 0 & -2\sqrt{3}i \end{bmatrix}$ where

$$
P =
\begin{bmatrix}
1 & 1 & 1 & 1 \\
\sqrt{2}i & -\sqrt{2}i & 2\sqrt{3}i & -2\sqrt{3}i \\
3 & 3 & -\frac{1}{3} & -\frac{1}{3} \\
3\sqrt{2}i & -3\sqrt{2}i & -\frac{2\sqrt{3}}{3}i & \frac{2\sqrt{3}}{3}i
\end{bmatrix}.
$$

The columns of P are the basis vectors for the eigenspaces. Two complex solutions to $V' = DV$ are

$$
Z_1 =
\begin{bmatrix} e^{\sqrt{2}it} \\ 0 \\ 0 \\ 0 \end{bmatrix}
=
\begin{bmatrix} \cos\sqrt{2}t + i\sin\sqrt{2}t \\ 0 \\ 0 \\ 0 \end{bmatrix}
$$

and

$$
Z_2 =
\begin{bmatrix} 0 \\ 0 \\ e^{2\sqrt{3}it} \\ 0 \end{bmatrix}
=
\begin{bmatrix} 0 \\ 0 \\ \cos 2\sqrt{3}t + i\sin 2\sqrt{3}t \\ 0 \end{bmatrix}.
$$

By the first part of Theorem 6.7,

$$
\begin{aligned}
PZ_1 &=
\begin{bmatrix}
1 & 1 & 1 & 1 \\
\sqrt{2}i & -\sqrt{2}i & 2\sqrt{3}i & -2\sqrt{3}i \\
3 & 3 & -\frac{1}{3} & -\frac{1}{3} \\
3\sqrt{2}i & -3\sqrt{2}i & -\frac{2\sqrt{3}}{3}i & \frac{2\sqrt{3}}{3}i
\end{bmatrix}
\begin{bmatrix} \cos\sqrt{2}t + i\sin\sqrt{2}t \\ 0 \\ 0 \\ 0 \end{bmatrix} \\[2mm]
&=
\begin{bmatrix}
\cos\sqrt{2}t + i\sin\sqrt{2}t \\
-\sqrt{2}\sin\sqrt{2}t + \sqrt{2}i\cos\sqrt{2}t \\
3\cos\sqrt{2}t + 3i\sin\sqrt{2}t \\
-3\sqrt{2}\sin\sqrt{2}t + 3\sqrt{2}i\cos\sqrt{2}t
\end{bmatrix}
\end{aligned}
$$

$$
= \begin{bmatrix} \cos\sqrt{2}t \\ -\sqrt{2}\sin\sqrt{2}t \\ 3\cos\sqrt{2}t \\ -3\sqrt{2}\sin\sqrt{2}t \end{bmatrix} + i \begin{bmatrix} \sin\sqrt{2}t \\ \sqrt{2}\cos\sqrt{2}t \\ 3\sin\sqrt{2}t \\ 3\sqrt{2}\cos\sqrt{2}t \end{bmatrix}
$$

is a complex valued solution to $V' = AV$. By Theorem 6.8

$$
\begin{bmatrix} \cos\sqrt{2}t \\ -\sqrt{2}\sin\sqrt{2}t \\ 3\cos\sqrt{2}t \\ -3\sqrt{2}\sin\sqrt{2}t \end{bmatrix} \text{ and } \begin{bmatrix} \sin\sqrt{2}t \\ \sqrt{2}\cos\sqrt{2}t \\ 3\sin\sqrt{2}t \\ 3\sqrt{2}\cos\sqrt{2}t \end{bmatrix}
$$

are two real valued solutions to $V' = AV$.

We obtain two more real valued solutions by considering

$$
PZ_2 = \begin{bmatrix} 1 & 1 & 1 & 1 \\ \sqrt{2}i & -\sqrt{2}i & 2\sqrt{3}i & -2\sqrt{3}i \\ 3 & 3 & -\frac{1}{3} & -\frac{1}{3} \\ 3\sqrt{2}i & -3\sqrt{2}i & -\frac{2\sqrt{3}}{3}i & \frac{2\sqrt{3}}{3}i \end{bmatrix} \begin{bmatrix} 0 \\ 0 \\ \cos 2\sqrt{3}t + i\sin 2\sqrt{3}t \\ 0 \end{bmatrix}
$$

$$
= \begin{bmatrix} \cos 2\sqrt{3}t + i\sin 2\sqrt{3}t \\ -2\sqrt{3}\sin 2\sqrt{3}t + 2\sqrt{3}i\cos 2\sqrt{3}t \\ -\frac{1}{3}\cos 2\sqrt{3}t - \frac{1}{3}i\sin 2\sqrt{3}t \\ \frac{2\sqrt{3}}{3}\sin 2\sqrt{3}t - \frac{2\sqrt{3}}{3}i\cos 2\sqrt{3}t \end{bmatrix}
$$

$$
= \begin{bmatrix} \cos 2\sqrt{3}t \\ -2\sqrt{3}\sin 2\sqrt{3}t \\ -\frac{1}{3}\cos 2\sqrt{3}t \\ \frac{2\sqrt{3}}{3}\sin 2\sqrt{3}t \end{bmatrix} + i \begin{bmatrix} \sin 2\sqrt{3}t \\ 2\sqrt{3}\cos 2\sqrt{3}t \\ -\frac{1}{3}\sin 2\sqrt{3}t \\ -\frac{2\sqrt{3}}{3}\cos 2\sqrt{3}t \end{bmatrix}.
$$

The two new real valued solutions are

$$
\begin{bmatrix} \cos 2\sqrt{3}t \\ -2\sqrt{3}\sin 2\sqrt{3}t \\ -\frac{1}{3}\cos 2\sqrt{3}t \\ \frac{2\sqrt{3}}{3}\sin 2\sqrt{3}t \end{bmatrix} \text{ and } \begin{bmatrix} \sin 2\sqrt{3}t \\ 2\sqrt{3}\cos 2\sqrt{3}t \\ -\frac{1}{3}\sin 2\sqrt{3}t \\ -\frac{2\sqrt{3}}{3}\cos 2\sqrt{3}t \end{bmatrix}.
$$

Therefore the general solution to $V' = AV$ is

$$
V = c_1 \begin{bmatrix} \cos\sqrt{2}t \\ -\sqrt{2}\sin\sqrt{2}t \\ 3\cos\sqrt{2}t \\ -3\sqrt{2}\sin\sqrt{2}t \end{bmatrix} + c_2 \begin{bmatrix} \sin\sqrt{2}t \\ \sqrt{2}\cos\sqrt{2}t \\ 3\sin\sqrt{2}t \\ 3\sqrt{2}\cos\sqrt{2}t \end{bmatrix} + c_3 \begin{bmatrix} \cos 2\sqrt{3}t \\ -2\sqrt{3}\sin 2\sqrt{3}t \\ -\frac{1}{3}\cos 2\sqrt{3}t \\ \frac{2\sqrt{3}}{3}\sin 2\sqrt{3}t \end{bmatrix} + c_4 \begin{bmatrix} \sin 2\sqrt{3}t \\ 2\sqrt{3}\cos 2\sqrt{3}t \\ -\frac{1}{3}\sin 2\sqrt{3}t \\ -\frac{2\sqrt{3}}{3}\cos 2\sqrt{3}t \end{bmatrix}
$$

$$
= \begin{bmatrix} c_1\cos\sqrt{2}t + c_2\sin\sqrt{2}t + c_3\cos 2\sqrt{3}t + c_4\sin 2\sqrt{3}t \\ -\sqrt{2}c_1\sin\sqrt{2}t + \sqrt{2}c_2\cos\sqrt{2}t - 2\sqrt{3}c_3\sin 2\sqrt{3}t + 2\sqrt{3}c_4\cos 2\sqrt{3}t \\ 3c_1\cos\sqrt{2}t + 3c_2\sin\sqrt{2}t - \frac{1}{3}c_3\cos 2\sqrt{3}t - \frac{1}{3}c_4\sin 2\sqrt{3}t \\ -3\sqrt{2}c_1\sin\sqrt{2}t + 3\sqrt{2}c_2\cos\sqrt{2}t + \frac{2\sqrt{3}}{3}c_3\sin 2\sqrt{3}t - \frac{2\sqrt{3}}{3}c_4\cos 2\sqrt{3}t \end{bmatrix}.
$$

Finally we use the initial positions and velocities of the masses. We have $v_1(0) = v_3(0) = 0.1$ and $v_2 = v_4(0) = 0$. We have the system of equations

$$
\frac{1}{10} = c_1 + c_3
$$

$$0 = \sqrt{2}c_2 + 2\sqrt{3}c_4$$
$$\frac{1}{10} = 3c_1 - \frac{1}{3}c_3$$
$$0 = 3\sqrt{2}c_2 - \frac{2\sqrt{3}}{3}c_4$$

which has the solution $c_1 = \frac{1}{25}$, $c_2 = 0$, $c_3 = \frac{3}{50}$, $c_4 = 0$. We have the solution

$$v_1 = x_1 = \frac{1}{25}\cos\sqrt{2}t + \frac{3}{50}\cos 2\sqrt{3}t$$

$$v_3 = x_2 = \frac{3}{25}\cos\sqrt{2}t - \frac{1}{50}\cos 2\sqrt{3}t.$$

15. We have the same system as in Exercise 11, with the exception that $f_1 = f_2 = 1$. Our matrix A now is

$$A = \begin{bmatrix} 0 & 1 & 0 & 0 \\ -5 & -1 & 2 & 0 \\ 0 & 0 & 0 & 1 \\ 2 & 0 & -2 & -1 \end{bmatrix}.$$

The characteristic equation of A is

$$\det(\lambda I - A) = \begin{vmatrix} \lambda & -1 & 0 & 0 \\ 5 & \lambda+1 & -2 & 0 \\ 0 & 0 & \lambda & -1 \\ -2 & 0 & 2 & \lambda+1 \end{vmatrix} = (\lambda^2+\lambda+1)(\lambda^2+\lambda+6) = 0.$$

Therefore, the eigenvalues of A are $\lambda = -\frac{1}{2} \pm \frac{\sqrt{3}}{2}i$ and $\lambda = -\frac{1}{2} \pm \frac{\sqrt{23}}{2}i$. We now wish to find bases for the corresponding eigenspaces.

For $\lambda = -\frac{1}{2} + \frac{\sqrt{3}}{2}i$ we need to find a basis for the null space of

$$\left(-\frac{1}{2} + \frac{\sqrt{3}}{2}i\right)I - A = \begin{bmatrix} -\frac{1}{2}+\frac{\sqrt{3}}{2}i & -1 & 0 & 0 \\ 5 & \frac{1}{2}+\frac{\sqrt{3}}{2}i & -2 & 0 \\ 0 & 0 & -\frac{1}{2}+\frac{\sqrt{3}}{2}i & -1 \\ -2 & 0 & 2 & \frac{1}{2}+\frac{\sqrt{3}}{2}i \end{bmatrix}.$$

Reducing the matrix for the associated homogeneous system,

$$\begin{bmatrix} -\frac{1}{2}+\frac{\sqrt{3}}{2}i & -1 & 0 & 0 & \vdots & 0 \\ 5 & \frac{1}{2}+\frac{\sqrt{3}}{2}i & -2 & 0 & \vdots & 0 \\ 0 & 0 & -\frac{1}{2}+\frac{\sqrt{3}}{2}i & -1 & \vdots & 0 \\ -2 & 0 & 2 & \frac{1}{2}+\frac{\sqrt{3}}{2}i & \vdots & 0 \end{bmatrix} \rightarrow \begin{bmatrix} -\frac{1}{2}+\frac{\sqrt{3}}{2}i & -1 & 0 & 0 & \vdots & 0 \\ 2 & 0 & -1 & 0 & \vdots & 0 \\ -1+\sqrt{3}i & 0 & 0 & -1 & \vdots & 0 \\ 0 & 0 & 0 & 0 & \vdots & 0 \end{bmatrix},$$

we see our solutions are

$$\begin{bmatrix} v_1 \\ v_2 \\ v_3 \\ v_4 \end{bmatrix} = \begin{bmatrix} v_1 \\ \left(-\frac{1}{2}+\frac{\sqrt{3}}{2}i\right)v_1 \\ 2v_1 \\ (-1+\sqrt{3}i)v_1 \end{bmatrix} = v_1 \begin{bmatrix} 1 \\ -\frac{1}{2}+\frac{\sqrt{3}}{2}i \\ 2 \\ -1+\sqrt{3}i \end{bmatrix}$$

and hence the column vector $\begin{bmatrix} 1 \\ -\frac{1}{2}+\frac{\sqrt{3}}{2}i \\ 2 \\ -1+\sqrt{3}i \end{bmatrix}$ forms a basis for $E_{-\frac{1}{2}+\frac{\sqrt{3}}{2}i}$.

For $\lambda = -\frac{1}{2} - \frac{\sqrt{3}}{2}i$ we need to find a basis for the null space of

$$\left(-\frac{1}{2} - \frac{\sqrt{3}}{2}i\right)I - A = \begin{bmatrix} -\frac{1}{2} - \frac{\sqrt{3}}{2}i & -1 & 0 & 0 \\ 5 & \frac{1}{2} - \frac{\sqrt{3}}{2}i & -2 & 0 \\ 0 & 0 & -\frac{1}{2} - \frac{\sqrt{3}}{2}i & -1 \\ -2 & 0 & 2 & \frac{1}{2} - \frac{\sqrt{3}}{2}i \end{bmatrix}.$$

Reducing the matrix for the associated homogeneous system,

$$\begin{bmatrix} -\frac{1}{2} - \frac{\sqrt{3}}{2}i & -1 & 0 & 0 & \vdots & 0 \\ 5 & \frac{1}{2} - \frac{\sqrt{3}}{2}i & -2 & 0 & \vdots & 0 \\ 0 & 0 & -\frac{1}{2} - \frac{\sqrt{3}}{2}i & -1 & \vdots & 0 \\ -2 & 0 & 2 & \frac{1}{2} - \frac{\sqrt{3}}{2}i & \vdots & 0 \end{bmatrix} \rightarrow \begin{bmatrix} -\frac{1}{2} - \frac{\sqrt{3}}{2}i & -1 & 0 & 0 & \vdots & 0 \\ 2 & 0 & -1 & 0 & \vdots & 0 \\ -1-\sqrt{3}i & 0 & 0 & -1 & \vdots & 0 \\ 0 & 0 & 0 & 0 & \vdots & 0 \end{bmatrix},$$

we see our solutions are

$$\begin{bmatrix} v_1 \\ v_2 \\ v_3 \\ v_4 \end{bmatrix} = \begin{bmatrix} v_1 \\ \left(-\frac{1}{2} - \frac{\sqrt{3}}{2}i\right)v_1 \\ 2v_1 \\ (-1-\sqrt{3}i)v_1 \end{bmatrix} = v_1 \begin{bmatrix} 1 \\ -\frac{1}{2} - \frac{\sqrt{3}}{2}i \\ 2 \\ -1-\sqrt{3}i \end{bmatrix}$$

and hence the column vector $\begin{bmatrix} 1 \\ -\frac{1}{2} - \frac{\sqrt{3}}{2}i \\ 2 \\ -1-\sqrt{3}i \end{bmatrix}$ forms a basis for $E_{-\frac{1}{2} - \frac{\sqrt{3}}{2}i}.$

For $\lambda = -\frac{1}{2} + \frac{\sqrt{23}}{2}i$ we need to find a basis for the null space of

$$\left(-\frac{1}{2} + \frac{\sqrt{23}}{2}i\right)I - A = \begin{bmatrix} -\frac{1}{2} + \frac{\sqrt{23}}{2}i & -1 & 0 & 0 \\ 5 & \frac{1}{2} + \frac{\sqrt{23}}{2}i & -2 & 0 \\ 0 & 0 & -\frac{1}{2} + \frac{\sqrt{23}}{2}i & -1 \\ -2 & 0 & 2 & \frac{1}{2} + \frac{\sqrt{23}}{2}i \end{bmatrix}.$$

Reducing the matrix for the associated homogeneous system,

$$\begin{bmatrix} -\frac{1}{2} + \frac{\sqrt{23}}{2}i & -1 & 0 & 0 & \vdots & 0 \\ 5 & \frac{1}{2} + \frac{\sqrt{23}}{2}i & -2 & 0 & \vdots & 0 \\ 0 & 0 & -\frac{1}{2} + \frac{\sqrt{23}}{2}i & -1 & \vdots & 0 \\ -2 & 0 & 2 & \frac{1}{2} + \frac{\sqrt{23}}{2}i & \vdots & 0 \end{bmatrix} \rightarrow \begin{bmatrix} -\frac{1}{2} + \frac{\sqrt{23}}{2}i & -1 & 0 & 0 & \vdots & 0 \\ \frac{1}{2} & 0 & 1 & 0 & \vdots & 0 \\ \frac{1}{4} - \frac{\sqrt{23}}{4}i & 0 & 0 & -1 & \vdots & 0 \\ 0 & 0 & 0 & 0 & \vdots & 0 \end{bmatrix},$$

we see our solutions are

$$\begin{bmatrix} v_1 \\ v_2 \\ v_3 \\ v_4 \end{bmatrix} = \begin{bmatrix} v_1 \\ \left(-\frac{1}{2} + \frac{\sqrt{23}}{2}i\right)v_1 \\ -\frac{1}{2}v_1 \\ \left(\frac{1}{4} - \frac{\sqrt{23}}{4}i\right)v_1 \end{bmatrix} = v_1 \begin{bmatrix} 1 \\ -\frac{1}{2} + \frac{\sqrt{23}}{2}i \\ -\frac{1}{2} \\ \frac{1}{4} - \frac{\sqrt{23}}{4}i \end{bmatrix}$$

and hence the column vector $\begin{bmatrix} 1 \\ -\frac{1}{2} + \frac{\sqrt{23}}{2}i \\ -\frac{1}{2} \\ \frac{1}{4} - \frac{\sqrt{23}}{4}i \end{bmatrix}$ forms a basis for $E_{-\frac{1}{2} + \frac{\sqrt{23}}{2}i}$.

For $\lambda = -\frac{1}{2} - \frac{\sqrt{23}}{2}i$ we need to find a basis for the null space of

$$\left(-\frac{1}{2} - \frac{\sqrt{23}}{2}i \right)I - A = \begin{bmatrix} -\frac{1}{2} - \frac{\sqrt{23}}{2}i & -1 & 0 & 0 \\ 5 & \frac{1}{2} - \frac{\sqrt{23}}{2}i & -2 & 0 \\ 0 & 0 & -\frac{1}{2} - \frac{\sqrt{23}}{2}i & -1 \\ -2 & 0 & 2 & \frac{1}{2} - \frac{\sqrt{23}}{2}i \end{bmatrix}.$$

Reducing the matrix for the associated homogeneous system,

$$\begin{bmatrix} -\frac{1}{2} - \frac{\sqrt{23}}{2}i & -1 & 0 & 0 & \vdots & 0 \\ 5 & \frac{1}{2} - \frac{\sqrt{23}}{2}i & -2 & 0 & \vdots & 0 \\ 0 & 0 & -\frac{1}{2} - \frac{\sqrt{23}}{2}i & -1 & \vdots & 0 \\ -2 & 0 & 2 & \frac{1}{2} - \frac{\sqrt{23}}{2}i & \vdots & 0 \end{bmatrix} \rightarrow \begin{bmatrix} -\frac{1}{2} - \frac{\sqrt{23}}{2}i & -1 & 0 & 0 & \vdots & 0 \\ \frac{1}{2} & 0 & 1 & 0 & \vdots & 0 \\ \frac{1}{4} + \frac{\sqrt{23}}{4}i & 0 & 0 & -1 & \vdots & 0 \\ 0 & 0 & 0 & 0 & \vdots & 0 \end{bmatrix},$$

we see our solutions are

$$\begin{bmatrix} v_1 \\ v_2 \\ v_3 \\ v_4 \end{bmatrix} = \begin{bmatrix} v_1 \\ \left(-\frac{1}{2} - \frac{\sqrt{23}}{2}i \right)v_1 \\ -\frac{1}{2}v_1 \\ \left(\frac{1}{4} + \frac{\sqrt{23}}{4}i \right)v_1 \end{bmatrix} = v_1 \begin{bmatrix} 1 \\ -\frac{1}{2} - \frac{\sqrt{23}}{2}i \\ -\frac{1}{2} \\ \frac{1}{4} + \frac{\sqrt{23}}{4}i \end{bmatrix}$$

and hence the column vector $\begin{bmatrix} 1 \\ -\frac{1}{2} - \frac{\sqrt{23}}{2}i \\ -\frac{1}{2} \\ \frac{1}{4} + \frac{\sqrt{23}}{4}i \end{bmatrix}$ forms a basis for $E_{-\frac{1}{2} - \frac{\sqrt{23}}{2}i}$.

Matrix A is similar to the diagonal matrix $D = P^{-1}AP = \begin{bmatrix} -\frac{1}{2} + \frac{\sqrt{3}}{2}i & 0 & 0 & 0 \\ 0 & -\frac{1}{2} - \frac{\sqrt{3}}{2}i & 0 & 0 \\ 0 & 0 & -\frac{1}{2} + \frac{\sqrt{23}}{2}i & 0 \\ 0 & 0 & 0 & -\frac{1}{2} - \frac{\sqrt{23}}{2}i \end{bmatrix}$ where

$$P = \begin{bmatrix} 1 & 1 & 1 & 1 \\ -\frac{1}{2} + \frac{\sqrt{3}}{2}i & -\frac{1}{2} - \frac{\sqrt{3}}{2}i & -\frac{1}{2} + \frac{\sqrt{23}}{2}i & -\frac{1}{2} - \frac{\sqrt{23}}{2}i \\ 2 & 2 & -\frac{1}{2} & -\frac{1}{2} \\ -1 + \sqrt{3}i & -1 - \sqrt{3}i & \frac{1}{4} - \frac{\sqrt{23}}{4}i & \frac{1}{4} + \frac{\sqrt{23}}{4}i \end{bmatrix}.$$

The columns of P are the basis vectors for the eigenspaces. Two complex solutions to $V' = DV$ are

$$Z_1 = \begin{bmatrix} e^{\left(-\frac{1}{2} + \frac{\sqrt{3}}{2}i \right)t} \\ 0 \\ 0 \\ 0 \end{bmatrix} = \begin{bmatrix} e^{-\frac{1}{2}t} \cos \frac{\sqrt{3}}{2}t + ie^{-\frac{1}{2}t} \sin \frac{\sqrt{3}}{2}t \\ 0 \\ 0 \\ 0 \end{bmatrix}$$

and

$$Z_2 = \begin{bmatrix} 0 \\ 0 \\ e^{\left(-\frac{1}{2}+\frac{\sqrt{23}}{2}i\right)t} \\ 0 \end{bmatrix} = \begin{bmatrix} 0 \\ 0 \\ e^{-\frac{1}{2}t}\cos\frac{\sqrt{23}}{2}t + ie^{-\frac{1}{2}t}\sin\frac{\sqrt{23}}{2}t \\ 0 \end{bmatrix}.$$

By the first part of Theorem 6.7,

$$PZ_1 = \begin{bmatrix} 1 & 1 & 1 & 1 \\ -\frac{1}{2}+\frac{\sqrt{3}}{2}i & -\frac{1}{2}-\frac{\sqrt{3}}{2}i & -\frac{1}{2}+\frac{\sqrt{23}}{2}i & -\frac{1}{2}-\frac{\sqrt{23}}{2}i \\ 2 & 2 & -\frac{1}{2} & -\frac{1}{2} \\ -1+\sqrt{3}i & -1-\sqrt{3}i & \frac{1}{4}-\frac{\sqrt{23}}{4}i & \frac{1}{4}+\frac{\sqrt{23}}{4}i \end{bmatrix} \begin{bmatrix} e^{-\frac{1}{2}t}\cos\frac{\sqrt{3}}{2}t + ie^{-\frac{1}{2}t}\sin\frac{\sqrt{3}}{2}t \\ 0 \\ 0 \\ 0 \end{bmatrix}$$

$$= \begin{bmatrix} e^{-\frac{1}{2}t}\left(\cos\frac{\sqrt{3}}{2}t + i\sin\frac{\sqrt{3}}{2}t\right) \\ \frac{1}{2}e^{-\frac{1}{2}t}\left(-\cos\frac{\sqrt{3}}{2}t - \sqrt{3}\sin\frac{\sqrt{3}}{2}t + i\left(-\sin\frac{\sqrt{3}}{2}t + \sqrt{3}\cos\frac{\sqrt{3}}{2}t\right)\right) \\ e^{-\frac{1}{2}t}\left(2\cos\frac{\sqrt{3}}{2}t + 2i\sin\frac{\sqrt{3}}{2}t\right) \\ e^{-\frac{1}{2}t}\left(-\cos\frac{\sqrt{3}}{2}t - \sqrt{3}\sin\frac{\sqrt{3}}{2}t + i\left(-\sin\frac{\sqrt{3}}{2}t + \sqrt{3}\cos\frac{\sqrt{3}}{2}t\right)\right) \end{bmatrix}$$

$$= \begin{bmatrix} e^{-\frac{1}{2}t}\cos\frac{\sqrt{3}}{2}t \\ \frac{1}{2}e^{-\frac{1}{2}t}\left(-\cos\frac{\sqrt{3}}{2}t - \sqrt{3}\sin\frac{\sqrt{3}}{2}t\right) \\ 2e^{-\frac{1}{2}t}\cos\frac{\sqrt{3}}{2}t \\ e^{-\frac{1}{2}t}\left(-\cos\frac{\sqrt{3}}{2}t - \sqrt{3}\sin\frac{\sqrt{3}}{2}t\right) \end{bmatrix} + i\begin{bmatrix} e^{-\frac{1}{2}t}\sin\frac{\sqrt{3}}{2}t \\ \frac{1}{2}e^{-\frac{1}{2}t}\left(-\sin\frac{\sqrt{3}}{2}t + \sqrt{3}\cos\frac{\sqrt{3}}{2}t\right) \\ 2e^{-\frac{1}{2}t}\sin\frac{\sqrt{3}}{2}t \\ e^{-\frac{1}{2}t}\left(-\sin\frac{\sqrt{3}}{2}t + \sqrt{3}\cos\frac{\sqrt{3}}{2}t\right) \end{bmatrix}$$

is a complex valued solution to $V' = AV$. By Theorem 6.8

$$\begin{bmatrix} e^{-\frac{1}{2}t}\cos\frac{\sqrt{3}}{2}t \\ \frac{1}{2}e^{-\frac{1}{2}t}\left(-\cos\frac{\sqrt{3}}{2}t - \sqrt{3}\sin\frac{\sqrt{3}}{2}t\right) \\ 2e^{-\frac{1}{2}t}\cos\frac{\sqrt{3}}{2}t \\ e^{-\frac{1}{2}t}\left(-\cos\frac{\sqrt{3}}{2}t - \sqrt{3}\sin\frac{\sqrt{3}}{2}t\right) \end{bmatrix} \text{ and } \begin{bmatrix} e^{-\frac{1}{2}t}\sin\frac{\sqrt{3}}{2}t \\ \frac{1}{2}e^{-\frac{1}{2}t}\left(-\sin\frac{\sqrt{3}}{2}t + \sqrt{3}\cos\frac{\sqrt{3}}{2}t\right) \\ 2e^{-\frac{1}{2}t}\sin\frac{\sqrt{3}}{2}t \\ e^{-\frac{1}{2}t}\left(-\sin\frac{\sqrt{3}}{2}t + \sqrt{3}\cos\frac{\sqrt{3}}{2}t\right) \end{bmatrix}$$

are two real valued solutions to $V' = AV$.

We obtain two more real valued solutions by considering

$$PZ_2 = \begin{bmatrix} 1 & 1 & 1 & 1 \\ -\frac{1}{2}+\frac{\sqrt{3}}{2}i & -\frac{1}{2}-\frac{\sqrt{3}}{2}i & -\frac{1}{2}+\frac{\sqrt{23}}{2}i & -\frac{1}{2}-\frac{\sqrt{23}}{2}i \\ 2 & 2 & -\frac{1}{2} & -\frac{1}{2} \\ -1+\sqrt{3}i & -1-\sqrt{3}i & \frac{1}{4}-\frac{\sqrt{23}}{4}i & \frac{1}{4}+\frac{\sqrt{23}}{4}i \end{bmatrix} \begin{bmatrix} 0 \\ 0 \\ e^{-\frac{1}{2}t}\cos\frac{\sqrt{23}}{2}t + ie^{-\frac{1}{2}t}\sin\frac{\sqrt{23}}{2}t \\ 0 \end{bmatrix}$$

$$= \begin{bmatrix} e^{-\frac{1}{2}t}\left(\cos\frac{\sqrt{23}}{2}t + i\sin\frac{\sqrt{23}}{2}t\right) \\ \frac{1}{2}e^{-\frac{1}{2}t}\left(-\cos\frac{\sqrt{23}}{2}t - \sqrt{23}\sin\frac{\sqrt{3}}{2}t + i\left(-\sin\frac{\sqrt{23}}{2}t + \sqrt{23}\cos\frac{\sqrt{23}}{2}t\right)\right) \\ -\frac{1}{2}e^{-\frac{1}{2}t}\left(\cos\frac{\sqrt{23}}{2}t + i\sin\frac{\sqrt{23}}{2}t\right) \\ \frac{1}{4}e^{-\frac{1}{2}t}\left(\cos\frac{\sqrt{23}}{2}t + \sqrt{23}\sin\frac{\sqrt{23}}{2}t + i\left(\sin\frac{\sqrt{23}}{2}t - \sqrt{23}\cos\frac{\sqrt{23}}{2}t\right)\right) \end{bmatrix}$$

$$= \begin{bmatrix} e^{-\frac{1}{2}t}\cos\frac{\sqrt{23}}{2}t \\ -\frac{1}{2}e^{-\frac{1}{2}t}\left(\cos\frac{\sqrt{23}}{2}t + \sqrt{23}\sin\frac{\sqrt{23}}{2}t\right) \\ -\frac{1}{2}e^{-\frac{1}{2}t}\cos\frac{\sqrt{23}}{2}t \\ \frac{1}{4}e^{-\frac{1}{2}t}\left(\cos\frac{\sqrt{23}}{2}t + \sqrt{23}\sin\frac{\sqrt{23}}{2}t\right) \end{bmatrix} + i\begin{bmatrix} e^{-\frac{1}{2}t}\sin\frac{\sqrt{23}}{2}t \\ \frac{1}{2}e^{-\frac{1}{2}t}\left(-\sin\frac{\sqrt{23}}{2}t + \sqrt{23}\cos\frac{\sqrt{23}}{2}t\right) \\ -\frac{1}{2}e^{-\frac{1}{2}t}\sin\frac{\sqrt{23}}{2}t \\ \frac{1}{4}e^{-\frac{1}{2}t}\left(\sin\frac{\sqrt{23}}{2}t - \sqrt{23}\cos\frac{\sqrt{23}}{2}t\right) \end{bmatrix}$$

The two new real valued solutions are

$$
\begin{bmatrix}
e^{-\frac{1}{2}t}\cos\frac{\sqrt{23}}{2}t \\
-\frac{1}{2}e^{-\frac{1}{2}t}\left(\cos\frac{\sqrt{23}}{2}t+\sqrt{23}\sin\frac{\sqrt{23}}{2}t\right) \\
-\frac{1}{2}e^{-\frac{1}{2}t}\cos\frac{\sqrt{23}}{2}t \\
\frac{1}{4}e^{-\frac{1}{2}t}\left(\cos\frac{\sqrt{23}}{2}t+\sqrt{23}\sin\frac{\sqrt{23}}{2}t\right)
\end{bmatrix}
\text{ and }
\begin{bmatrix}
e^{-\frac{1}{2}t}\sin\frac{\sqrt{23}}{2}t \\
\frac{1}{2}e^{-\frac{1}{2}t}\left(-\sin\frac{\sqrt{23}}{2}t+\sqrt{23}\cos\frac{\sqrt{23}}{2}t\right) \\
-\frac{1}{2}e^{-\frac{1}{2}t}\sin\frac{\sqrt{23}}{2}t \\
\frac{1}{4}e^{-\frac{1}{2}t}\left(\sin\frac{\sqrt{23}}{2}t-\sqrt{23}\cos\frac{\sqrt{23}}{2}t\right)
\end{bmatrix}.
$$

Therefore the general solution to $V' = AV$ is

$$
V = c_1
\begin{bmatrix}
e^{-\frac{1}{2}t}\cos\frac{\sqrt{3}}{2}t \\
\frac{1}{2}e^{-\frac{1}{2}t}\left(-\cos\frac{\sqrt{3}}{2}t-\sqrt{3}\sin\frac{\sqrt{3}}{2}t\right) \\
2e^{-\frac{1}{2}t}\cos\frac{\sqrt{3}}{2}t \\
e^{-\frac{1}{2}t}\left(-\cos\frac{\sqrt{3}}{2}t-\sqrt{3}\sin\frac{\sqrt{3}}{2}t\right)
\end{bmatrix}
+ c_2
\begin{bmatrix}
e^{-\frac{1}{2}t}\sin\frac{\sqrt{3}}{2}t \\
\frac{1}{2}e^{-\frac{1}{2}t}\left(-\sin\frac{\sqrt{3}}{2}t+\sqrt{3}\cos\frac{\sqrt{3}}{2}t\right) \\
2e^{-\frac{1}{2}t}\sin\frac{\sqrt{3}}{2}t \\
e^{-\frac{1}{2}t}\left(-\sin\frac{\sqrt{3}}{2}t+\sqrt{3}\cos\frac{\sqrt{3}}{2}t\right)
\end{bmatrix}
$$

$$
+ c_3
\begin{bmatrix}
e^{-\frac{1}{2}t}\cos\frac{\sqrt{23}}{2}t \\
-\frac{1}{2}e^{-\frac{1}{2}t}\left(\cos\frac{\sqrt{23}}{2}t+\sqrt{23}\sin\frac{\sqrt{23}}{2}t\right) \\
-\frac{1}{2}e^{-\frac{1}{2}t}\cos\frac{\sqrt{23}}{2}t \\
\frac{1}{4}e^{-\frac{1}{2}t}\left(\cos\frac{\sqrt{23}}{2}t+\sqrt{23}\sin\frac{\sqrt{23}}{2}t\right)
\end{bmatrix}
+ c_4
\begin{bmatrix}
e^{-\frac{1}{2}t}\sin\frac{\sqrt{23}}{2}t \\
\frac{1}{2}e^{-\frac{1}{2}t}\left(-\sin\frac{\sqrt{23}}{2}t+\sqrt{23}\cos\frac{\sqrt{23}}{2}t\right) \\
-\frac{1}{2}e^{-\frac{1}{2}t}\sin\frac{\sqrt{23}}{2}t \\
\frac{1}{4}e^{-\frac{1}{2}t}\left(\sin\frac{\sqrt{23}}{2}t-\sqrt{23}\cos\frac{\sqrt{23}}{2}t\right)
\end{bmatrix}.
$$

Finally we use the initial positions and velocities of the masses. As in Exercise 11 we have

$$v_1(0) = v_3(0) = 0 \text{ and } v_2 = v_4(0) = 1.$$

We have the system of equations

$$
\begin{aligned}
0 &= c_1 + c_3 \\
1 &= -\frac{1}{2}c_1 + \frac{\sqrt{3}}{2}c_2 - \frac{1}{2}c_3 + \frac{\sqrt{23}}{2}c_4 \\
0 &= 2c_1 - \frac{1}{2}c_3 \\
1 &= -c_1 + \sqrt{3}c_2 + \frac{1}{4}c_3 - \frac{\sqrt{23}}{4}c_4
\end{aligned}
$$

which has the solution $c_1 = 0$, $c_2 = \frac{2\sqrt{3}}{5}$, $c_3 = 0$, $c_4 = \frac{4\sqrt{23}}{115}$. We have the solution

$$
v_1 = x_1 = \frac{2\sqrt{3}}{5}\sin\frac{\sqrt{3}}{2}t + \frac{4\sqrt{23}}{115}\sin\frac{\sqrt{23}}{2}t
$$

$$
v_3 = x_2 = \frac{4\sqrt{3}}{5}\sin\frac{\sqrt{3}}{2}t - \frac{2\sqrt{23}}{115}\sin\frac{\sqrt{23}}{2}t.
$$

17. We have the same system as in Exercise 11 with the first mass subject to the external force of $\cos 2t$ kg m/sec^2. Thus, we have the system $V' = AV + G(t)$, where A as in Exercise 11 and $G(t) = \begin{bmatrix} 0 \\ \cos 2t \\ 0 \\ 0 \end{bmatrix}$.

From Exercise 11 we already have the general solution

$$
V_H =
\begin{bmatrix}
c_1\cos t + c_2\sin t + c_3\cos\sqrt{6}t + c_4\sin\sqrt{6}t \\
-c_1\sin t + c_2\cos t - \sqrt{6}c_3\sin\sqrt{6}t + \sqrt{6}c_4\sqrt{6}\cos\sqrt{6}t \\
2c_1\cos t + 2c_2\sin t - \frac{1}{2}c_3\cos\sqrt{6}t - \frac{1}{2}c_4\sin\sqrt{6}t \\
-2c_1\sin t + 2c_2\cos t + \frac{\sqrt{6}}{2}c_3\sin\sqrt{6}t - \frac{\sqrt{6}}{2}c_4\cos\sqrt{6}t
\end{bmatrix}.
$$

To find a particular solution we will use the formula $V_P = M \int M^{-1} G(t)\, dt$, where M is a matrix of fundamental solutions.

$$\text{Here } M = \begin{bmatrix} \cos t & \sin t & \cos\sqrt{6}t & \sin\sqrt{6}t \\ -\sin t & \cos t & -\sqrt{6}\sin\sqrt{6}t & \sqrt{6}\cos\sqrt{6}t \\ 2\cos t & 2\sin t & -\frac{1}{2}\cos\sqrt{6}t & -\frac{1}{2}\sin\sqrt{6}t \\ -2\sin t & 2\cos t & \frac{\sqrt{6}}{2}\sin\sqrt{6}t & -\frac{\sqrt{6}}{2}\cos\sqrt{6}t \end{bmatrix}.$$

We use Maple to find M^{-1} using the commands

```
with(linalg):
M:=matrix(4,4,[cos(t),sin(t),cos(sqrt(6)*t),sin(sqrt(6)*t),-sin(t),cos(t),
    -sqrt(6)*sin(sqrt(6)*t),sqrt(6)*cos(sqrt(6)*t),2*cos(t),2*sin(t),
    -(1/2)*cos(sqrt(6)*t),-(1/2)*sin(sqrt(6)*t),-2*sin(t),2*cos(t),
    (sqrt(6)/2)*sin(sqrt(6)*t),-(sqrt(6)/2)*cos(sqrt(6)*t)]);
inverse(M);
simplify(%);
```

We find

$$M^{-1} = \begin{bmatrix} \frac{1}{5}\cos t & -\frac{1}{5}\sin t & \frac{2}{5}\cos t & -\frac{2}{5}\sin t \\ \frac{1}{5}\sin t & \frac{1}{5}\cos t & \frac{2}{5}\sin t & \frac{2}{5}\cos t \\ \frac{4}{5}\cos\sqrt{6}t & -\frac{2\sqrt{6}}{15}\sin\sqrt{6}t & -\frac{2}{5}\cos\sqrt{6}t & \frac{\sqrt{6}}{15}\sin\sqrt{6}t \\ \frac{4}{5}\sin\sqrt{6}t & \frac{2\sqrt{6}}{15}\cos\sqrt{6}t & -\frac{2}{5}\sin\sqrt{6}t & -\frac{\sqrt{6}}{15}\cos\sqrt{6}t \end{bmatrix}.$$

The product inside the integral is then

$$M^{-1}G(t) = \begin{bmatrix} \frac{1}{5}\cos t & -\frac{1}{5}\sin t & \frac{2}{5}\cos t & -\frac{2}{5}\sin t \\ \frac{1}{5}\sin t & \frac{1}{5}\cos t & \frac{2}{5}\sin t & \frac{2}{5}\cos t \\ \frac{4}{5}\cos\sqrt{6}t & -\frac{2\sqrt{6}}{15}\sin\sqrt{6}t & -\frac{2}{5}\cos\sqrt{6}t & \frac{\sqrt{6}}{15}\sin\sqrt{6}t \\ \frac{4}{5}\sin\sqrt{6}t & \frac{2\sqrt{6}}{15}\cos\sqrt{6}t & -\frac{2}{5}\sin\sqrt{6}t & -\frac{\sqrt{6}}{15}\cos\sqrt{6}t \end{bmatrix} \begin{bmatrix} 0 \\ \cos 2t \\ 0 \\ 0 \end{bmatrix} = \begin{bmatrix} -\frac{1}{5}\sin t\cos 2t \\ \frac{1}{5}\cos t\cos 2t \\ -\frac{2\sqrt{6}}{15}\sin\sqrt{6}t\cos 2t \\ \frac{2\sqrt{6}}{15}\cos\sqrt{6}t\cos 2t \end{bmatrix}.$$

Integrating each component of this column vector gives us

$$\int M^{-1}G(t)\, dt = \begin{bmatrix} -\int \frac{1}{5}\sin t\cos 2t\, dt \\ \int \frac{1}{5}\cos t\cos 2t\, dt \\ -\int \frac{2\sqrt{6}}{15}\sin\sqrt{6}t\cos 2t\, dt \\ \int \frac{2\sqrt{6}}{15}\cos\sqrt{6}t\cos 2t\, dt \end{bmatrix} = \begin{bmatrix} \frac{2}{15}\cos^3 t - \frac{1}{5}\cos t \\ \frac{1}{15}\sin t + \frac{2}{15}\sin t\cos^2 t \\ \frac{1}{15}\left(12\cos\sqrt{6}t\cos^2 t - 6\cos\sqrt{6}t + 2\sqrt{6}\sin 2t\sin\sqrt{6}t\right) \\ \frac{1}{15}\left(12\sin\sqrt{6}t\cos^2 t - 6\sin\sqrt{6}t - 2\sqrt{6}\sin 2t\cos\sqrt{6}t\right) \end{bmatrix}.$$

Thus we have

$$V_P = M\int M^{-1}G(t)\, dt$$

$$= \begin{bmatrix} \cos t & \sin t & \cos\sqrt{6}t & \sin\sqrt{6}t \\ -\sin t & \cos t & -\sqrt{6}\sin\sqrt{6}t & \sqrt{6}\cos\sqrt{6}t \\ 2\cos t & 2\sin t & -\frac{1}{2}\cos\sqrt{6}t & -\frac{1}{2}\sin\sqrt{6}t \\ -2\sin t & 2\cos t & \frac{\sqrt{6}}{2}\sin\sqrt{6}t & -\frac{\sqrt{6}}{2}\cos\sqrt{6}t \end{bmatrix} \begin{bmatrix} \frac{2}{15}\cos^3 t - \frac{1}{5}\cos t \\ \frac{1}{15}\sin t + \frac{2}{15}\sin t\cos^2 t \\ \frac{1}{15}\left(\cos\sqrt{6}t\left(12\cos^2 t - 6\right) + 2\sqrt{6}\sin 2t\sin\sqrt{6}t\right) \\ \frac{1}{15}\left(\sin\sqrt{6}t\left(12\cos^2 t - 6\right) - 2\sqrt{6}\sin 2t\cos\sqrt{6}t\right) \end{bmatrix}$$

$$= \begin{bmatrix} \frac{1}{3}\cos 2t \\ -\frac{2}{3}\sin 2t \\ -\frac{1}{3}\cos 2t \\ \frac{2}{3}\sin 2t \end{bmatrix}.$$

The general solution is then

$$
V = V_H + V_P =
\begin{bmatrix}
c_1\cos t + c_2\sin t + c_3\cos\sqrt{6}t + c_4\sin\sqrt{6}t \\
-c_1\sin t + c_2\cos t - \sqrt{6}c_3\sin\sqrt{6}t + \sqrt{6}c_4\sqrt{6}\cos\sqrt{6}t \\
2c_1\cos t + 2c_2\sin t - \tfrac{1}{2}c_3\cos\sqrt{6}t - \tfrac{1}{2}c_4\sin\sqrt{6}t \\
-2c_1\sin t + 2c_2\cos t + \tfrac{\sqrt{6}}{2}c_3\sin\sqrt{6}t - \tfrac{\sqrt{6}}{2}c_4\cos\sqrt{6}t
\end{bmatrix}
+
\begin{bmatrix}
\tfrac{1}{3}\cos 2t \\
-\tfrac{2}{3}\sin 2t \\
-\tfrac{1}{3}\cos 2t \\
\tfrac{2}{3}\sin 2t
\end{bmatrix}
$$

$$
=
\begin{bmatrix}
c_1\cos t + c_2\sin t + c_3\cos\sqrt{6}t + c_4\sin\sqrt{6}t + \tfrac{1}{3}\cos 2t \\
-c_1\sin t + c_2\cos t - \sqrt{6}c_3\sin\sqrt{6}t + \sqrt{6}c_4\sqrt{6}\cos\sqrt{6}t - \tfrac{2}{3}\sin 2t \\
2c_1\cos t + 2c_2\sin t - \tfrac{1}{2}c_3\cos\sqrt{6}t - \tfrac{1}{2}c_4\sin\sqrt{6}t - \tfrac{1}{3}\cos 2t \\
-2c_1\sin t + 2c_2\cos t + \tfrac{\sqrt{6}}{2}c_3\sin\sqrt{6}t - \tfrac{\sqrt{6}}{2}c_4\cos\sqrt{6}t + \tfrac{2}{3}\sin 2t
\end{bmatrix}.
$$

Finally we use the initial positions and velocities of the masses. As in Exercise 11 we have

$$v_1(0) = v_3(0) = 0 \text{ and } v_2 = v_4(0) = 1.$$

We have the system of equations

$$
\begin{aligned}
0 &= c_1 + c_3 + \frac{1}{3} \\
1 &= c_2 + \sqrt{6}c_4 \\
0 &= 2c_1 - \frac{1}{2}c_3 - \frac{1}{3} \\
1 &= 2c_2 - \frac{\sqrt{6}}{2}c_4
\end{aligned}
$$

which has the solution $c_1 = \tfrac{1}{15}$, $c_2 = \tfrac{3}{5}$, $c_3 = -\tfrac{2}{5}$, $c_4 = \tfrac{\sqrt{6}}{15}$. We have the solution

$$v_1 = x_1 = \frac{1}{15}\cos t + \frac{3}{5}\sin t - \frac{2}{5}\cos\sqrt{6}t + \frac{\sqrt{6}}{15}\sin\sqrt{6}t + \frac{1}{3}\cos 2t$$

$$v_3 = x_2 = \frac{2}{15}\cos t + \frac{6}{5}\sin t + \frac{1}{5}\cos\sqrt{6}t - \frac{\sqrt{6}}{30}\sin\sqrt{6}t - \frac{1}{3}\cos 2t.$$

21. We wish to compute

$$\int_a^b \left(f(x(t),y(t))x'(t) + (g(x(t),y(t))y'(t) \right) dt.$$

From Exercise 9 we have $f(x,y) = 500x - 250y$ and $g(x,y) = 125y$. Furthermore we know that

$$x = 2e^{-5t} + 2e^{5t} - e^{-10t} - e^{10t} \text{ and } y = 3e^{-5t} + 3e^{5t}.$$

We are also given $a = 0$ and $b = 1$. Thus we evaluate

$$
\int_0^1 \big((500(2e^{-5t} + 2e^{5t} - e^{-10t} - e^{10t}) - 250(3e^{-5t} + 3e^{5t})(-10e^{-5t} + 10e^{5t} + 10e^{-10t} - 10e^{10t})
$$
$$
+ 125(3e^{-5t} + 3e^{5t})(-15e^{-5t} + 15e^{5t})\big)\, dt
$$
$$
= \int_0^1 (-5000e^{-20t} + 7500e^{-15t} - 8125e^{-10t} - 2500e^{-5t} + 2500e^{5t} + 8125e^{10t} - 7500e^{15t} + 5000e^{20t})\, dt
$$
$$
= \left(250e^{-20t} - 500e^{-15t} + \frac{1625}{2}e^{-10t} + 500e^{-5t} + 500e^{5t} + \frac{1625}{2}e^{10t} - 500e^{15t} + 250e^{20t} \right)\Big|_0^1
$$
$$
= 250e^{-20} - 500e^{-15} + \frac{1625}{2}e^{-10} + 500e^{-5} + 500e^5 + \frac{1625}{2}e^{10} - 500e^{15} + 250e^{20} - 2125.
$$

23. If the elevation at a point is $f(x,y) = x^2 + 4xy + y^2$ then

$$x' = 2x + 4y \text{ and } y' = 4x + 2y.$$

Thus we have the initial value problem $V' = AV$, where $V = \begin{bmatrix} x \\ y \end{bmatrix}$ and $A = \begin{bmatrix} 2 & 4 \\ 4 & 2 \end{bmatrix}$.

The characteristic equation of A is

$$\begin{aligned} \det(\lambda I - A) &= \begin{vmatrix} \lambda - 2 & -4 \\ -4 & \lambda - 2 \end{vmatrix} \\ &= (\lambda - 2)^2 - 16 = \lambda^2 - 4\lambda - 12 = (\lambda - 6)(\lambda + 2) = 0. \end{aligned}$$

Therefore, the eigenvalues of A are $\lambda = 6$ and $\lambda = -2$. We now wish to find bases for the corresponding eigenspaces.

For $\lambda = 6$ we need to find a basis for the null space of $6I - A = \begin{bmatrix} 4 & -4 \\ -4 & 4 \end{bmatrix}$. Reducing the matrix for the associated

homogeneous system, $\begin{bmatrix} 4 & -4 & \vdots & 0 \\ -4 & 4 & \vdots & 0 \end{bmatrix} \rightarrow \begin{bmatrix} 1 & -1 & \vdots & 0 \\ 0 & 0 & \vdots & 0 \end{bmatrix}$, we see our solutions are

$$\begin{bmatrix} x \\ y \end{bmatrix} = \begin{bmatrix} x \\ x \end{bmatrix} = x \begin{bmatrix} 1 \\ 1 \end{bmatrix}$$

and hence the column vector $\begin{bmatrix} 1 \\ 1 \end{bmatrix}$ forms a basis for E_6.

For $\lambda = -2$ we need to find a basis for the null space of $-2I - A = \begin{bmatrix} -4 & -4 \\ -4 & -4 \end{bmatrix}$. Reducing the matrix for the associated

homogeneous system, $\begin{bmatrix} -4 & -4 & \vdots & 0 \\ -4 & -4 & \vdots & 0 \end{bmatrix} \rightarrow \begin{bmatrix} 1 & 1 & \vdots & 0 \\ 0 & 0 & \vdots & 0 \end{bmatrix}$, we see our solutions are

$$\begin{bmatrix} x \\ y \end{bmatrix} = \begin{bmatrix} x \\ -x \end{bmatrix} = x \begin{bmatrix} 1 \\ -1 \end{bmatrix}$$

and hence the column vector $\begin{bmatrix} 1 \\ -1 \end{bmatrix}$ forms a basis for E_{-2}.

Matrix A is similar to the diagonal matrix $D = P^{-1}AP = \begin{bmatrix} 6 & 0 \\ 0 & -2 \end{bmatrix}$ where $P = \begin{bmatrix} 1 & 1 \\ 1 & -1 \end{bmatrix}$. The columns of P are the

basis vectors for the eigenspaces. We also know that $\begin{bmatrix} c_1 e^{6t} \\ c_2 e^{-2t} \end{bmatrix}$ is the general solution to $V' = DV$. Therefore the general

solution to $V' = AV$ is

$$V = P \begin{bmatrix} c_1 e^{6t} \\ c_2 e^{-2t} \end{bmatrix} = \begin{bmatrix} 1 & 1 \\ 1 & -1 \end{bmatrix} \begin{bmatrix} c_1 e^{6t} \\ c_2 e^{-2t} \end{bmatrix} = \begin{bmatrix} c_1 e^{6t} + c_2 e^{-2t} \\ c_1 e^{6t} - c_2 e^{-2t} \end{bmatrix}.$$

Since the hiker starts at the point $x = 10$ and $y = 15$ we have the initial conditions $x(0) = 10$, $y(0) = 15$. We have the system of equations

$$x(0) = 10 = c_1 + c_2, \ y(0) = 15 = c_1 - c_2.$$

This system has the solution $c_1 = \frac{25}{2}$, $c_2 = -\frac{5}{2}$. Thus the solution to the initial value problem is

$$x = \frac{25}{2} e^{6t} - \frac{5}{2} e^{-2t}, \ y = \frac{25}{2} e^{6t} + \frac{5}{2} e^{-2t}.$$

6.7 2×2 Systems of Nonlinear Differential Equations

Exercises 6.7, pp. 342-343

1. We let
$$F(x,y) = x + xy = x(1+y) \text{ and } G(x,y) = 2y - xy = y(2-x).$$

(a) We see that $F(x,y) = 0$ if $x = 0$ or $y = -1$, and that $G(x,y) = 0$ if $y = 0$ or $x = 2$. We can now see that $(0,0)$ and $(2,-1)$ are the equilibrium solutions.

(b) To obtain our Taylor polynomials, we note that
$$F_x(x,y) = 1 + y, \ F_y(x,y) = x, \ G_x(x,y) = -y, \ G_y(x,y) = 2 - x.$$

   i. We first consider the equilibrium solution $(0,0)$. At $(0,0)$ our first degree Taylor polynomial representation is
$$F(x,y) = F(0,0) + F_x(0,0)(x-0) + F_y(0,0)(y-0) = x$$

   and
$$G(x,y) = G(0,0) + G_x(0,0)(x-0) + G_y(0,0)(y-0) = 2y.$$

   Ignoring $R_F(x,y)$ and $R_G(x,y)$ we obtain the linear system:
$$x' = x, \ y' = 2y.$$

   ii. We now consider the equilibrium solution $(2,-1)$. At $(2,-1)$ our first degree Taylor polynomial representation is
$$F(x,y) = F(2,-1) + F_x(2,-1)(x-2) + F_y(2,-1)(y+1) = 2(y+1)$$

   and
$$G(x,y) = G(2,-1) + G_x(2,-1)(x-2) + G_y(2,-1)(y+1) = x-2.$$

   Again ignoring $R_F(x,y)$ and $R_G(x,y)$ and now using the substitutions $u = x - 2$ and $v = y + 1$, we obtain the linear system:
$$u' = 2v, \ v' = u.$$

(c) i. For $(0,0)$ we have the linear system:
$$x' = x, \ y' = 2y.$$

   The solution to this system is $x = c_1 e^t$, $y = c_2 e^{2t}$.

   ii. For $(2,-1)$ we have the linear system:
$$u' = 2v, \ v' = u.$$

   We let $A = \begin{bmatrix} 0 & 2 \\ 1 & 0 \end{bmatrix}$. The characteristic equation for A is
$$\det(\lambda I - A) = \begin{vmatrix} \lambda & -2 \\ -1 & \lambda \end{vmatrix} = \lambda^2 - 2 = 0.$$

   Therefore the eigenvalues of A are $\lambda = \pm\sqrt{2}$. We now wish to find bases for the corresponding eigenspaces. For $\lambda = \sqrt{2}$ we need to find a basis for the null space of $\sqrt{2}I - A = \begin{bmatrix} \sqrt{2} & -2 \\ -1 & \sqrt{2} \end{bmatrix}$. Reducing the matrix for the associated homogeneous system, $\begin{bmatrix} \sqrt{2} & -2 & \vdots & 0 \\ -1 & \sqrt{2} & \vdots & 0 \end{bmatrix} \rightarrow \begin{bmatrix} \frac{\sqrt{2}}{2} & -1 & \vdots & 0 \\ 0 & 0 & \vdots & 0 \end{bmatrix}$, we see our solutions are
$$\begin{bmatrix} u \\ v \end{bmatrix} = \begin{bmatrix} u \\ \frac{\sqrt{2}}{2}u \end{bmatrix} = u \begin{bmatrix} 1 \\ \frac{\sqrt{2}}{2} \end{bmatrix}$$

   and hence the column vector $\begin{bmatrix} 1 \\ \frac{\sqrt{2}}{2} \end{bmatrix}$ forms a basis for $E_{\sqrt{2}}$.

For $\lambda = -\sqrt{2}$ we need to find a basis for the null space of $-\sqrt{2}I - A = \begin{bmatrix} -\sqrt{2} & -2 \\ -1 & -\sqrt{2} \end{bmatrix}$. Reducing the matrix for

the associated homogeneous system, $\begin{bmatrix} -\sqrt{2} & -2 & \vdots & 0 \\ -1 & -\sqrt{2} & \vdots & 0 \end{bmatrix} \rightarrow \begin{bmatrix} \frac{\sqrt{2}}{2} & 1 & \vdots & 0 \\ 0 & 0 & \vdots & 0 \end{bmatrix}$, we see our solutions are

$$\begin{bmatrix} u \\ v \end{bmatrix} = \begin{bmatrix} u \\ -\frac{\sqrt{2}}{2}u \end{bmatrix} = u \begin{bmatrix} 1 \\ -\frac{\sqrt{2}}{2} \end{bmatrix}$$

and hence the column vector $\begin{bmatrix} 1 \\ -\frac{\sqrt{2}}{2} \end{bmatrix}$ forms a basis for $E_{-\sqrt{2}}$.

Matrix A is similar to the diagonal matrix $D = P^{-1}AP = \begin{bmatrix} \sqrt{2} & 0 \\ 0 & -\sqrt{2} \end{bmatrix}$ where $P = \begin{bmatrix} 1 & 1 \\ \frac{\sqrt{2}}{2} & -\frac{\sqrt{2}}{2} \end{bmatrix}$. The columns

of P are the basis vectors for the eigenspaces. We also know that $\begin{bmatrix} c_1 e^{\sqrt{2}t} \\ c_2 e^{-\sqrt{2}t} \end{bmatrix}$ is the general solution to

$$\begin{bmatrix} u' \\ v' \end{bmatrix} = D \begin{bmatrix} u \\ v \end{bmatrix}.$$

Therefore the general solution to $\begin{bmatrix} u' \\ v' \end{bmatrix} = A \begin{bmatrix} u \\ v \end{bmatrix}$ is

$$\begin{bmatrix} u \\ v \end{bmatrix} = P \begin{bmatrix} c_1 e^{\sqrt{2}t} \\ c_2 e^{-\sqrt{2}t} \end{bmatrix} = \begin{bmatrix} 1 & 1 \\ \frac{\sqrt{2}}{2} & -\frac{\sqrt{2}}{2} \end{bmatrix} \begin{bmatrix} c_1 e^{\sqrt{2}t} \\ c_2 e^{-\sqrt{2}t} \end{bmatrix} = \begin{bmatrix} c_1 e^{\sqrt{2}t} + c_2 e^{-\sqrt{2}t} \\ \frac{\sqrt{2}}{2} c_1 e^{\sqrt{2}t} - \frac{\sqrt{2}}{2} c_2 e^{-\sqrt{2}t} \end{bmatrix}.$$

(d)   i. The system for $(0,0)$ is unstable since

$$\lim_{t \to \infty} x(t) = \lim_{t \to \infty} c_1 e^t = \infty$$

and

$$\lim_{t \to \infty} y(t) = \lim_{t \to \infty} c_2 e^{2t} = \infty.$$

ii. The system for $(2,-1)$ is unstable since

$$\lim_{t \to \infty} u(t) = \lim_{t \to \infty} \left(c_1 e^{\sqrt{2}t} + c_2 e^{-\sqrt{2}t} \right) = \infty$$

and

$$\lim_{t \to \infty} v(t) = \lim_{t \to \infty} \left(c_1 e^{\sqrt{2}t} - \frac{\sqrt{2}}{2} c_2 e^{-\sqrt{2}t} \right) = \infty.$$

3. We let

$$F(x,y) = x - 2xy + xy^2 = x(y-1)^2 \text{ and } G(x,y) = y + xy = y(1+x).$$

(a) We see that $F(x,y) = 0$ if $x = 0$ or $y = 1$, and that $G(x,y) = 0$ if $y = 0$ or $x = -1$. We can now see that $(0,0)$ and $(-1,1)$ are the equilibrium solutions.

(b) To obtain our Taylor polynomials, we note that

$$F_x(x,y) = 1 - 2y + y^2, \ F_y(x,y) = -2x + 2xy, \ G_x(x,y) = y, \ G_y(x,y) = 1 + x.$$

i. We first consider the equilibrium solution $(0,0)$. At $(0,0)$ our first degree Taylor polynomial representation is

$$F(x,y) = F(0,0) + F_x(0,0)(x-0) + F_y(0,0)(y-0) = x$$

and

$$G(x,y) = G(0,0) + G_x(0,0)(x-0) + G_y(0,0)(y-0) = y.$$

Ignoring $R_F(x,y)$ and $R_G(x,y)$ we obtain the linear system:

$$x' = x, \ y' = y.$$

ii. We now consider the equilibrium solution $(-1,1)$. At $(-1,1)$ our first degree Taylor polynomial representation is

$$F(x,y) = F(-1,1) + F_x(-1,1)(x+1) + F_y(-1,1)(y-1) = 0$$

and

$$G(x,y) = G(-1,1) + G_x(-1,1)(x+1) + G_y(-1,1)(y-1) = x+1.$$

Again ignoring $R_F(x,y)$ and $R_G(x,y)$ and now using the substitutions $u = x+1$ and $v = y-1$, we obtain the linear system:

$$u' = 0,\ v' = u.$$

(c)   i. For $(0,0)$ we have the linear system:

$$x' = x,\ y' = y.$$

The solution to this system is $x = c_1 e^t$, $y = c_2 e^t$.

ii. For $(-1,1)$ we have the linear system:

$$u' = 0,\ v' = u.$$

We immediately see that $u = c_1$. Substituting this into the equation for v' we see $v' = c_1$. Therefore, $v = c_1 t + c_2$.

(d)   i. The system for $(0,0)$ is unstable since

$$\lim_{t\to\infty} x(t) = \lim_{t\to\infty} c_1 e^t = \infty$$

and

$$\lim_{t\to\infty} y(t) = \lim_{t\to\infty} c_2 e^t = \infty.$$

ii. The system for $(-1,1)$ is unstable since

$$\lim_{t\to\infty} u(t) = \lim_{t\to\infty} c_1 = c_1$$

and

$$\lim_{t\to\infty} v(t) = \lim_{t\to\infty} (c_1 t + c_2) = \infty.$$

5. We let

$$F(x,y) = 2 + x - 2e^{-2y} \text{ and } G(x,y) = x - \sin y.$$

(a) We see $G(x,y) = 0$ if $x = \sin y$. Substituting this into the equation $2 + x - 2e^{-2y} = 0$ we have

$$2 + \sin y = 2e^{-2y}.$$

The only solution to this equation is $y = 0$. Thus $(0,0)$ is the only equilibrium solution.

(b) To obtain our Taylor polynomial, we note that

$$F_x(x,y) = 1,\ F_y(x,y) = 4e^{-2y},\ G_x(x,y) = 1,\ G_y(x,y) = -\cos y.$$

At $(0,0)$ our first degree Taylor polynomial representation is

$$F(x,y) = F(0,0) + F_x(0,0)(x-0) + F_y(0,0)(y-0) = x + 4y$$

and

$$G(x,y) = G(0,0) + G_x(0,0)(x-0) + G_y(0,0)(y-0) = x - y.$$

Ignoring $R_F(x,y)$ and $R_G(x,y)$ we obtain the linear system:

$$x' = x + 4y,\ y' = x - y.$$

(c) We let $A = \begin{bmatrix} 1 & 4 \\ 1 & -1 \end{bmatrix}$. The characteristic equation for A is

$$\det(\lambda I - A) = \begin{vmatrix} \lambda-1 & -4 \\ -1 & \lambda+1 \end{vmatrix} = \lambda^2 - 5 = 0.$$

Therefore the eigenvalues of A are $\lambda = \pm\sqrt{5}$. We now wish to find bases for the corresponding eigenspaces.

For $\lambda = \sqrt{5}$ we need to find a basis for the null space of $\sqrt{5}I - A = \begin{bmatrix} \sqrt{5}-1 & -4 \\ -1 & \sqrt{5}+1 \end{bmatrix}$. Reducing the matrix for the

associated homogeneous system, $\begin{bmatrix} \sqrt{5}-1 & -4 & \vdots & 0 \\ -1 & \sqrt{5}+1 & \vdots & 0 \end{bmatrix} \rightarrow \begin{bmatrix} -\frac{1}{4}+\frac{\sqrt{5}}{4} & -1 & \vdots & 0 \\ 0 & 0 & \vdots & 0 \end{bmatrix}$, we see our solutions

are

$$\begin{bmatrix} u \\ v \end{bmatrix} = \begin{bmatrix} u \\ -\frac{1}{4}+\frac{\sqrt{5}}{4}u \end{bmatrix} = u \begin{bmatrix} 1 \\ -\frac{1}{4}+\frac{\sqrt{5}}{4} \end{bmatrix}$$

and hence the column vector $\begin{bmatrix} 1 \\ -\frac{1}{4}+\frac{\sqrt{5}}{4} \end{bmatrix}$ forms a basis for $E_{\sqrt{5}}$.

For $\lambda = -\sqrt{5}$ we need to find a basis for the null space of $-\sqrt{5}I - A = \begin{bmatrix} -\sqrt{5}-1 & -4 \\ -1 & -\sqrt{5}+1 \end{bmatrix}$. Reducing the matrix

for the associated homogeneous system, $\begin{bmatrix} -\sqrt{5}-1 & -4 & \vdots & 0 \\ -1 & -\sqrt{5}+1 & \vdots & 0 \end{bmatrix} \rightarrow \begin{bmatrix} -\frac{1}{4}-\frac{\sqrt{5}}{4} & -1 & \vdots & 0 \\ 0 & 0 & \vdots & 0 \end{bmatrix}$, we see our

solutions are

$$\begin{bmatrix} u \\ v \end{bmatrix} = \begin{bmatrix} u \\ -\frac{1}{4}-\frac{\sqrt{5}}{4}u \end{bmatrix} = u \begin{bmatrix} 1 \\ -\frac{1}{4}-\frac{\sqrt{5}}{4} \end{bmatrix}$$

and hence the column vector $\begin{bmatrix} 1 \\ -\frac{1}{4}-\frac{\sqrt{5}}{4} \end{bmatrix}$ forms a basis for $E_{-\sqrt{5}}$.

Matrix A is similar to the diagonal matrix $D = P^{-1}AP = \begin{bmatrix} \sqrt{5} & 0 \\ 0 & -\sqrt{5} \end{bmatrix}$ where

$$P = \begin{bmatrix} 1 & 1 \\ -\frac{1}{4}+\frac{\sqrt{5}}{4} & -\frac{1}{4}-\frac{\sqrt{5}}{4} \end{bmatrix}.$$

The columns of P are the basis vectors for the eigenspaces. We also know that $\begin{bmatrix} c_1 e^{\sqrt{5}t} \\ c_2 e^{-\sqrt{5}t} \end{bmatrix}$ is the general solution to

$\begin{bmatrix} x' \\ y' \end{bmatrix} = D \begin{bmatrix} x \\ y \end{bmatrix}$. Therefore the general solution to $\begin{bmatrix} x' \\ y' \end{bmatrix} = A \begin{bmatrix} x \\ y \end{bmatrix}$ is

$$\begin{bmatrix} x \\ y \end{bmatrix} = P \begin{bmatrix} c_1 e^{\sqrt{5}t} \\ c_2 e^{-\sqrt{5}t} \end{bmatrix} = \begin{bmatrix} 1 & 1 \\ -\frac{1}{4}+\frac{\sqrt{5}}{4} & -\frac{1}{4}-\frac{\sqrt{5}}{4} \end{bmatrix} \begin{bmatrix} c_1 e^{\sqrt{5}t} \\ c_2 e^{-\sqrt{5}t} \end{bmatrix}$$

$$= \begin{bmatrix} c_1 e^{\sqrt{5}t} + c_2 e^{-\sqrt{5}t} \\ \left(-\frac{1}{4}+\frac{\sqrt{5}}{4}\right)c_1 e^{\sqrt{5}t} - \left(\frac{1}{4}+\frac{\sqrt{5}}{4}\right)c_2 e^{-\sqrt{5}t} \end{bmatrix}.$$

(d) The system for $(0,0)$ is unstable since

$$\lim_{t \to \infty} x(t) = \lim_{t \to \infty} \left(c_1 e^{\sqrt{5}t} + c_2 e^{-\sqrt{5}t} \right) = \infty$$

and

$$\lim_{t \to \infty} y(t) = \lim_{t \to \infty} \left(\left(-\frac{1}{4}+\frac{\sqrt{5}}{4}\right)c_1 e^{\sqrt{5}t} - \left(\frac{1}{4}+\frac{\sqrt{5}}{4}\right)c_2 e^{-\sqrt{5}t} \right) = \infty.$$

9. We have the 2×2 system

$$m'(t) = m(t)\left(\frac{1}{2} - \frac{1}{4}w(t)\right)$$

$$w'(t) = w(t)\left(-\frac{1}{4}w(t) + \frac{3}{4}m(t)\right).$$

(a) There are two equilibrium solutions $(0,0)$ and $\left(\frac{\gamma\alpha}{\delta\beta}, \frac{\alpha}{\beta}\right) = \left(\frac{2}{3}, 2\right)$.

(b)   i. The linear part of the problem at the equilibrium $(0,0)$ is $A(0,0) = \begin{bmatrix} \frac{1}{2} & 0 \\ 0 & 0 \end{bmatrix}$. The eigenvalues of A are 0 and $\frac{1}{2}$.

   ii. The linear part of the problem at the equilibrium $\left(\frac{2}{3}, 2\right)$ is $A\left(\frac{2}{3}, 2\right) = \begin{bmatrix} 0 & -\frac{1}{6} \\ \frac{3}{2} & -\frac{1}{2} \end{bmatrix}$.

(c)   i. We first consider the equilibrium solution $(0,0)$. The general solution to the system is $m = c_1 e^{\frac{1}{2}t}$, $w = c_2$.

   ii. We now consider the equilibrium solution $\left(\frac{2}{3}, 2\right)$. The characteristic equation for $A\left(\frac{2}{3}, 2\right)$ is

$$\det(\lambda I - A) = \begin{vmatrix} \lambda & \frac{1}{6} \\ -\frac{3}{2} & \lambda + \frac{1}{2} \end{vmatrix} = \lambda^2 + \frac{1}{2}\lambda + \frac{1}{4} = 0.$$

Therefore the eigenvalues of A are $\lambda = \frac{-1 \pm \sqrt{3}i}{4}$. We now wish to find bases for the corresponding eigenspaces.

For $\lambda = \frac{-1+\sqrt{3}i}{4}$ we need to find a basis for the null space of $\frac{-1+\sqrt{3}i}{4}I - A = \begin{bmatrix} \frac{-1+\sqrt{3}i}{4} & \frac{1}{6} \\ -\frac{3}{2} & \frac{1+\sqrt{3}i}{4} \end{bmatrix}$. Reducing the

matrix for the associated homogeneous system,

$$\begin{bmatrix} \frac{-1+\sqrt{3}i}{4} & \frac{1}{6} & : & 0 \\ -\frac{3}{2} & \frac{1+\sqrt{3}i}{4} & : & 0 \end{bmatrix} \rightarrow \begin{bmatrix} \frac{3-3\sqrt{3}i}{2} & -1 & : & 0 \\ 0 & 0 & : & 0 \end{bmatrix},$$

we see our solutions are

$$\begin{bmatrix} u \\ v \end{bmatrix} = \begin{bmatrix} u \\ \frac{3-3\sqrt{3}i}{2}u \end{bmatrix} = u\begin{bmatrix} 1 \\ \frac{3-3\sqrt{3}i}{2} \end{bmatrix}$$

and hence the column vector $\begin{bmatrix} 1 \\ \frac{3-3\sqrt{3}i}{2} \end{bmatrix}$ forms a basis for $E_{\frac{-1+\sqrt{3}i}{4}}$.

For $\lambda = \frac{-1-\sqrt{3}i}{4}$ we need to find a basis for the null space of $\frac{-1-\sqrt{3}i}{4}I - A = \begin{bmatrix} \frac{-1-\sqrt{3}i}{4} & \frac{1}{6} \\ -\frac{3}{2} & \frac{1-\sqrt{3}i}{4} \end{bmatrix}$. Reducing the

matrix for the associated homogeneous system,

$$\begin{bmatrix} \frac{-1-\sqrt{3}i}{4} & \frac{1}{6} & : & 0 \\ -\frac{3}{2} & \frac{1-\sqrt{3}i}{4} & : & 0 \end{bmatrix} \rightarrow \begin{bmatrix} \frac{3+3\sqrt{3}i}{2} & -1 & : & 0 \\ 0 & 0 & : & 0 \end{bmatrix},$$

we see our solutions are

$$\begin{bmatrix} u \\ v \end{bmatrix} = \begin{bmatrix} u \\ \frac{3+3\sqrt{3}i}{2}u \end{bmatrix} = u\begin{bmatrix} 1 \\ \frac{3+3\sqrt{3}i}{2} \end{bmatrix}$$

and hence the column vector $\begin{bmatrix} 1 \\ \frac{3+3\sqrt{3}i}{2} \end{bmatrix}$ forms a basis for $E_{\frac{-1-\sqrt{3}i}{4}}$.

Matrix A is similar to the diagonal matrix $D = P^{-1}AP = \begin{bmatrix} \frac{-1+\sqrt{3}i}{4} & 0 \\ 0 & \frac{-1-\sqrt{3}i}{4} \end{bmatrix}$ where

$$P = \begin{bmatrix} 1 & 1 \\ \frac{3-3\sqrt{3}i}{2} & \frac{3+3\sqrt{3}i}{2} \end{bmatrix}.$$

The columns of P are the basis vectors for the eigenspaces.

One complex valued solution to $\begin{bmatrix} m' \\ w' \end{bmatrix} = D\begin{bmatrix} m \\ w \end{bmatrix}$ is

$$Z = \begin{bmatrix} e^{\frac{-1+\sqrt{3}i}{4}t} \\ 0 \end{bmatrix} = \begin{bmatrix} e^{-\frac{1}{4}t}\cos\frac{\sqrt{3}}{4}t + ie^{-\frac{1}{4}t}\sin\frac{\sqrt{3}}{4}t \\ 0 \end{bmatrix}.$$

By the first part of Theorem 6.7,

$$PZ = \begin{bmatrix} 1 & 1 \\ \frac{3-3\sqrt{3}i}{2} & \frac{3+3\sqrt{3}i}{2} \end{bmatrix} \begin{bmatrix} e^{-\frac{1}{4}t}\cos\frac{\sqrt{3}}{4}t + ie^{-\frac{1}{4}t}\sin\frac{\sqrt{3}}{4}t \\ 0 \end{bmatrix}$$

$$= \begin{bmatrix} e^{-\frac{1}{4}t}\cos\frac{\sqrt{3}}{4}t + ie^{-\frac{1}{4}t}\sin\frac{\sqrt{3}}{4}t \\ \frac{3}{2}e^{-\frac{1}{4}t}\cos\frac{\sqrt{3}}{4}t + \frac{3\sqrt{3}}{2}e^{-\frac{1}{4}t}\sin\frac{\sqrt{3}}{4}t + i\left(\frac{3}{2}e^{-\frac{1}{4}t}\sin\frac{\sqrt{3}}{4}t - \frac{3\sqrt{3}}{2}e^{-\frac{1}{4}t}\cos\frac{\sqrt{3}}{4}t\right) \end{bmatrix}$$

$$= \begin{bmatrix} e^{-\frac{1}{4}t}\cos\frac{\sqrt{3}}{4}t \\ \frac{3}{2}e^{-\frac{1}{4}t}\cos\frac{\sqrt{3}}{4}t + \frac{3\sqrt{3}}{2}e^{-\frac{1}{4}t}\sin\frac{\sqrt{3}}{4}t \end{bmatrix} + i\begin{bmatrix} e^{-\frac{1}{4}t}\sin\frac{\sqrt{3}}{4}t \\ \frac{3}{2}e^{-\frac{1}{4}t}\sin\frac{\sqrt{3}}{4}t - \frac{3\sqrt{3}}{2}e^{-\frac{1}{4}t}\cos\frac{\sqrt{3}}{4}t \end{bmatrix}$$

is a complex valued solution to $\begin{bmatrix} m' \\ w' \end{bmatrix} = A\begin{bmatrix} m \\ w \end{bmatrix}$. By Theorem 6.8

$$\begin{bmatrix} e^{-\frac{1}{4}t}\cos\frac{\sqrt{3}}{4}t \\ \frac{3}{2}e^{-\frac{1}{4}t}\cos\frac{\sqrt{3}}{4}t + \frac{3\sqrt{3}}{2}e^{-\frac{1}{4}t}\sin\frac{\sqrt{3}}{4}t \end{bmatrix} \quad \text{and} \quad \begin{bmatrix} e^{-\frac{1}{4}t}\sin\frac{\sqrt{3}}{4}t \\ \frac{3}{2}e^{-\frac{1}{4}t}\sin\frac{\sqrt{3}}{4}t - \frac{3\sqrt{3}}{2}e^{-\frac{1}{4}t}\cos\frac{\sqrt{3}}{4}t \end{bmatrix}$$

are real valued solutions to $\begin{bmatrix} m' \\ w' \end{bmatrix} = A\begin{bmatrix} m \\ w \end{bmatrix}$.

The general solution to $\begin{bmatrix} m' \\ w' \end{bmatrix} = A\begin{bmatrix} m \\ w \end{bmatrix}$ is then

$$c_1\begin{bmatrix} e^{-\frac{1}{4}t}\cos\frac{\sqrt{3}}{4}t \\ \frac{3}{2}e^{-\frac{1}{4}t}\cos\frac{\sqrt{3}}{4}t + \frac{3\sqrt{3}}{2}e^{-\frac{1}{4}t}\sin\frac{\sqrt{3}}{4}t \end{bmatrix} + c_2\begin{bmatrix} e^{-\frac{1}{4}t}\sin\frac{\sqrt{3}}{4}t \\ \frac{3}{2}e^{-\frac{1}{4}t}\sin\frac{\sqrt{3}}{4}t - \frac{3\sqrt{3}}{2}e^{-\frac{1}{4}t}\cos\frac{\sqrt{3}}{4}t \end{bmatrix} =$$

$$\begin{bmatrix} c_1 e^{-\frac{1}{4}t}\cos\frac{\sqrt{3}}{4}t + c_2 e^{-\frac{1}{4}t}\sin\frac{\sqrt{3}}{4}t \\ c_1\left(\frac{3}{2}e^{-\frac{1}{4}t}\cos\frac{\sqrt{3}}{4}t + \frac{3\sqrt{3}}{2}e^{-\frac{1}{4}t}\sin\frac{\sqrt{3}}{4}t\right) + c_2\left(\frac{3}{2}e^{-\frac{1}{4}t}\sin\frac{\sqrt{3}}{4}t - \frac{3\sqrt{3}}{2}e^{-\frac{1}{4}t}\cos\frac{\sqrt{3}}{4}t\right) \end{bmatrix}.$$

(d) i. The system for $(0,0)$ is unstable since

$$\lim_{t\to\infty} m(t) = \lim_{t\to\infty} c_1 e^{\frac{1}{2}t} = \infty$$

and

$$\lim_{t\to\infty} w(t) = \lim_{t\to\infty} c_2 = c_2.$$

ii. The system for $\left(\frac{2}{3}, 2\right)$ is stable since

$$\lim_{t\to\infty} u(t) = \lim_{t\to\infty} e^{-\frac{1}{4}t}\left(c_1\cos\frac{\sqrt{3}}{4}t + c_2\sin\frac{\sqrt{3}}{4}t\right) = 0$$

and

$$\lim_{t\to\infty} v(t) = \lim_{t\to\infty} \frac{1}{2}e^{-\frac{1}{4}t}\left(c_1\left(3\cos\frac{\sqrt{3}}{4}t + 3\sqrt{3}\sin\frac{\sqrt{3}}{4}t\right) + c_2\left(3\sin\frac{\sqrt{3}}{4}t - 3\sqrt{3}\cos\frac{\sqrt{3}}{4}t\right)\right) = 0.$$

11. We have the 2 × 2 system

$$m'(t) = m(t)\left(\frac{3}{4} - \frac{1}{4}w(t)\right)$$

$$w'(t) = w(t)\left(-\frac{1}{4}w(t) + \frac{1}{4}m(t)\right).$$

(a) There are two equilibrium solutions $(0,0)$ and $\left(\frac{\gamma\alpha}{\delta\beta}, \frac{\alpha}{\beta}\right) = (3,3)$.

(b) i. The linear part of the problem at the equilibrium $(0,0)$ is $A(0,0) = \begin{bmatrix} \frac{3}{4} & 0 \\ 0 & 0 \end{bmatrix}$. The eigenvalues of A are 0 and $\frac{3}{4}$.

ii. The linear part of the problem at the equilibrium $(3,3)$ is $A(3,3) = \begin{bmatrix} 0 & -\frac{3}{4} \\ \frac{3}{4} & -\frac{3}{4} \end{bmatrix}$.

(c) i. We first consider the equilibrium solution $(0,0)$. The general solution to the system is $m = c_1 e^{\frac{3}{4}t}$, $w = c_2$.

ii. We now consider the equilibrium solution $\left(\frac{2}{3}, 2\right)$. The characteristic equation for $A(3,3)$ is

$$\det(\lambda I - A) = \begin{vmatrix} \lambda & \frac{3}{4} \\ -\frac{3}{4} & \lambda + \frac{3}{4} \end{vmatrix} = \lambda^2 + \frac{3}{4}\lambda + \frac{9}{16} = 0.$$

Therefore the eigenvalues of A are $\lambda = \frac{-3 \pm 3\sqrt{3}i}{8}$. We now wish to find bases for the corresponding eigenspaces.

For $\lambda = \frac{-3+3\sqrt{3}i}{8}$ we need to find a basis for the null space of $\frac{-3+3\sqrt{3}i}{8}I - A = \begin{bmatrix} \frac{-3+3\sqrt{3}i}{8} & \frac{3}{4} \\ -\frac{3}{4} & \frac{3+3\sqrt{3}i}{8} \end{bmatrix}$. Reducing

the matrix for the associated homogeneous system,

$$\begin{bmatrix} \frac{-3+3\sqrt{3}i}{8} & \frac{3}{4} & \vdots & 0 \\ -\frac{3}{4} & \frac{3+3\sqrt{3}i}{8} & \vdots & 0 \end{bmatrix} \rightarrow \begin{bmatrix} \frac{1-\sqrt{3}i}{2} & -1 & \vdots & 0 \\ 0 & 0 & \vdots & 0 \end{bmatrix},$$

we see our solutions are

$$\begin{bmatrix} u \\ v \end{bmatrix} = \begin{bmatrix} u \\ \frac{1-\sqrt{3}i}{2}u \end{bmatrix} = u \begin{bmatrix} 1 \\ \frac{1-\sqrt{3}i}{2} \end{bmatrix}$$

and hence the column vector $\begin{bmatrix} 1 \\ \frac{1-\sqrt{3}i}{2} \end{bmatrix}$ forms a basis for $E_{\frac{-3+3\sqrt{3}i}{8}}$.

For $\lambda = \frac{-3-3\sqrt{3}i}{8}$ we need to find a basis for the null space of $\frac{-3-3\sqrt{3}i}{8}I - A = \begin{bmatrix} \frac{-3-3\sqrt{3}i}{8} & \frac{3}{4} \\ -\frac{3}{4} & \frac{3-3\sqrt{3}i}{8} \end{bmatrix}$. Reducing

the matrix for the associated homogeneous system,

$$\begin{bmatrix} \frac{-3-3\sqrt{3}i}{8} & \frac{3}{4} & \vdots & 0 \\ -\frac{3}{4} & \frac{3-3\sqrt{3}i}{8} & \vdots & 0 \end{bmatrix} \rightarrow \begin{bmatrix} \frac{1+\sqrt{3}i}{2} & -1 & \vdots & 0 \\ 0 & 0 & \vdots & 0 \end{bmatrix},$$

we see our solutions are

$$\begin{bmatrix} u \\ v \end{bmatrix} = \begin{bmatrix} u \\ \frac{1+\sqrt{3}i}{2}u \end{bmatrix} = u \begin{bmatrix} 1 \\ \frac{1+\sqrt{3}i}{2} \end{bmatrix}$$

and hence the column vector $\begin{bmatrix} 1 \\ \frac{1+\sqrt{3}i}{2} \end{bmatrix}$ forms a basis for $E_{\frac{-3-3\sqrt{3}i}{8}}$.

Matrix A is similar to the diagonal matrix $D = P^{-1}AP = \begin{bmatrix} \frac{-3+3\sqrt{3}i}{8} & 0 \\ 0 & \frac{-3-3\sqrt{3}i}{8} \end{bmatrix}$ where

$$P = \begin{bmatrix} 1 & 1 \\ \frac{1-\sqrt{3}i}{2} & \frac{1+\sqrt{3}i}{2} \end{bmatrix}.$$

The columns of P are the basis vectors for the eigenspaces.

One complex valued solution to $\begin{bmatrix} m' \\ w' \end{bmatrix} = D \begin{bmatrix} m \\ w \end{bmatrix}$ is

$$Z = \begin{bmatrix} e^{\frac{-3+3\sqrt{3}i}{8}t} \\ 0 \end{bmatrix} = \begin{bmatrix} e^{-\frac{3}{8}t} \cos \frac{3\sqrt{3}}{8}t + i e^{-\frac{3}{8}t} \sin \frac{3\sqrt{3}}{8}t \\ 0 \end{bmatrix}.$$

By the first part of Theorem 6.7,

$$
PZ = \begin{bmatrix} 1 & 1 \\ \frac{1-\sqrt{3}i}{2} & \frac{1+\sqrt{3}i}{2} \end{bmatrix} \begin{bmatrix} e^{-\frac{3}{8}t}\cos\frac{3\sqrt{3}}{8}t + ie^{-\frac{3}{8}t}\sin\frac{3\sqrt{3}}{8}t \\ 0 \end{bmatrix}
$$

$$
= \begin{bmatrix} e^{-\frac{3}{8}t}\cos\frac{3\sqrt{3}}{8}t + ie^{-\frac{3}{8}t}\sin\frac{3\sqrt{3}}{8}t \\ \frac{1}{2}e^{-\frac{3}{8}t}\left(\cos\frac{3\sqrt{3}}{8}t + \sqrt{3}\sin\frac{3\sqrt{3}}{8}t + i\left(\sin\frac{3\sqrt{3}}{8}t - \sqrt{3}\cos\frac{3\sqrt{3}}{8}t\right)\right) \end{bmatrix}
$$

$$
= \begin{bmatrix} e^{-\frac{3}{8}t}\cos\frac{3\sqrt{3}}{8}t \\ \frac{1}{2}e^{-\frac{3}{8}t}\left(\cos\frac{3\sqrt{3}}{8}t + \sqrt{3}\sin\frac{3\sqrt{3}}{8}t\right) \end{bmatrix} + i\begin{bmatrix} e^{-\frac{3}{8}t}\sin\frac{3\sqrt{3}}{8}t \\ \frac{1}{2}e^{-\frac{3}{8}t}\left(\sin\frac{3\sqrt{3}}{8}t - \sqrt{3}\cos\frac{3\sqrt{3}}{8}t\right) \end{bmatrix}
$$

is a complex valued solution to $\begin{bmatrix} m' \\ w' \end{bmatrix} = A\begin{bmatrix} m \\ w \end{bmatrix}$. By Theorem 6.8

$$
\begin{bmatrix} e^{-\frac{3}{8}t}\cos\frac{3\sqrt{3}}{8}t \\ \frac{1}{2}e^{-\frac{3}{8}t}\left(\cos\frac{3\sqrt{3}}{8}t + \sqrt{3}\sin\frac{3\sqrt{3}}{8}t\right) \end{bmatrix} \text{ and } \begin{bmatrix} e^{-\frac{3}{8}t}\sin\frac{3\sqrt{3}}{8}t \\ \frac{1}{2}e^{-\frac{3}{8}t}\left(\sin\frac{3\sqrt{3}}{8}t - \sqrt{3}\cos\frac{3\sqrt{3}}{8}t\right) \end{bmatrix}
$$

are real valued solutions to $\begin{bmatrix} m' \\ w' \end{bmatrix} = A\begin{bmatrix} m \\ w \end{bmatrix}$.

The general solution to $\begin{bmatrix} m' \\ w' \end{bmatrix} = A\begin{bmatrix} m \\ w \end{bmatrix}$ is then

$$
c_1 \begin{bmatrix} e^{-\frac{3}{8}t}\cos\frac{3\sqrt{3}}{8}t \\ \frac{1}{2}e^{-\frac{3}{8}t}\left(\cos\frac{3\sqrt{3}}{8}t + \sqrt{3}\sin\frac{3\sqrt{3}}{8}t\right) \end{bmatrix} + c_2 \begin{bmatrix} e^{-\frac{3}{8}t}\sin\frac{3\sqrt{3}}{8}t \\ \frac{1}{2}e^{-\frac{3}{8}t}\left(\sin\frac{3\sqrt{3}}{8}t - \sqrt{3}\cos\frac{3\sqrt{3}}{8}t\right) \end{bmatrix} =
$$

$$
\begin{bmatrix} c_1 e^{-\frac{3}{8}t}\cos\frac{3\sqrt{3}}{8}t + c_2 e^{-\frac{3}{8}t}\sin\frac{3\sqrt{3}}{8}t \\ \frac{1}{2}e^{-\frac{3}{8}t}\left(c_1\left(\cos\frac{3\sqrt{3}}{8}t + \sqrt{3}\sin\frac{3\sqrt{3}}{8}t\right) + c_2\left(\sin\frac{3\sqrt{3}}{8}t - \sqrt{3}\cos\frac{3\sqrt{3}}{8}t\right)\right) \end{bmatrix}.
$$

(d) $\quad$ i. The system for $(0,0)$ is unstable since

$$
\lim_{t\to\infty} m(t) = \lim_{t\to\infty} c_1 e^{\frac{3}{4}t} = \infty
$$

and

$$
\lim_{t\to\infty} w(t) = \lim_{t\to\infty} c_2 = c_2.
$$

ii. The system for $(3,3)$ is stable since

$$
\lim_{t\to\infty} u(t) = \lim_{t\to\infty} e^{-\frac{3}{8}t}\left(c_1\cos\frac{3\sqrt{3}}{8}t + c_2\sin\frac{3\sqrt{3}}{8}t\right) = 0
$$

and

$$
\lim_{t\to\infty} v(t) = \lim_{t\to\infty} \frac{1}{2}e^{-\frac{3}{8}t}\left(c_1\left(\cos\frac{3\sqrt{3}}{8}t + \sqrt{3}\sin\frac{3\sqrt{3}}{8}t\right) + c_2\left(\sin\frac{3\sqrt{3}}{8}t - \sqrt{3}\cos\frac{3\sqrt{3}}{8}t\right)\right) = 0.
$$

13. We have the system $x' = y$, $y' = -\frac{g}{l}\sin x$. Thus,

$$
\frac{dy}{dx} = \frac{dy/dt}{dx/dt} = \frac{y'}{x'} = \frac{-\frac{g}{l}\sin x}{y}.
$$

This is a separable differential equation. We have

$$
y\,dy = -\frac{g}{l}\sin x\,dx.
$$

Integrating each side of the equation we obtain $\frac{1}{2}y^2 = \frac{g}{l}\cos x + C$ or equivalently,

$$
y^2 - \frac{2g}{l}\cos x = c.
$$

Using the initial conditions $x(0) = \frac{\pi}{4}$, $y(0) = 0$ with $g = 32$ ft/sec^2 and $l = \frac{1}{2}$ foot, we have $0 - \frac{2 \cdot 32}{\frac{1}{2}} \cos \frac{\pi}{4} = c$. Thus, $c = -64\sqrt{2}$.

15. We have the differential equation $ml\theta'' + c\theta' + mg\sin\theta = 0$. We let $x = \theta$ and $y = \theta'$ thus we have the system

$$x' = \theta' = y, \; y' = \theta'' = -\frac{c}{ml}\theta' - \frac{mg}{ml}\sin\theta$$

or more simply,

$$x' = y, \; y' = -\frac{g}{l}\sin x - \frac{c}{ml}y.$$

We let

$$F(x,y) = y \text{ and } G(x,y) = -\frac{g}{l}\sin x - \frac{c}{ml}y.$$

We see that $F(x,y) = 0$ if $y = 0$, and that when $y = 0$, $G(x,y) = 0$ if $x = n\pi$ for n any integer. We can now see that $(n\pi, 0)$ are the equilibrium solutions for n any integer. To obtain our Taylor polynomials, we note that

$$F_x(x,y) = 0, \; F_y(x,y) = 1, \; G_x(x,y) = -\frac{g}{l}\cos x, \; G_y(x,y) = -\frac{c}{ml}.$$

For $(n\pi, 0)$ our first degree Taylor polynomial representation is

$$F(x,y) = F(n\pi, 0) + F_x(n\pi, 0)(x - n\pi) + F_y(n\pi, 0)(y - 0) = y$$

and

$$G(x,y) = G(n\pi, 0) + G_x(n\pi, 0)(x - n\pi) + G_y(n\pi, 0)(y - 0) = \pm\frac{g}{l}(x - n\pi) - \frac{c}{ml}y.$$

where the sign of the first term in the second equation depends upon the parity of n. Ignoring $R_F(x,y)$ and $R_G(x,y)$ and using the substitutions $u = x - n\pi$ and $v = y$, we obtain the linear system:

$$u' = v, \; v' = -\frac{g}{l}u - \frac{c}{ml}v$$

if $n = 2k$ and

$$u' = v, \; v' = \frac{g}{l}u - \frac{c}{ml}v$$

if $n = 2k + 1$.

For $(2k\pi, 0)$ we have the linear system

$$A(2k\pi, 0) = \begin{bmatrix} 0 & 1 \\ -\frac{g}{l} & -\frac{c}{ml} \end{bmatrix}.$$

For $((2k+1)\pi, 0)$ we have the linear system:

$$A((2k+1)\pi, 0) = \begin{bmatrix} 0 & 1 \\ \frac{g}{l} & -\frac{c}{ml} \end{bmatrix}.$$

17. We have the 2×2 system

$$x' = x\left(-\frac{3}{4} + \frac{1}{4}y\right)$$

$$y' = y\left(1 - \frac{1}{4}x\right).$$

We let

$$F(x,y) = x\left(-\frac{3}{4} + \frac{1}{4}y\right) \text{ and } G(x,y) = y\left(1 - \frac{1}{4}x\right).$$

We see that $F(x,y) = 0$ if $x = 0$ or $y = 3$, and that $G(x,y) = 0$ if $y = 0$ or $x = 4$. We can now see that $(0,0)$ and $(4,3)$ are the equilibrium solutions.

To obtain our Taylor polynomials, we note that

$$F_x(x,y) = -\frac{3}{4} + \frac{1}{4}y, \; F_y(x,y) = \frac{1}{4}x, \; G_x(x,y) = -\frac{1}{4}y, \; G_y(x,y) = 1 - \frac{1}{4}x.$$

We first consider the equilibrium solution $(0,0)$. At $(0,0)$ our first degree Taylor polynomial representation is

$$F(x,y) = F(0,0) + F_x(0,0)(x-0) + F_y(0,0)(y-0) = -\frac{3}{4}x$$

and

$$G(x,y) = G(0,0) + G_x(0,0)(x-0) + G_y(0,0)(y-0) = y.$$

Ignoring $R_F(x,y)$ and $R_G(x,y)$ we obtain the linear system:

$$x' = -\frac{3}{4}x, \ y' = y.$$

The solution to this system is $x = c_1 e^{-\frac{3}{4}t}, \ y = c_2 e^t$.

The system for $(0,0)$ is unstable since

$$\lim_{t\to\infty} x(t) = \lim_{t\to\infty} c_1 e^{-\frac{3}{4}t} = 0$$

and

$$\lim_{t\to\infty} y(t) = \lim_{t\to\infty} c_2 e^t = \infty.$$

We now consider the equilibrium solution $(4,3)$. At $(4,3)$ our first degree Taylor polynomial representation is

$$F(x,y) = F(4,3) + F_x(4,3)(x-4) + F_y(4,3)(y-3) = y-3$$

and

$$G(x,y) = G(4,3) + G_x(4,3)(x-4) + G_y(4,3)(y-3) = -\frac{3}{4}(x-4).$$

Again ignoring $R_F(x,y)$ and $R_G(x,y)$ and now using the substitutions $u = x-4$ and $v = y-3$, we obtain the linear system:

$$u' = v, \ v' = -\frac{3}{4}u.$$

We let $A = \begin{bmatrix} 0 & 1 \\ -\frac{3}{4} & 0 \end{bmatrix}$. The characteristic equation for A is

$$\det(\lambda I - A) = \begin{vmatrix} \lambda & -1 \\ \frac{3}{4} & \lambda \end{vmatrix} = \lambda^2 + \frac{3}{4} = 0.$$

Therefore the eigenvalues of A are $\lambda = \pm\frac{\sqrt{3}}{2}i$. We now wish to find bases for the corresponding eigenspaces.

For $\lambda = \frac{\sqrt{3}}{2}i$ we need to find a basis for the null space of $\frac{\sqrt{3}}{2}iI - A = \begin{bmatrix} \frac{\sqrt{3}}{2}i & -1 \\ \frac{3}{4} & \frac{\sqrt{3}}{2}i \end{bmatrix}$. Reducing the matrix for the associated

homogeneous system, $\begin{bmatrix} \frac{\sqrt{3}}{2}i & -1 & \vdots & 0 \\ \frac{3}{4} & \frac{\sqrt{3}}{2}i & \vdots & 0 \end{bmatrix} \rightarrow \begin{bmatrix} \frac{\sqrt{3}}{2}i & -1 & \vdots & 0 \\ 0 & 0 & \vdots & 0 \end{bmatrix}$, we see our solutions are

$$\begin{bmatrix} u \\ v \end{bmatrix} = \begin{bmatrix} u \\ \frac{\sqrt{3}}{2}iu \end{bmatrix} = u\begin{bmatrix} 1 \\ \frac{\sqrt{3}}{2}i \end{bmatrix}$$

and hence the column vector $\begin{bmatrix} 1 \\ \frac{\sqrt{3}}{2}i \end{bmatrix}$ forms a basis for $E_{\frac{\sqrt{3}}{2}i}$.

For $\lambda = -\frac{\sqrt{3}}{2}i$ we need to find a basis for the null space of $-\frac{\sqrt{3}}{2}iI - A = \begin{bmatrix} -\frac{\sqrt{3}}{2}i & -1 \\ \frac{3}{4} & -\frac{\sqrt{3}}{2}i \end{bmatrix}$. Reducing the matrix for the

associated homogeneous system, $\begin{bmatrix} -\frac{\sqrt{3}}{2}i & -1 & \vdots & 0 \\ \frac{3}{4} & -\frac{\sqrt{3}}{2}i & \vdots & 0 \end{bmatrix} \rightarrow \begin{bmatrix} \frac{\sqrt{3}}{2}i & 1 & \vdots & 0 \\ 0 & 0 & \vdots & 0 \end{bmatrix}$, we see our solutions are

$$\begin{bmatrix} u \\ v \end{bmatrix} = \begin{bmatrix} u \\ -\frac{\sqrt{3}}{2}iu \end{bmatrix} = u\begin{bmatrix} 1 \\ -\frac{\sqrt{3}}{2}i \end{bmatrix}$$

and hence the column vector $\begin{bmatrix} 1 \\ -\frac{\sqrt{3}}{2}i \end{bmatrix}$ forms a basis for $E_{-\frac{\sqrt{3}}{2}i}$.

Matrix A is similar to the diagonal matrix $D = P^{-1}AP = \begin{bmatrix} \frac{\sqrt{3}}{2}i & 0 \\ 0 & -\frac{\sqrt{3}}{2}i \end{bmatrix}$ where $P = \begin{bmatrix} 1 & 1 \\ \frac{\sqrt{3}}{2}i & -\frac{\sqrt{3}}{2}i \end{bmatrix}$. The columns of P are the basis vectors for the eigenspaces.

One complex valued solution to $\begin{bmatrix} u' \\ v' \end{bmatrix} = D \begin{bmatrix} u \\ v \end{bmatrix}$ is

$$Z = \begin{bmatrix} e^{\frac{\sqrt{3}}{2}it} \\ 0 \end{bmatrix} = \begin{bmatrix} \cos\frac{\sqrt{3}}{2}t + i\sin\frac{\sqrt{3}}{2}t \\ 0 \end{bmatrix}.$$

By the first part of Theorem 6.7,

$$\begin{aligned}
PZ &= \begin{bmatrix} 1 & 1 \\ \frac{\sqrt{3}}{2}i & -\frac{\sqrt{3}}{2}i \end{bmatrix} \begin{bmatrix} \cos\frac{\sqrt{3}}{2}t + i\sin\frac{\sqrt{3}}{2}t \\ 0 \end{bmatrix} \\
&= \begin{bmatrix} \cos\frac{\sqrt{3}}{2}t + i\sin\frac{\sqrt{3}}{2}t \\ -\frac{\sqrt{3}}{2}\sin\frac{\sqrt{3}}{2}t + \frac{\sqrt{3}}{2}i\cos\frac{\sqrt{3}}{2}t \end{bmatrix} \\
&= \begin{bmatrix} \cos\frac{\sqrt{3}}{2}t \\ -\frac{\sqrt{3}}{2}\sin\frac{\sqrt{3}}{2}t \end{bmatrix} + i \begin{bmatrix} \sin\frac{\sqrt{3}}{2}t \\ \frac{\sqrt{3}}{2}\cos\frac{\sqrt{3}}{2}t \end{bmatrix}
\end{aligned}$$

is a complex valued solution to $\begin{bmatrix} u' \\ v' \end{bmatrix} = A \begin{bmatrix} u \\ v \end{bmatrix}$. By Theorem 6.8

$$\begin{bmatrix} \cos\frac{\sqrt{3}}{2}t \\ -\frac{\sqrt{3}}{2}\sin\frac{\sqrt{3}}{2}t \end{bmatrix} \quad \text{and} \quad \begin{bmatrix} \sin\frac{\sqrt{3}}{2}t \\ \frac{\sqrt{3}}{2}\cos\frac{\sqrt{3}}{2}t \end{bmatrix}$$

are real valued solutions to $\begin{bmatrix} u' \\ v' \end{bmatrix} = A \begin{bmatrix} u \\ v \end{bmatrix}$.

The general solution to $\begin{bmatrix} u' \\ v' \end{bmatrix} = A \begin{bmatrix} u \\ v \end{bmatrix}$ is then

$$c_1 \begin{bmatrix} \cos\frac{\sqrt{3}}{2}t \\ -\frac{\sqrt{3}}{2}\sin\frac{\sqrt{3}}{2}t \end{bmatrix} + c_2 \begin{bmatrix} \sin\frac{\sqrt{3}}{2}t \\ \frac{\sqrt{3}}{2}\cos\frac{\sqrt{3}}{2}t \end{bmatrix} = \begin{bmatrix} c_1\cos\frac{\sqrt{3}}{2}t + c_2\sin\frac{\sqrt{3}}{2}t \\ -\frac{\sqrt{3}}{2}c_1\sin\frac{\sqrt{3}}{2}t + \frac{\sqrt{3}}{2}c_2\cos\frac{\sqrt{3}}{2}t \end{bmatrix}.$$

The system for $(4, 3)$ is unstable since neither of the limits

$$\lim_{t\to\infty} u(t) = \lim_{t\to\infty} \left(c_1\cos\frac{\sqrt{3}}{2}t + c_2\sin\frac{\sqrt{3}}{2}t \right)$$

nor

$$\lim_{t\to\infty} v(t) = \lim_{t\to\infty} \frac{\sqrt{3}}{2} \left(c_1\sin\frac{\sqrt{3}}{2}t + c_2\cos\frac{\sqrt{3}}{2}t \right)$$

exists.

We also wish to solve the differential equation

$$\frac{dy}{dx} = \frac{dy/dt}{dx/dt} = \frac{y'}{x'} = \frac{x\left(-\frac{3}{4} + \frac{1}{4}y\right)}{y\left(1 - \frac{1}{4}x\right)}.$$

This is a separable equation. After separating the variables we have

$$\left(\frac{1}{4} - \frac{3}{4}\frac{1}{y}\right) dy = \left(\frac{1}{x} - \frac{1}{4}\right) dx.$$

Integrating each side we obtain

$$\frac{1}{4}y - \frac{3}{4}\ln y = \ln x - \frac{1}{4}x + c.$$

Note, since x and y are both positive, absolute values are not necessary in the logarithms above.

19. We have the 2×2 system

$$x' = \left(1 - \frac{1}{4}x - \frac{1}{2}y\right)x$$

$$y' = \left(1 - \frac{1}{2}x - \frac{1}{4}y\right)y.$$

We let

$$F(x,y) = \left(1 - \frac{1}{4}x - \frac{1}{2}y\right)x \text{ and } G(x,y) = \left(1 - \frac{1}{2}x - \frac{1}{4}y\right)y.$$

We see that $F(x,y) = 0$ if $x = 0$ or $1 - \frac{1}{4}x - \frac{1}{2}y = 0$, and that $G(x,y) = 0$ if $y = 0$ or $1 - \frac{1}{2}x - \frac{1}{4}y$. We can now see that $(0,0)$, $(4,0)$, $(0,4)$, and $\left(\frac{4}{3},\frac{4}{3}\right)$ are the equilibrium solutions.

To obtain our Taylor polynomials, we note that

$$F_x(x,y) = 1 - \frac{1}{2}x - \frac{1}{2}y, \ F_y(x,y) = -\frac{1}{2}x, \ G_x(x,y) = -\frac{1}{2}y, \ G_y(x,y) = 1 - \frac{1}{2}x - \frac{1}{2}y.$$

We first consider the equilibrium solution $(0,0)$. At $(0,0)$ our first degree Taylor polynomial representation is

$$F(x,y) = F(0,0) + F_x(0,0)(x-0) + F_y(0,0)(y-0) = x$$

and

$$G(x,y) = G(0,0) + G_x(0,0)(x-0) + G_y(0,0)(y-0) = y.$$

Ignoring $R_F(x,y)$ and $R_G(x,y)$ we obtain the linear system:

$$x' = x, \ y' = y.$$

The solution to this system is $x = c_1 e^t$, $y = c_2 e^t$.

The system for $(0,0)$ is unstable since

$$\lim_{t\to\infty} x(t) = \lim_{t\to\infty} c_1 e^t = \infty$$

and

$$\lim_{t\to\infty} y(t) = \lim_{t\to\infty} c_2 e^t = \infty.$$

We now consider the equilibrium solution $(4,0)$. At $(4,0)$ our first degree Taylor polynomial representation is

$$F(x,y) = F(4,0) + F_x(4,0)(x-4) + F_y(4,0)(y-0) = -(x-4) - 2y$$

and

$$G(x,y) = G(4,0) + G_x(4,0)(x-4) + G_y(4,0)(y-0) = -y.$$

Again ignoring $R_F(x,y)$ and $R_G(x,y)$ and now using the substitutions $u = x - 4$ and $v = y$, we obtain the linear system:

$$u' = -u - 2v, \ v' = -v.$$

We let $A = \begin{bmatrix} -1 & -2 \\ 0 & -1 \end{bmatrix}$. The second equation $v' = -v$ has the general solution $v = c_2 e^{-t}$. Substituting this into the first equation we have $u' = -u - 2c_2 e^{-t}$. The related homogeneous equation $u' + u = 0$ has the solution $u_H = c_1 e^{-t}$. We may then use the method of undetermined coefficients to find the particular solution of the form $u_P = Ate^{-t}$. Here, $u_P' = Ae^{-t} - Ate^{-t}$. Substituting these into the equation $u' = -u - 2c_2 e^{-t}$ we obtain

$$Ae^{-t} - Ate^{-t} = -(Ate^{-t}) - 2c_2 e^{-t}.$$

Simplifying this we see that $Ae^{-t} = -2c_2e^{-t}$. Thus $A = -2c_2$. Adding the particular solution to the general solution of the related homogeneous equation we have $u = c_1e^{-t} - 2c_2te^{-t}$. Thus the general solution to the linear system at $(4,0)$ is

$$u = c_1e^{-t} - 2c_2te^{-t}, \ v = c_2e^{-t}.$$

The system for $(4,0)$ is stable since

$$\lim_{t\to\infty} u(t) = \lim_{t\to\infty} c_1e^{-t} - 2c_2te^{-t} = 0$$

and

$$\lim_{t\to\infty} v(t) = \lim_{t\to\infty} c_2e^{-t} = 0.$$

We now consider the equilibrium solution $(0,4)$. At $(0,4)$ our first degree Taylor polynomial representation is

$$F(x,y) = F(0,4) + F_x(0,4)(x-0) + F_y(0,4)(y-4) = -x$$

and

$$G(x,y) = G(0,4) + G_x(0,4)(x-0) + G_y(0,4)(y-4) = -2x - (y-4).$$

Again ignoring $R_F(x,y)$ and $R_G(x,y)$ and now using the substitutions $u = x$ and $v = y - 4$, we obtain the linear system:

$$u' = -u, \ v' = -2u - v.$$

We let $A = \begin{bmatrix} -1 & 0 \\ -2 & -1 \end{bmatrix}$. Please note that this system is virtually identical to the system we solved above, with the roles of the variables u and v reversed. Therefore, the solution to the system at $(0,4)$ is

$$u = c_1e^{-t}, \ v = -2c_1e^{-t} + c_2e^{-t}.$$

The system for $(0,4)$ is stable since

$$\lim_{t\to\infty} x(t) = \lim_{t\to\infty} c_1e^{-t} = 0$$

and

$$\lim_{t\to\infty} y(t) = \lim_{t\to\infty} -2c_1e^{-t} + c_2e^{-t} = 0.$$

Finally we consider the equilibrium solution $\left(\frac{4}{3}, \frac{4}{3}\right)$. At $\left(\frac{4}{3}, \frac{4}{3}\right)$ our first degree Taylor polynomial representation is

$$F(x,y) = F\left(\frac{4}{3}, \frac{4}{3}\right) + F_x\left(\frac{4}{3}, \frac{4}{3}\right)\left(x - \frac{4}{3}\right) + F_y\left(\frac{4}{3}, \frac{4}{3}\right)\left(y - \frac{4}{3}\right) = -\frac{1}{3}\left(x - \frac{4}{3}\right) - \frac{2}{3}\left(y - \frac{4}{3}\right)$$

and

$$G(x,y) = G\left(\frac{4}{3}, \frac{4}{3}\right) + G_x\left(\frac{4}{3}, \frac{4}{3}\right)\left(x - \frac{4}{3}\right) + G_y\left(\frac{4}{3}, \frac{4}{3}\right)\left(y - \frac{4}{3}\right) = -\frac{2}{3}\left(x - \frac{4}{3}\right) - \frac{1}{3}\left(y - \frac{4}{3}\right).$$

Again ignoring $R_F(x,y)$ and $R_G(x,y)$ and now using the substitutions $u = x - \frac{4}{3}$ and $v = y - \frac{4}{3}$, we obtain the linear system:

$$u' = -\frac{1}{3}u - \frac{2}{3}v, \ v' = -\frac{2}{3}u - \frac{1}{3}v.$$

We let $A = \begin{bmatrix} -\frac{1}{3} & -\frac{2}{3} \\ -\frac{2}{3} & -\frac{1}{3} \end{bmatrix}$. The characteristic equation for A is

$$\det(\lambda I - A) = \begin{vmatrix} \lambda + \frac{1}{3} & \frac{2}{3} \\ \frac{2}{3} & \lambda + \frac{1}{3} \end{vmatrix} = \left(\lambda + \frac{1}{3}\right)^2 - \frac{4}{9} = \lambda^2 + \frac{2}{3}\lambda - \frac{1}{3} = (\lambda + 1)\left(\lambda - \frac{1}{3}\right) = 0.$$

Therefore the eigenvalues of A are $\lambda = -1$ and $\lambda = \frac{1}{3}$. We now wish to find bases for the corresponding eigenspaces.

For $\lambda = -1$ we need to find a basis for the null space of $-I - A = \begin{bmatrix} -\frac{2}{3} & \frac{2}{3} \\ \frac{2}{3} & -\frac{2}{3} \end{bmatrix}$. Reducing the matrix for the associated

homogeneous system, $\begin{bmatrix} -\frac{2}{3} & \frac{2}{3} & \vdots & 0 \\ \frac{2}{3} & -\frac{2}{3} & \vdots & 0 \end{bmatrix} \rightarrow \begin{bmatrix} 1 & -1 & \vdots & 0 \\ 0 & 0 & \vdots & 0 \end{bmatrix}$, we see our solutions are

$$\begin{bmatrix} u \\ v \end{bmatrix} = \begin{bmatrix} u \\ u \end{bmatrix} = u \begin{bmatrix} 1 \\ 1 \end{bmatrix}$$

and hence the column vector $\begin{bmatrix} 1 \\ 1 \end{bmatrix}$ forms a basis for E_{-1}.

For $\lambda = \frac{1}{3}$ we need to find a basis for the null space of $\frac{1}{3}I - A = \begin{bmatrix} \frac{2}{3} & \frac{2}{3} \\ \frac{2}{3} & \frac{2}{3} \end{bmatrix}$. Reducing the matrix for the associated homo-

geneous system, $\begin{bmatrix} \frac{2}{3} & \frac{2}{3} & \vdots & 0 \\ \frac{2}{3} & \frac{2}{3} & \vdots & 0 \end{bmatrix} \rightarrow \begin{bmatrix} 1 & 1 & \vdots & 0 \\ 0 & 0 & \vdots & 0 \end{bmatrix}$, we see our solutions are

$$\begin{bmatrix} u \\ v \end{bmatrix} = \begin{bmatrix} u \\ -u \end{bmatrix} = u \begin{bmatrix} 1 \\ -1 \end{bmatrix}$$

and hence the column vector $\begin{bmatrix} 1 \\ -1 \end{bmatrix}$ forms a basis for $E_{\frac{1}{3}}$.

Matrix A is similar to the diagonal matrix $D = P^{-1}AP = \begin{bmatrix} -1 & 0 \\ 0 & \frac{1}{3} \end{bmatrix}$ where $P = \begin{bmatrix} 1 & 1 \\ 1 & -1 \end{bmatrix}$. The columns of P are the

basis vectors for the eigenspaces.

We also know that $\begin{bmatrix} c_1 e^{-t} \\ c_2 e^{\frac{1}{3}t} \end{bmatrix}$ is the general solution to $\begin{bmatrix} u' \\ v' \end{bmatrix} = D \begin{bmatrix} u \\ v \end{bmatrix}$. Therefore the general solution to

$$\begin{bmatrix} u' \\ v' \end{bmatrix} = A \begin{bmatrix} u \\ v \end{bmatrix}$$

is

$$\begin{bmatrix} u \\ v \end{bmatrix} = P \begin{bmatrix} c_1 e^{-t} \\ c_2 e^{\frac{1}{3}t} \end{bmatrix} = \begin{bmatrix} 1 & 1 \\ 1 & -1 \end{bmatrix} \begin{bmatrix} c_1 e^{-t} \\ c_2 e^{\frac{1}{3}t} \end{bmatrix} = \begin{bmatrix} c_1 e^{-t} + c_2 e^{\frac{1}{3}t} \\ c_1 e^{-t} - c_2 e^{\frac{1}{3}t} \end{bmatrix}.$$

The system for $\left(\frac{4}{3}, \frac{4}{3}\right)$ is unstable since neither of the limits

$$\lim_{t \to \infty} u(t) = \lim_{t \to \infty} \left(c_1 e^{-t} + c_2 e^{\frac{1}{3}t} \right)$$

nor

$$\lim_{t \to \infty} v(t) = \lim_{t \to \infty} \left(c_1 e^{-t} - c_2 e^{\frac{1}{3}t} \right)$$

is finite.

Chapter 7

The Laplace Transform

7.1 Definition and Properties of the Laplace Transform

Exercises 7.1, pp. 351-352

1.
$$L(4) = \int_0^\infty e^{-st} 4 \, dt = -\lim_{b \to \infty} \left(\frac{4}{s} e^{-st} \Big|_0^b \right) = -\frac{4}{s} \lim_{b \to \infty} \left(e^{-sb} - 1 \right) = \frac{4}{s}.$$

3.
$$
\begin{aligned}
L(2t^2) &= \int_0^\infty e^{-st}(2t^2) \, dt = -2 \lim_{b \to \infty} \left(\left(\frac{t^2 e^{-st}}{s} + \frac{2t e^{-st}}{s^2} + \frac{2 e^{-st}}{s^3} \right) \Big|_0^b \right) \\
&= -2 \lim_{b \to \infty} \left(\frac{b^2 e^{-sb}}{s} + \frac{2b e^{-sb}}{s^2} + \frac{2 e^{-sb}}{s^3} - \frac{2}{s^3} \right) = \frac{4}{s^3}.
\end{aligned}
$$

5.
$$
\begin{aligned}
L(e^{-4t}) &= \int_0^\infty e^{-st} e^{-4t} \, dt = \lim_{b \to \infty} \int_0^\infty e^{-(s+4)t} \, dt = -\frac{1}{s+4} \lim_{b \to \infty} \left(e^{-(s+4)t} \Big|_0^b \right) \\
&= -\frac{1}{s+4} \lim_{b \to \infty} \left(e^{-(s+4)b} - 1 \right) = \frac{1}{s+4}.
\end{aligned}
$$

$$\left(\frac{f}{g} \right)' = \frac{f'g - fg'}{g^2}$$

7. $L(\cos 3t) = \dfrac{s}{s^2 + 9}.$

9. By property (3) of Theorem 7.1

$$L(t \sin t) = -\frac{d}{ds} \left(L(\sin t) \right) = -\frac{d}{ds} \left(\frac{1}{s^2 + 1} \right) = \frac{2s}{s^2 + 1}$$

wrong
should be squared

11. By property (3) of Theorem 7.1

$$L(t^2 e^{3t}) = -\frac{d}{ds} \left(L(t e^{3t}) \right) = \frac{d^2}{ds^2} \left(L(e^{3t}) \right) = \frac{d^2}{ds^2} \left(\frac{1}{s-3} \right) = \frac{d}{ds} \left(-\frac{1}{(s-3)^2} \right) = \frac{2}{(s-3)^3}.$$

13. Let $F(s) = L(\sin 3t) = \dfrac{3}{s^2 + 9}$, then by property (1) of Theorem 7.1, $L(e^{2t} \sin 3t) = F(s-2) = \dfrac{3}{(s-2)^2 + 9}.$

15. Let $F(s) = L(\cos bt) = \dfrac{s}{s^2 + b^2}$, then by property (1) of Theorem 7.1, $L(e^{at} \cos bt) = F(s-a) = \dfrac{s-a}{(s-a)^2 + b^2}.$

17. Since $\cosh 3t = \dfrac{e^{3t} + e^{-3t}}{2}$, we have

$$\mathcal{L}(\cosh 3t) = \frac{1}{2}\left(\mathcal{L}(e^{3t}) + \mathcal{L}(e^{-3t})\right) = \frac{1}{2}\left(\frac{1}{s-3} + \frac{1}{s+3}\right) = \frac{1}{2}\left(\frac{2s}{s^2-9}\right) = \frac{s}{s^2-9}.$$

19. By property (3) of Theorem 7.1

$$\mathcal{L}(2t\sinh t) = -2\frac{d}{ds}\left(\mathcal{L}(\sinh t)\right) = -2\frac{d}{ds}\left(\mathcal{L}(\frac{e^t - e^{-t}}{2})\right) = -\frac{d}{ds}\left(\frac{1}{s-1} - \frac{1}{s+1}\right) = -\frac{d}{ds}\left(\frac{2}{s^2-1}\right) = \frac{4s}{(s^2-1)^2}.$$

21. First

$$\mathcal{L}(\cosh bt) = \frac{1}{2}\left(\mathcal{L}(e^{bt}) + \mathcal{L}(e^{-bt})\right) = \frac{1}{2}\left(\frac{1}{s-b} + \frac{1}{s+b}\right) = \frac{1}{2}\left(\frac{2s}{s^2-b^2}\right) = \frac{s}{s^2-b^2}.$$

Now let $F(s) = \mathcal{L}(\cosh bt) = \dfrac{s}{s^2 - b^2}$, then by property (1) of Theorem 7.1,

$$\mathcal{L}(e^{at}\cosh bt) = F(s-a) = \frac{s-a}{(s-a)^2 - b^2}.$$

25. First $4t^3$ is continuous on $[0,\infty)$. In addition, $\lim\limits_{t\to\infty}\dfrac{4t^3}{e^t} = \lim\limits_{t\to\infty}\dfrac{12t^2}{e^t} = \lim\limits_{t\to\infty}\dfrac{24t}{e^t} = \lim\limits_{t\to\infty}\dfrac{24}{e^t} = 0$ by 3 applications of l'Hôpital's rule, so $4t^3$ is exponentially bounded on $[0,\infty)$.

27. First $t^2\sin 3t$ is continuous on $[0,\infty)$. In addition, $-\dfrac{t^2}{e^t} \le \dfrac{t^2\sin 3t}{e^t} \le \dfrac{t^2}{e^t}$ and each of the limits $\lim\limits_{t\to\infty}\left(\pm\dfrac{t^2}{e^t}\right) = 0$ by l'Hôpital's rule. Therefore by the Squeeze Theorem $\lim\limits_{t\to\infty}\dfrac{t^2\sin 3t}{e^t} = 0$. So $t^2\sin 3t$ is exponentially bounded on $[0,\infty)$.

29. First, $t^2\sinh 3t = \dfrac{t^2}{2}(e^{3t} - e^{-3t})$ is continuous on $[0,\infty)$. In addition, $\lim\limits_{t\to\infty}\dfrac{\frac{t^2}{2}(e^{3t} - e^{-3t})}{e^{4t}} = \lim\limits_{t\to\infty}\left(\dfrac{t^2}{2e^t} - \dfrac{t^2}{2e^{7t}}\right) = 0$ by 2 applications of l'Hôpital's rule, so $t^2\sinh 3t$ is exponentially bounded on $[0,\infty)$.

31. Any function which is continuous on $[0,\infty)$ which also has a finite limit as $t\to\infty$ is exponentially bounded on $[0,\infty)$. That is the case here since $\lim\limits_{t\to\infty}2t^3e^{-7t} = \lim\limits_{t\to\infty}\dfrac{2t^3}{e^{7t}} = 0$ by 3 applications of l'Hôpital's rule, so $t^2\sinh 3t$ is exponentially bounded on $[0,\infty)$.

33. First $t^n e^{at}\cos bt$ is continuous on $[0,\infty)$. In addition,

$$-\frac{t^n e^{at}}{e^{(a+1)t}} \le \frac{t^n e^{at}\cos bt}{e^{(a+1)t}} \le \frac{t^n e^{at}}{e^{(a+1)t}}.$$

Since $\pm\dfrac{t^n e^{at}}{e^{(a+1)t}} = \pm\dfrac{t^n}{e^t}$ and each of the limits $\lim\limits_{t\to\infty}\left(\pm\dfrac{t^n}{e^t}\right) = 0$ with n applications of l'Hôpital's rule, by the Squeeze Theorem $\lim\limits_{t\to\infty}\dfrac{t^n e^{at}\cos bt}{e^t} = 0$. So $t^n e^{at}\cos bt$ is exponentially bounded on $[0,\infty)$.

35. $\mathcal{L}^{-1}\left(\dfrac{4}{s}\right) = 4\mathcal{L}^{-1}\left(\dfrac{1}{s}\right) = 4.$

37. $\mathcal{L}^{-1}\left(\dfrac{4}{s-6}\right) = 4\mathcal{L}^{-1}\left(\dfrac{1}{s-6}\right) = 4e^{6t}.$

39. We first find the partial fraction decomposition $\dfrac{4}{s^2+2s-8} = \dfrac{2}{3}\dfrac{1}{s-2} - \dfrac{2}{3}\dfrac{1}{s+4}$. Therefore

$$\mathcal{L}^{-1}\left(\frac{4}{s^2+2s-8}\right) = \frac{2}{3}\left(\mathcal{L}^{-1}\left(\frac{1}{s-2}\right) - \mathcal{L}^{-1}\left(\frac{1}{s+4}\right)\right) = \frac{2}{3}\left(e^{2t} - e^{-4t}\right).$$

41. $\mathcal{L}^{-1}\left(\dfrac{2s}{s^2+9}\right) = 2\mathcal{L}^{-1}\left(\dfrac{s}{s^2+9}\right) = 2\cos 3t.$

7.2   Solving Constant Coefficient Linear Initial Value Problems with Laplace Transforms

Exercises 7.2, pp. 355-356

1. We have the differential equation $y' - 2y = 0$, $y(0) = 1$. Applying the Laplace transform to each side of the differential equation gives us

$$L(y') - 2L(y) = L(0).$$

Using equation (1) of this section, this equation becomes

$$-y(0) + sL(y) - 2L(y) = -1 + (s-2)L(y) = 0.$$

Solving for $L(y)$ gives us

$$L(y) = \frac{1}{s-2}.$$

Finally using the inverse Laplace transform, we find our solution is

$$y = e^{2t}.$$

3. We have the differential equation $y'' + 8y' = 0$, $y(0) = -1$, $y'(0) = 0$. Applying the Laplace transform to each side of the differential equation gives us

$$L(y'') + 8L(y') = L(0).$$

Using equations (1) and (2) of this section, this equation becomes

$$-y'(0) - sy(0) + s^2 L(y) + 8(-y(0) + sL(y)) = s + 8 + (s^2 + 8s)L(y) = 0.$$

Solving for $L(y)$ gives us

$$L(y) = -\frac{1}{s}.$$

Finally using the inverse Laplace transform, we find our solution is

$$y = -1.$$

5. We have the differential equation $y'' + 6y' + 9y = 0$, $y(0) = 1$, $y'(0) = 2$. Applying the Laplace transform to each side of the differential equation gives us

$$L(y'') + 6L(y') + 9L(y) = L(0).$$

Using equations (1) and (2) of this section, this equation becomes

$$-y'(0) - sy(0) + s^2 L(y) + 6(-y(0) + sL(y)) + 9L(y) = -8 - s + (s^2 + 6s + 9)L(y) = 0.$$

Solving for $L(y)$ gives us

$$L(y) = \frac{s+8}{(s+3)^2} = \frac{1}{s+3} + \frac{5}{(s+3)^2}.$$

Finally using the inverse Laplace transform, we find our solution is

$$y = e^{-3t} + 5te^{-3t}.$$

7. We have the differential equation $4y'' + 16y = 0$, $y(0) = 0$, $y'(0) = 1$. Applying the Laplace transform to each side of the differential equation gives us

$$L(y'') + 16L(y) = L(0).$$

Using equation (2) of this section, this equation becomes

$$-y'(0) - sy(0) + s^2 L(y) + 16L(y) = -1 + (s^2 + 16)L(y) = 0.$$

Solving for $L(y)$ gives us

$$L(y) = \frac{1}{s^2 + 16} = \frac{1}{2}\frac{2}{s^2 + 16}.$$

Finally using the inverse Laplace transform, we find our solution is

$$y = \frac{1}{2}\sin 2t.$$

9. We have the differential equation $y''' + 4y'' - y' - 4y = 0$, $y(0) = -1$, $y'(0) = 0$, $y''(0) = 1$. Applying the Laplace transform to each side of the differential equation gives us

$$L(y''') + 4L(y'') - L(y') - 4L(y) = L(0).$$

Using equations (1), (2) and (3) of this section, this equation becomes

$$-y''(0) - sy'(0) - s^2 y(0) + s^3 L(y) + 4(-y'(0) - sy(0) + s^2 L(y)) - (-y(0) + sL(y)) - 4L(y)$$

$$= s^2 + 4s - 2 + (s^3 + 4s^2 - s - 4)L(y) = 0.$$

Solving for $L(y)$ gives us

$$L(y) = \frac{-s^2 - 4s + 2}{s^3 + 4s^2 - s - 4} = -\frac{3}{10}\frac{1}{s-1} + \frac{2}{15}\frac{1}{s+4} - \frac{5}{6}\frac{1}{s+1}.$$

Finally using the inverse Laplace transform, we find our solution is

$$y = -\frac{3}{10}e^t + \frac{2}{15}e^{-4t} - \frac{5}{6}e^{-t}.$$

11. We have the differential equation $y' + 3y = 2$, $y(0) = 0$. Applying the Laplace transform to each side of the differential equation gives us

$$L(y') + 3L(y) = L(2).$$

Using equation (1) of this section and Table 7.1, this equation becomes

$$-y(0) + sL(y) + 3L(y) = (s+3)L(y) = \frac{2}{s}.$$

Solving for $L(y)$ gives us

$$L(y) = \frac{2}{s(s+3)} = \frac{2}{3}\left(\frac{1}{s} - \frac{1}{s+3}\right).$$

Finally using the inverse Laplace transform, we find our solution is

$$y = \frac{2}{3}\left(1 - e^{-3t}\right).$$

13. We have the differential equation $y'' + 4y' + 3y = 2 - e^{3t}$, $y(0) = 0, y'(0) = 0$. Applying the Laplace transform to each side of the differential equation gives us

$$L(y'') + 4L(y') + 3L(y) = L(2) - L(e^{3t}).$$

Using equations (1) and (2) of this section and Table 7.1, this equation becomes

$$-y'(0) - sy(0) + s^2 L(y) + 4(-y(0) + sL(y)) + 3L(y) = (s^2 + 4s + 3)L(y) = \frac{2}{s} - \frac{1}{s-3}.$$

Solving for $L(y)$ gives us

$$L(y) = \frac{s-6}{s(s-3)(s+3)(s+1)} = \frac{2}{3}\frac{1}{s} - \frac{1}{24}\frac{1}{s-3} + \frac{1}{4}\frac{1}{s+3} - \frac{7}{8}\frac{1}{s+1}.$$

Finally using the inverse Laplace transform, we find our solution is

$$y = \frac{2}{3} - \frac{1}{24}e^{3t} + \frac{1}{4}e^{-3t} - \frac{7}{8}e^{-t}.$$

15. We have the differential equation $y''' - 4y'' + 5y' = 7 - 2\cos t$, $y(0) = 1, y'(0) = 0, y''(0) = 2$. Applying the Laplace transform to each side of the differential equation gives us

$$L(y''') - 4L(y'') + 5L(y') = L(7) - 2L(\cos t).$$

Using equations (1), (2) and (3) of this section and Table 7.1, this equation becomes

$$-y''(0) - sy'(0) - s^2 y(0) + s^3 L(y) - 4(-y'(0) - sy(0) + s^2 L(y)) + 5(-y(0) + sL(y)) = \frac{7}{s} - \frac{2s}{s^2+1},$$

which simplifies to

$$-s^2 + 4s - 7 + (s^2 + 4s + 3)L(y) = \frac{7}{s} - \frac{2s}{s^2+1}.$$

Solving for $L(y)$ gives us

$$L(y) = \frac{s^5 - 4s^4 + 8s^3 + s^2 + 7s + 7}{s^2(s^2+1)(s^2-4s+5)}.$$

Finding the partial fraction decomposition of this we have

$$
\begin{aligned}
L(y) &= \frac{7}{5}\frac{1}{s^2} + \frac{63}{25}\frac{1}{s} - \frac{1}{4}\frac{s}{(s^2+1)^2} - \frac{1}{4}\frac{1}{s^2+1} - \frac{1}{100}\frac{127s - 393}{s^2 - 4s + 5} \\
&= \frac{7}{5}\frac{1}{s^2} + \frac{63}{25}\frac{1}{s} - \frac{1}{4}\frac{(s^2+1)^2}{-}\frac{1}{4}\frac{1}{s^2+1} - \frac{127}{100}\frac{s-2}{(s-2)^2+1} + \frac{139}{100}\frac{1}{(s-2)^2+1}.
\end{aligned}
$$

Finally using the inverse Laplace transform, we find our solution is

$$y = \frac{7}{5}t + \frac{63}{25} - \frac{1}{4}\cos t - \frac{1}{4}\sin t - \frac{127}{100}e^{2t}\cos t + \frac{139}{100}e^{2t}\sin t.$$

17. Since the growth is exponential we have the model $P(t) = ce^{kt}$ where k is the growth constant. Initially we have a population of 1000 amoeba which doubles in 30 minutes. Solving the equation $2000 = 1000e^{\frac{1}{2}k}$ we have $k = \ln 4$. After the population reaches 2000, 1000 amoeba per hour are removed continuously from the jar. Assuming this has no effect on the growth constant, we have the initial value problem

$$P' = (\ln 4)P - 1000, \quad P(0) = 2000.$$

Applying the Laplace transform to each side of the differential equation gives us

$$L(P') = (\ln 4)L(P) - L(1000).$$

Using equation (1) of this section, and the initial condition, this equation becomes

$$-2000 + sL(P)) = (\ln 4)L(P) - \frac{1000}{s}.$$

Solving for $L(P)$ gives us

$$L(P) = \frac{2000s - 1000}{s(s - \ln 4)} = \frac{1000}{\ln 4}\frac{1}{s} + \left(2000 - \frac{1000}{\ln 4}\right)\frac{1}{s - \ln 4}.$$

Using the inverse Laplace transform, we find the population is given by the formula

$$P = \frac{1000}{\ln 4} + \left(2000 - \frac{1000}{\ln 4}\right)e^{(\ln 4)t}.$$

Finally we are asked to find the population in the jar after 4 more hours. This is

$$P(4) = \frac{1000}{\ln 4} + \left(2000 - \frac{1000}{\ln 4}\right)e^{4\ln 4} \approx 328,056.$$

19. Let θ be the temperature of the liquid at time t. By Newton's law of cooling

$$\theta' = k(\theta - 70), \quad \theta(0) = 180.$$

Applying the Laplace transform to each side of the differential equation gives us

$$L(\theta') = kL(\theta) - L(70k).$$

Using equation (1) of this section, this equation becomes

$$-\theta(0) + sL(\theta)) = kL(\theta) - \frac{70k}{s}.$$

Solving for $L(\theta)$ gives us

$$L(\theta) = \frac{180s - 70k}{s(s-k)} = \frac{70}{s} + \frac{110}{s-k}.$$

Using the inverse Laplace transform, we find the temperature is given by the formula

$$\theta = 70 + 110e^{kt}.$$

We are also told that $\theta(2) = 160$. Using this in the equation above we have $160 = 70 + 110e^{2k}$. Solving this equation we have $k = \frac{1}{2} \ln \frac{9}{11}$. Thus, in general the temperature is

$$\theta = 70 + 110e^{\left(\frac{1}{2} \ln \frac{9}{11}\right)t}.$$

Finally we find the temperature after 10 minutes

$$\theta(10) = 70 + 110e^{5 \ln \frac{9}{11}} \approx 110°\,\mathrm{F}.$$

21. We standardize our units to meters. Since $k = \frac{mg}{l} = \frac{5 \cdot 10}{0.25} = 200$ as in Section 4.5 we have the initial value problem

$$5u'' + 20u' + 200u = 3e^{-\frac{1}{10}t}; \; u(0) = 0.25, \; u'(0) = 0.5.$$

Applying the Laplace transform to each side of the differential equation gives us

$$5L(u'') + 20L(u') + 200L(u) = L(e^{-\frac{1}{10}t}).$$

Using equations (1) and (2) of this section and Table 7.1, this equation becomes

$$5(-u'(0) - su(0) + s^2 L(u)) + 20(-u(0) + sL(u)) + 200L(u) = 7.5 + 1.25s + (5s^2 + 20s + 200)L(u) = \frac{3}{s + \frac{1}{10}}.$$

Solving for $L(u)$ gives us

$$\begin{aligned}
L(u) &= \frac{1}{40} \frac{10s^2 + 61s + 30}{\left(s + \frac{1}{10}\right)(s^2 + 4s + 40)} = \frac{60}{3961} \cdot \frac{1}{s + \frac{1}{10}} + \frac{1}{15844} \frac{3721s + 22830}{s^2 + 4s + 40} \\
&= \frac{60}{3961} \frac{1}{s + \frac{1}{10}} + \frac{3721}{15844} \frac{s+2}{(s+2)^2 + 6^2} + \frac{3847}{23766} \frac{6}{(s+2)^2 + 6^2}.
\end{aligned}$$

Finally using the inverse Laplace transform, we find our solution is

$$u = \frac{60}{3961}e^{-\frac{1}{10}t} + \frac{3721}{15844}e^{-2t}\cos 6t + \frac{3847}{23766}e^{-2t}\sin 6t.$$

23. As in Section 4.5 we have the initial value problem

$$\frac{1}{2}I'' + 2I' + \frac{3}{2}I = 3\sin 2t; \; I(0) = 0, \; I'(0) = 0.$$

First simplifying the equation we have

$$I'' + 4I' + 3I = 6\sin 2t.$$

Applying the Laplace transform to each side of the differential equation gives us

$$L(I'') + 4L(I') + 3L(I) = 6L(\sin 2t).$$

Using equations (1) and (2) of this section and Table 7.1, this equation becomes

$$-I'(0) - sI(0) + s^2 \mathcal{L}(I) + 4(-I(0) + s\mathcal{L}(I)) + 3\mathcal{L}(I) = (s^2 + 4s + 3)\mathcal{L}(I) = \frac{12s}{s^2 + 4}.$$

Solving for $\mathcal{L}(I)$ gives us

$$
\begin{aligned}
\mathcal{L}(I) &= \frac{12s}{(s^2 + 4)(s + 3)(s + 1)} = -\frac{12}{65}\frac{s - 16}{s^2 + 4} + \frac{18}{13}\frac{1}{s + 3} - \frac{6}{5}\frac{1}{s + 1} \\
&= \frac{96}{65}\frac{2}{s^2 + 4} - \frac{12}{65}\frac{s}{s^2 + 4} + \frac{18}{13}\frac{1}{s + 3} - \frac{6}{5}\frac{1}{s + 1}.
\end{aligned}
$$

Finally using the inverse Laplace transform, we find our solution is

$$I = \frac{96}{65}\sin 2t - \frac{12}{65}\cos 2t + \frac{18}{13}e^{-3t} - \frac{6}{5}e^{-t}.$$

7.3 Step Functions, Impulse Functions, and the Delta Function

Exercises 7.3, pp. 364-366

1. (a) $f(t) = t - tu_2(t)$

    (c) Since $tu_2(t) = (t - 2)u_2(t) + 2u_2(t)$, we have $f(t) = t - (t - 2)u_2(t) - 2u_2(t)$. Thus,

    $$\mathcal{L}(f(t)) = \frac{1}{s^2} - e^{-2s}\frac{1}{s^2} - e^{-2s}\frac{2}{s}.$$

3. (a) $\sigma(t) = e^{2t}u_2(t) - e^{2t}u_4(t)$

    (c) Since $e^{2t}u_2(t) = e^4 \cdot e^{2(t-2)}u_2(t)$ and $e^{2t}u_4(t) = e^8 \cdot e^{8(t-4)}u_4(t)$ we have

    $$\sigma(t) = e^4 \cdot e^{2(t-2)}u_2(t) - e^8 \cdot e^{8(t-4)}u_4(t).$$

    Thus,

    $$\mathcal{L}(\sigma(t)) = \frac{e^4 \cdot e^{-2s}}{s - 2} - \frac{e^8 \cdot e^{-4s}}{s - 4} = \frac{e^{-2(s-2)}}{s - 2} - \frac{e^{-4(s-2)}}{s - 4}.$$

5. (a) $h(t) = \cos 2t\, u_\pi(t) - \cos 2t\, u_{2\pi}(t)$

    (c) Since $\cos 2t = \cos 2(t - \pi) = \cos 2(t - 2\pi)$ we have $h(t) = \cos 2(t - \pi)u_\pi(t) - \cos 2(t - 2\pi)u_{2\pi}(t)$. Thus,

    $$\mathcal{L}(h(t)) = e^{-\pi s}\frac{s}{s^2 + 4} - e^{-2\pi s}\frac{s}{s^2 + 4}.$$

7. (b) Since

    $$t\,u_2(t) - 2t\,u_4(t) = (t - 2)u_2(t) + 2u_2(t) - 2(t - 4)u_4(t) - 8u_4(t)$$

    the Laplace transform of the function is

    $$e^{-2s}\frac{1}{s^2} + e^{-2s}\frac{2}{s} - e^{-4s}\frac{2}{s^2} - e^{-4s}\frac{8}{s} = e^{-2s}\left(\frac{1}{s^2} + \frac{2}{s}\right) - e^{-4s}\left(\frac{2}{s^2} + \frac{8}{s}\right).$$

9. (b) Since

    $$u_\pi(t)\cos t = -u_\pi(t)\cos(t - \pi)$$

    the Laplace transform of the function is

    $$-\frac{se^{-\pi s}}{s^2 + 1}.$$

11. $\mathcal{L}^{-1}\left(\frac{e^{-2s}}{s}\right) = u_2(t)$

13. Since

$$\frac{e^{-s}}{s^2 - 4s} = \frac{e^{-s}}{4}\left(\frac{1}{s-4} - \frac{1}{s}\right),$$

we have

$$\mathcal{L}^{-1}\left(\frac{e^{-s}}{s^2 - 4s}\right) = \frac{1}{4}u_1(t)e^{4(t-1)} - \frac{1}{4}u_1(t) = \frac{1}{4}u_1(t)\left(e^{4(t-1)} - 1\right).$$

15. We are given the initial value problem

$$y' + 3y = t - t\,u_2(t),\ y(0) = 0.$$

Since $t - t\,u_2(t) = t - (t-2)u_2(t) - 2u_2(t)$ and $\mathcal{L}(y') = -y(0) + s\mathcal{L}(y)$ when we apply the Laplace transform we have

$$s\mathcal{L}(y) + 3\mathcal{L}(y) = \frac{1}{s^2} - \frac{e^{-2s}}{s^2} - \frac{2e^{-2s}}{s}.$$

Solving for $\mathcal{L}(y)$ we obtain

$$\mathcal{L}(y) = \frac{1}{s^2(s+3)} - e^{-2s}\left(\frac{1+2s}{s^2(s+3)}\right).$$

Finding the partial fraction decompositions of these expressions we have

$$\mathcal{L}(y) = \frac{1}{3}\frac{1}{s^2} - \frac{1}{9}\frac{1}{s} + \frac{1}{9}\frac{1}{s+3} + e^{-2s}\left(-\frac{1}{3}\frac{1}{s^2} - \frac{5}{9}\frac{1}{s} + \frac{5}{9}\frac{1}{s+3}\right).$$

Applying the inverse Laplace transform we obtain

$$y = \frac{1}{3}t - \frac{1}{9} + \frac{1}{9}e^{-3t} + u_2(t)\left(-\frac{1}{3}(t-2) - \frac{5}{9} + \frac{5}{9}e^{-3(t-2)}\right) = \frac{1}{3}t - \frac{1}{9} + \frac{1}{9}e^{-3t} + u_2(t)\left(-\frac{1}{3}t + \frac{1}{9} + \frac{5}{9}e^{-3(t-2)}\right).$$

17. We are given the initial value problem

$$y'' + 4y' + 3y = e^{3t-3}u_2(t),\ y(0) = 0,\ y'(0) = 0.$$

Since $\mathcal{L}(y'') = -y'(0) - sy(0) + s^2\mathcal{L}(y)$ and $\mathcal{L}(y') = -y(0) + s\mathcal{L}(y)$ when we apply the Laplace transform we have

$$s^2\mathcal{L}(y) + 4s\mathcal{L}(y) + 3\mathcal{L}(y) = \frac{e^{-s}}{s-3}.$$

Solving for $\mathcal{L}(y)$ we obtain

$$\mathcal{L}(y) = \frac{e^{-s}}{(s-3)(s+1)(s+3)}.$$

Finding the partial fraction decompositions of this expression we have

$$\mathcal{L}(y) = e^{-s}\left(\frac{1}{24}\frac{1}{s-3} - \frac{1}{8}\frac{1}{s+1} + \frac{1}{12}\frac{1}{s+3}\right).$$

Applying the inverse Laplace transform we obtain

$$y = u_1(t)\left(\frac{1}{24}e^{3(t-1)} - \frac{1}{8}e^{-t+1} + \frac{1}{12}e^{-3(t-1)}\right).$$

19. We are given the initial value problem

$$y'' - 4y' + 5y = 7 - 3\cos 2t - 7u_{2\pi}(t) + 3u_{2\pi}(t)\cos 2t,\ y(0) = 1,\ y'(0) = 0.$$

Since $\mathcal{L}(y'') = -y'(0) - sy(0) + s^2\mathcal{L}(y)$ and $\mathcal{L}(y') = -y(0) + s\mathcal{L}(y)$ when we apply the Laplace transform we have

$$-s + s^2\mathcal{L}(y) + 4 - 4s\mathcal{L}(y) + 5\mathcal{L}(y) = \frac{7}{s} - \frac{3s}{s^2+4} + e^{-2\pi s}\left(\frac{3s}{s^2+4} - \frac{7}{s}\right).$$

Solving for $\mathcal{L}(y)$ we obtain

$$\mathcal{L}(y) = \frac{s^4 - 4s^3 + 8s^2 - 16s + 28}{s(s^2+4)(s^2-4s+5)} - e^{-2\pi s}\frac{4s^2+28}{s(s^2+4)(s^2-4s+5)}.$$

Finding the partial fraction decompositions of these expressions we have

$$
\begin{aligned}
\mathcal{L}(y) &= \frac{7}{5}\frac{1}{s} - \frac{3}{65}\frac{s-16}{s^2+4} - \frac{1}{65}\frac{23s-44}{s^2-4s+5} + e^{-2\pi s}\left(-\frac{7}{5}\frac{1}{s} + \frac{3}{65}\frac{s-16}{s^2+4} + \frac{8}{65}\frac{11s-38}{s^2-4s+5}\right)\\
&= \frac{7}{5}\frac{1}{s} - \frac{3}{65}\frac{s}{s^2+4} + \frac{24}{65}\frac{2}{s^2+4} - \frac{2}{65}\frac{2}{(s-2)^2+1} - \frac{23}{65}\frac{s-2}{(s-2)^2+1}\\
&\quad + e^{-2\pi s}\left(-\frac{7}{5}\frac{1}{s} + \frac{3}{65}\frac{s}{s^2+4} - \frac{24}{65}\frac{2}{s^2+4} + \frac{88}{65}\frac{s}{(s-2)^2+1} - \frac{128}{65}\frac{1}{(s-2)^2+1}\right).
\end{aligned}
$$

Applying the inverse Laplace transform we obtain

$$
\begin{aligned}
y &= \frac{7}{5} - \frac{3}{65}\cos 2t + \frac{24}{65}\sin 2t - \frac{2}{65}e^{2t}\sin t - \frac{23}{65}e^{2t}\cos t\\
&\quad + u_{2\pi}(t)\left(-\frac{7}{5} + \frac{3}{65}\cos 2t - \frac{24}{65}\sin 2t + \frac{88}{65}e^{2(t-2\pi)}\cos t - \frac{128}{65}e^{2(t-2\pi)}\sin t\right).
\end{aligned}
$$

21. We are given the initial value problem

$$y'' + 2y' - 8y = \delta(t),\ y(0) = 0,\ y'(0) = 0.$$

Since $\mathcal{L}(y'') = -y'(0) - sy(0) + s^2\mathcal{L}(y)$ and $\mathcal{L}(y') = -y(0) + s\mathcal{L}(y)$ when we apply the Laplace transform we have

$$s^2\mathcal{L}(y) + 2s\mathcal{L}(y) - 8\mathcal{L}(y) = 1.$$

Solving for $\mathcal{L}(y)$ we obtain

$$\mathcal{L}(y) = \frac{1}{s^2+2s-8} = \frac{1}{6}\frac{1}{s-2} - \frac{1}{6}\frac{1}{s+4}.$$

Applying the inverse Laplace transform we obtain

$$y = \frac{1}{6}e^{2t} - \frac{1}{6}e^{-4t}.$$

23. We are given two initial value problems

$$y'' + y = 0,\ y(0) = 0,\ y'(0) = 1.$$

Since $\mathcal{L}(y'') = -y'(0) - sy(0) + s^2\mathcal{L}(y)$ when we apply the Laplace transform we have

$$-1 + s^2\mathcal{L}(y) + \mathcal{L}(y) = 0.$$

Solving for $\mathcal{L}(y)$ we obtain

$$\mathcal{L}(y) = \frac{1}{s^2+1}.$$

Applying the inverse Laplace transform we obtain

$$y = \sin t.$$

Next, we have

$$y'' + y = \delta(t),\ y(0) = 0,\ y'(0) = 0.$$

Now when we apply the Laplace transform we have

$$s^2\mathcal{L}(y) + \mathcal{L}(y) = 1.$$

Solving for $\mathcal{L}(y)$ we obtain

$$\mathcal{L}(y) = \frac{1}{s^2+1}.$$

Applying the inverse Laplace transform we obtain

$$y = \sin t.$$

25. As in Exercise 17 of Section 7.2 we have the growth constant $k = \ln 4$. Thus the initial value problem is

$$y' = (\ln 4)y + 1000(u_2(t) - 1), \quad y(0) = 2000.$$

Since $L(y') = -y(0) + sL(y)$ when we apply the Laplace transform we have

$$-2000 + sL(y) = (\ln 4)L(y) + \frac{1000}{s}e^{-2s} - \frac{1000}{s}.$$

Solving for $L(y)$ we obtain

$$L(y) = \frac{2000s - 1000}{s(s - \ln 4)} + \frac{1000}{s(s - \ln 4)}e^{-2s}.$$

Finding the partial fraction decompositions of this expression we have

$$L(y) = \left(2000 - \frac{1000}{\ln 4}\right)\frac{1}{s - \ln 4} + \frac{1000}{\ln 4}\frac{1}{s} + e^{-2s}\left(\frac{1000}{\ln 4}\frac{1}{s - \ln 4} - \frac{1000}{\ln 4}\frac{1}{s}\right).$$

Applying the inverse Laplace transform we obtain

$$y = \left(2000 - \frac{1000}{\ln 4}\right)e^{(\ln 4)t} + \frac{1000}{\ln 4} + u_2(t)\left(\frac{1000}{\ln 4}e^{\ln 4(t-2)} - \frac{1000}{\ln 4}\right).$$

Finally we evaluate the function at $t = 3.5$ (this is 4 hours after the start of the experiment) to obtain

$$
\begin{aligned}
y(3.5) &= \left(2000 - \frac{1000}{\ln 4}\right)e^{(\ln 4)3.5} + \frac{1000}{\ln 4} + u_2(3.5)\left(\frac{1000}{\ln 4}e^{\ln 4(3.5-2)} - \frac{1000}{\ln 4}\right)\\
&= 128\left(2000 - \frac{1000}{\ln 4}\right) + \frac{8000}{\ln 4} \approx 169{,}438.
\end{aligned}
$$

27. This problem is similar to Exercise 21 of Section 7.2, only the external force is different. Here we have the initial value problem

$$5y'' + 20y' + 200y = 3e^{\frac{1}{10}t}(1 - u_3(t)); \quad y(0) = \frac{1}{4}, \quad y'(0) = \frac{1}{2}.$$

Since $L(y'') = -y'(0) - sy(0) + s^2L(y)$ and $L(y') = -y(0) + sL(y)$ when we apply the Laplace transform we have

$$-\frac{5}{2} - \frac{5}{4}s + s^2L(y) - 5 + 20sL(y) + 200L(y) = \frac{3}{s + \frac{1}{10}} - e^{-3s - \frac{3}{10}}\frac{3}{s + \frac{1}{10}}.$$

Solving for $L(y)$ we obtain

$$L(y) = \frac{1}{40}\frac{10s^2 + 61s + 30}{\left(s + \frac{1}{10}\right)(s^2 + 4s + 40)} - e^{-3s - \frac{3}{10}}\frac{3}{\left(s + \frac{1}{10}\right)(s^2 + 4s + 40)}.$$

Finding the partial fraction decompositions of this expression we have

$$
\begin{aligned}
L(y) &= \frac{60}{3961}\frac{1}{s + \frac{1}{10}} + \frac{3721}{15844}\frac{s + 2}{(s+2)^2 + 6^2} + \frac{3847}{23766}\frac{6}{(s+2)^2 + 6^2}\\
&\quad + e^{-\frac{3}{10}}e^{-3s}\left(-\frac{60}{3961}\frac{1}{s + \frac{1}{10}} + \frac{60}{3961}\frac{s + 2}{(s+2)^2 + 6^2} + \frac{19}{3961}\frac{6}{(s+2)^2 + 6^2}\right).
\end{aligned}
$$

Applying the inverse Laplace transform we obtain

$$
\begin{aligned}
y &= \frac{60}{3961}e^{-\frac{1}{10}t} + \frac{3721}{15844}e^{-2t}\cos 6t + \frac{3847}{23766}e^{-2t}\cos 6t\\
&\quad + \frac{1}{3961}e^{-\frac{3}{10}}u_3(t)\left(-60e^{-\frac{1}{10}(t-3)} + 60e^{-2(t-3)}\cos 6(t-3) + 19e^{-2(t-3)}\sin 6(t-3)\right).
\end{aligned}
$$

29. We have the initial value problem

$$\frac{1}{2}I'' + 2I' + \frac{3}{2}I = 6(u_0(t) - u_2(t)); \ I(0) = 0, \ I'(0) = 0.$$

Since $L(I'') = -I'(0) - sI(0) + s^2 L(I)$ and $L(I') = -I(0) + sL(I)$ when we apply the Laplace transform we have

$$\frac{1}{2}s^2 L(I) + 2sL(I) + \frac{3}{2}L(I) = \frac{6}{s} - e^{-2s}\frac{6}{s}.$$

Solving for $L(I)$ we obtain

$$L(I) = \frac{12}{s(s+1)(s+3)} - e^{-2s}\frac{12}{s(s+1)(s+3)}.$$

Finding the partial fraction decompositions of this expression we have

$$L(I) = \frac{4}{s} - \frac{6}{s+1} + \frac{2}{s+3} + e^{-2s}\left(-\frac{4}{s} + \frac{6}{s+1} - \frac{2}{s+3}\right).$$

Applying the inverse Laplace transform we obtain

$$y = 4 - 6e^{-t} + 2e^{-3t} + u_2(t)\left(-4 + 6e^{-t+2} - 2e^{-3(t-2)}\right).$$

31. We have the initial value problem

$$\frac{1}{2}I'' + 2I' + \frac{3}{2}I = 6\delta(t - \pi); \ I(0) = 0, \ I'(0) = 0.$$

As in Exercise 29 when we apply the Laplace transform we have

$$\frac{1}{2}s^2 L(I) + 2sL(I) + \frac{3}{2}L(I) = 6e^{-\pi s}.$$

Solving for $L(I)$ we obtain

$$L(I) = e^{-\pi s}\frac{12}{s^2 + 4s + 3} = e^{-\pi s}\left(\frac{6}{s+1} - \frac{6}{s+3}\right).$$

Applying the inverse Laplace transform we obtain

$$y = 6u_\pi(t)\left(e^{-t+\pi} - e^{-3(t-\pi)}\right).$$

7.4   Convolution Integrals

Exercises 7.4, pp. 369

1. We rewrite $\frac{1}{s^2 - s}$ as $\frac{1}{s-1} \cdot \frac{1}{s}$. Since $L(e^t) = \frac{1}{s-1}$ and $L(1) = \frac{1}{s}$ we have that

$$L^{-1}\left(\frac{1}{s^2 - s}\right) = e^t * 1 = \int_0^t e^\tau \, d\tau = e^\tau|_0^t = e^t - 1.$$

3. We rewrite $\frac{1}{(s+1)(s^2+1)}$ as $\frac{1}{s^2+1} \cdot \frac{1}{s+1}$. Since $L(\sin t) = \frac{1}{s^2+1}$ and $L(e^{-t}) = \frac{1}{s+1}$ we have that

$$
\begin{aligned}
L^{-1}\left(\frac{1}{(s+1)(s^2+1)}\right) &= \sin t * e^{-t} = \int_0^t \sin(\tau)e^{-(t-\tau)} \, d\tau = e^{-t}\int_0^t \sin(\tau)e^\tau \, d\tau \\
&= \frac{1}{2}e^{-t}\left((e^\tau \sin\tau - e^\tau \cos\tau)|_0^t\right) = \frac{1}{2}e^{-t}\left(e^t \sin t - e^t \cos t + 1\right) \\
&= \frac{1}{2}\left(e^{-t} + \sin t - \cos t\right).
\end{aligned}
$$

5. The integral $\int_0^t (t-\tau)e^\tau \, d\tau = e^t * t$. Therefore,

$$\mathcal{L}\left(\int_0^t (t-\tau)e^\tau \, d\tau\right) = \mathcal{L}(e^t * t) = \mathcal{L}(e^t) \cdot \mathcal{L}(t) = \frac{1}{s-1} \cdot \frac{1}{s^2} = \frac{1}{s^2(s-1)}.$$

7. When we apply the Laplace transform to the differential equation we have

$$-y'(0) - sy(0) + s^2 \mathcal{L}(y) + \mathcal{L}(y) = \frac{1}{s^2}.$$

Since $y(0) = 0$ and $y'(0) = 1$ this is

$$-1 + (s^2 + 1)\mathcal{L}(y) = \frac{1}{s^2}.$$

Solving for $\mathcal{L}(y)$ we have $\mathcal{L}(y) = \frac{1}{s^2}$. Here, we then see that $y = \mathcal{L}^{-1}\left(\frac{1}{s^2}\right) = t$. However, we are asked to use a convolution integral to find the solution. Since $\mathcal{L}(1) = \frac{1}{s}$ we have that

$$\mathcal{L}^{-1}\left(\frac{1}{s^2}\right) = 1 * 1 = \int_0^t d\tau = \tau|_0^t = t.$$

9. When we apply the Laplace transform to the differential equation we have

$$-y'(0) - sy(0) + s^2 \mathcal{L}(y) + 4\mathcal{L}(y) = \frac{2}{s^2 + 4}.$$

Since $y(0) = 1$ and $y'(0) = 0$ this is

$$-s + (s^2 + 4)\mathcal{L}(y) = \frac{2}{s^2 + 4}.$$

Solving for $\mathcal{L}(y)$ we have

$$\mathcal{L}(y) = \frac{s^3 + 4s + 2}{(s^2 + 4)^2} = \frac{s}{s^2 + 4} + \frac{2}{(s^2 + 4)^2}.$$

We find the inverse Laplace transform of the first expression above right using Table 7.1 and the inverse Laplace transform of the second expression using a convolution integral. Thus, the first expression is

$$\mathcal{L}^{-1}\left(\frac{s}{s^2 + 4}\right) = \cos 2t.$$

For the second

$$\frac{2}{(s^2 + 4)^2} = \frac{2}{s^2 + 4} \cdot \frac{1}{s^2 + 4}.$$

Since $\mathcal{L}(\sin 2t) = \frac{2}{s^2+4}$ and $\mathcal{L}\left(\frac{1}{2}\sin 2t\right) = \frac{1}{s^2+4}$ we have

$$\mathcal{L}^{-1}\left(\frac{2}{(s^2+4)^2}\right) = \sin 2t * \frac{1}{2}\sin 2t = \frac{1}{2}\int_0^t \sin 2(t-\tau)\sin 2\tau \, d\tau = \frac{1}{8}\sin 2t - \frac{1}{4}t\cos 2t,$$

the solution to the differential equation is

$$y = \cos 2t + \frac{1}{8}\sin 2t - \frac{1}{4}t\cos 2t.$$

11. To find the convolution product $1 * (t * t)$ we first evaluate

$$t * t = \int_0^t (t-\tau)\tau \, d\tau = \left(\frac{1}{2}t\tau^2 - \frac{1}{3}\tau^3\right)\Big|_0^t = \frac{1}{6}t^3.$$

Next we find

$$1 * \frac{1}{6}t^3 = \int_0^t 1 \cdot \frac{1}{6}\tau^3 \, d\tau = \frac{1}{24}\tau^4\Big|_0^t = \frac{1}{24}t^4.$$

We now evaluate $(1 * t) * t$. First

$$1 * t = \int_0^t 1 \cdot \tau \, d\tau = \frac{1}{2}\tau^2 \Big|_0^t = \frac{1}{2}t^2.$$

Next we find

$$\frac{1}{2}t^2 * t = \int_0^t \frac{1}{2}(t-\tau)^2 \cdot \tau \, d\tau = \frac{1}{2}\int_0^t (t^2\tau - 2t\tau^2 + \tau^3) \, d\tau = \frac{1}{2}\left(\frac{1}{2}t^2\tau^2 - \frac{2}{3}t\tau^3 + \frac{1}{4}\tau^4\right)\Big|_0^t = \frac{1}{24}t^4.$$

We see that the two convolution products share the value $\frac{1}{24}t^4$.

13. We are given

$$y(t) = 3 + \int_0^t e^{t-\tau}y(\tau) \, d\tau = 3 + e^t * y(t).$$

Using the Laplace transform we have

$$L(y) = L(3) + L(e^t) \cdot L(y).$$

or

$$L(y) = \frac{3}{s} + \frac{1}{s-1} \cdot L(y).$$

Solving for $L(y)$ we have

$$L(y) = \frac{3(s-1)}{s(s-2)} = \frac{3}{2}\cdot\frac{1}{s} + \frac{3}{2}\cdot\frac{1}{s-2}.$$

Applying the inverse Laplace transform we have

$$y(t) = \frac{3}{2}\left(1 + e^{2t}\right).$$

15. We are given

$$y(t) = e^t + 2\int_0^t e^{2(t-\tau)}y(\tau) \, d\tau = e^t + 2(e^{2t} * y(t)).$$

Using the Laplace transform we have

$$L(y) = L(e^t) + 2L(e^{2t}) \cdot L(y).$$

or

$$L(y) = \frac{1}{s-1} + 2\frac{1}{s-2} \cdot L(y).$$

Solving for $L(y)$ we have

$$L(y) = \frac{s-2}{(s-1)(s-4)} = \frac{1}{3}\cdot\frac{1}{s-1} + \frac{2}{3}\cdot\frac{1}{s-4}.$$

Applying the inverse Laplace transform we have

$$y(t) = \frac{1}{3}e^t + \frac{2}{3}e^{4t}.$$

17. We are given

$$y'(t) = 4 + \int_0^t (t-\tau)y(\tau) \, d\tau = 4 + t * y(t).$$

Using the Laplace transform we have

$$L(y') = L(4) + L(t) \cdot L(y).$$

or

$$-y(0) + sL(y) = \frac{4}{s} + \frac{1}{s^2} \cdot L(y).$$

Since $y(0) = 1$ when we solve for $L(y)$ we have

$$L(y) = \frac{s^2+4s}{s^3-1} = \frac{5}{3}\cdot\frac{1}{s-1} - \frac{1}{3}\cdot\frac{2s-5}{s^2+s+1} = \frac{5}{3}\cdot\frac{1}{s-1} - \frac{2}{3}\cdot\frac{s+\frac{1}{2}}{\left(s+\frac{1}{2}\right)^2+\left(\frac{\sqrt{3}}{2}\right)^2} + \frac{2}{\left(s+\frac{1}{2}\right)^2+\left(\frac{\sqrt{3}}{2}\right)^2}.$$

Applying the inverse Laplace transform we have

$$y(t) = \frac{5}{3}e^t - \frac{2}{3}e^{-\frac{1}{2}t}\cos\frac{\sqrt{3}}{2}t + \frac{4\sqrt{3}}{3}e^{-\frac{1}{2}t}\sin\frac{\sqrt{3}}{2}t.$$

7.5   Systems of Linear Differential Equations

Exercises 7.5, pp. 372-373

1. We use the unit step function to first write the system as:

$$\begin{aligned}
y_1' &= y_1 - y_2 \\
y_2' &= 2y_1 + 4y_2 + 4u_1(t).
\end{aligned}$$

Now applying the Laplace transform to each equation in the system gives us

$$\begin{aligned}
\mathcal{L}(y_1') &= \mathcal{L}(y_1 - y_2) \\
\mathcal{L}(y_2') &= \mathcal{L}(2y_1 + 4y_2 + 4u_1(t))
\end{aligned}$$

or

$$\begin{aligned}
\mathcal{L}(y_1') &= \mathcal{L}(y_1) - \mathcal{L}(y_2) \\
\mathcal{L}(y_2') &= 2\mathcal{L}(y_1) + 4\mathcal{L}(y_2) + \frac{4}{s}e^{-2s}.
\end{aligned}$$

Using the property $\mathcal{L}(y') = -y(0) + s\mathcal{L}(y)$ and the initial conditions $y_1(0) = 0$, $y_2(0) = 1$, we have

$$\begin{aligned}
s\mathcal{L}(y_1) &= \mathcal{L}(y_1) - \mathcal{L}(y_2) \\
-1 + s\mathcal{L}(y_2) &= \mathcal{L}(2y_1) + 4\mathcal{L}(y_2) + \frac{4}{s}e^{-2s}.
\end{aligned}$$

or

$$\begin{aligned}
(s-1)\mathcal{L}(y_1) + \mathcal{L}(y_2) &= 0 \\
-2\mathcal{L}(y_1) + (s-4)\mathcal{L}(y_2) &= \frac{4}{s}e^{-2s} + 1.
\end{aligned}$$

Solving this system for $\mathcal{L}(y_1)$ and $\mathcal{L}(y_2)$, we find

$$\begin{aligned}
\mathcal{L}(y_1) &= -\frac{1}{s^2 - 5s + 6} - \frac{4}{s(s^2 - 5s + 6)} + \frac{4}{s}e^{-2s} \\
\mathcal{L}(y_2) &= \frac{s-1}{s^2 - 5s + 6} + \frac{4(s-1)}{s(s^2 - 5s + 6)} + \frac{4}{s}e^{-2s}.
\end{aligned}$$

Finding the partial fraction decomposition of each of the expressions above we have

$$\begin{aligned}
\mathcal{L}(y_1) &= \frac{1}{s-2} - \frac{1}{s-3} + e^{-2s}\left(-\frac{2}{3}\cdot\frac{1}{s} + \frac{2}{s-2} - \frac{4}{3}\cdot\frac{1}{s-3}\right) \\
\mathcal{L}(y_2) &= -\frac{1}{s-2} + \frac{2}{s-3} + e^{-2s}\left(-\frac{2}{3}\cdot\frac{1}{s} - \frac{2}{s-2} + \frac{8}{3}\cdot\frac{1}{s-3}\right).
\end{aligned}$$

Applying the inverse Laplace transform leads to

$$\begin{aligned}
y_1 &= e^{2t} - e^{3t} + u_2(t)\left(-\frac{2}{3} + 2e^{2(t-2)} - \frac{4}{3}e^{3(t-2)}\right) \\
y_2 &= -e^{2t} + 2e^{3t} + u_2(t)\left(-\frac{2}{3} - 2e^{2(t-2)} + \frac{8}{3}e^{3(t-2)}\right).
\end{aligned}$$

3. We have the system:

$$\begin{aligned}
y_1' &= y_1 + y_2 \\
y_2' &= 4y_1 - 2y_2 + e^t u_2(t).
\end{aligned}$$

Now applying the Laplace transform to each equation in the system gives us

$$\begin{aligned} L(y_1') &= L(y_1 + y_2) \\ L(y_2') &= L(4y_1 - 2y_2 + e^t u_2(t)) \end{aligned}$$

or

$$\begin{aligned} L(y_1') &= L(y_1) + L(y_2) \\ L(y_2') &= 4L(y_1) - 2L(y_2) + e^{2-2s}\frac{1}{s-1}. \end{aligned}$$

Using the property $L(y') = -y(0) + sL(y)$ and the initial conditions $y_1(0) = 1$, $y_2(0) = 0$, we have

$$\begin{aligned} -1 + sL(y_1) &= L(y_1) + L(y_2) \\ sL(y_2) &= L(4y_1) - 2L(y_2) + e^{2-2s}\frac{1}{s-1}. \end{aligned}$$

or

$$\begin{aligned} (s-1)L(y_1) - L(y_2) &= 1 \\ -4L(y_1) + (s+2)L(y_2) &= e^{2-2s}\frac{1}{s-1}. \end{aligned}$$

Solving this system for $L(y_1)$ and $L(y_2)$, we find

$$\begin{aligned} L(y_1) &= \frac{s+2}{s^2+s-6} + \frac{1}{(s-1)(s-2)(s+3)}e^{2-2s} \\ L(y_2) &= \frac{4}{s^2+s-6} + \frac{1}{s^2+s-6}e^{2-2s}. \end{aligned}$$

Finding the partial fraction decomposition of each of the expressions above we have

$$\begin{aligned} L(y_1) &= \frac{4}{5}\cdot\frac{1}{s-2} + \frac{1}{5}\cdot\frac{1}{s+3} + e^{2-2s}\left(-\frac{1}{4}\cdot\frac{1}{s-1} + \frac{1}{5}\cdot\frac{1}{s-2} + \frac{1}{20}\cdot\frac{1}{s+3}\right) \\ L(y_2) &= \frac{4}{5}\cdot\frac{1}{s-2} - \frac{4}{5}\cdot\frac{1}{s+3} + e^{2-2s}\left(\frac{1}{5}\cdot\frac{1}{s-2} - \frac{1}{5}\cdot\frac{1}{s+3}\right). \end{aligned}$$

Applying the inverse Laplace transform leads to

$$\begin{aligned} y_1 &= \frac{4}{5}e^{2t} + \frac{1}{5}e^{-3t} + u_2(t)\left(-\frac{1}{4}e^t + \frac{1}{5}e^{2(t-2)+2} + \frac{1}{20}e^{-3(t-2)+2}\right) \\ y_2 &= \frac{4}{5}e^{2t} - \frac{4}{5}e^{-3t} + u_2(t)\left(\frac{1}{5}e^{2(t-2)+2} - \frac{1}{5}e^{3(t-2)+2}\right). \end{aligned}$$

Equivalently,

$$\begin{aligned} y_1 &= \frac{4}{5}e^{2t} + \frac{1}{5}e^{-3t} + u_2(t)\left(-\frac{1}{4}e^t + \frac{1}{5}e^{2t-2} + \frac{1}{20}e^{-3t+8}\right) \\ y_2 &= \frac{4}{5}e^{2t} - \frac{4}{5}e^{-3t} + u_2(t)\left(\frac{1}{5}e^{2t-2} - \frac{1}{5}e^{-3t+8}\right). \end{aligned}$$

5. We have the system:

$$\begin{aligned} x_1'' &= 10x_1 - 5x_2 \\ x_2'' &= -12x_1 + 6x_2 + tu_1(t). \end{aligned}$$

Now applying the Laplace transform to each equation in the system gives us

$$\begin{aligned} L(x_1'') &= L(10x_1 - 5x_2) \\ L(x_2'') &= L(-12x_1 + 6x_2 + tu_1(t)) \end{aligned}$$

or

$$L(x_1'') = 10L(x_1) - 5L(x_2)$$
$$L(x_2'') = -12L(x_1) + 6L(x_2) + e^{-s}\left(\frac{1}{s} - \frac{1}{s^2}\right).$$

Using the property $L(x'') = -x'(0) - sx(0) + s^2 L(x)$ and the initial conditions $x_1(0) = 0$, $x_1'(0) = 0$, $x_2(0) = 0$ and $x_2'(0) = 0$, we have

$$s^2 L(x_1) = 10L(x_1) - 5L(x_2)$$
$$s2L(x_2) = -12L(x_1) + 6L(x_2) + e^{-s}\left(\frac{1}{s} - \frac{1}{s^2}\right).$$

or

$$(s^2 - 10)L(x_1) + 5L(x_2) = 0$$
$$12L(x_1) + (s^2 - 6)L(x_2) = e^{-s}\left(\frac{1}{s} - \frac{1}{s^2}\right).$$

Solving this system for $L(x_1)$ and $L(x_2)$, we find

$$L(x_1) = \frac{-5(s+1)}{s^4(s^2 - 16)}e^{-s}$$
$$L(x_2) = \frac{(s+1)(s^2 - 10)}{s^4(s^2 - 16)}e^{-s}.$$

Finding the partial fraction decomposition of each of the expressions above we have

$$L(x_1) = \left(\frac{5}{256}\cdot\frac{1}{s} + \frac{5}{256}\cdot\frac{1}{s^2} + \frac{5}{16}\cdot\frac{1}{s^3} + \frac{5}{16}\cdot\frac{1}{s^4} - \frac{15}{2048}\cdot\frac{1}{s+4} + \frac{25}{2048}\cdot\frac{1}{s-4}\right)e^{-s}$$
$$L(x_2) = \left(-\frac{3}{128}\cdot\frac{1}{s} + \frac{3}{128}\cdot\frac{1}{s^2} + \frac{5}{8}\cdot\frac{1}{s^3} + \frac{5}{8}\cdot\frac{1}{s^4} + \frac{9}{1024}\cdot\frac{1}{s+4} + \frac{15}{1024}\cdot\frac{1}{s-4}\right)e^{-s}.$$

Applying the inverse Laplace transform leads to

$$x_1 = u_1(t)\left(\frac{5}{48} - \frac{35}{256}t + \frac{5}{96}t^3 - \frac{15}{2048}e^{-4(t-1)} - \frac{25}{2048}e^{4(t-1)}\right)$$
$$x_2 = u_1(t)\left(\frac{5}{24} - \frac{43}{128}t + \frac{5}{48}t^3 + \frac{9}{1024}e^{-4(t-1)} + \frac{15}{1024}e^{4(t-1)}\right).$$

7. We have the system:

$$x_1'' = -3x_1 + 2x_2 + \delta(t)$$
$$x_2'' = x_1 - 4x_2 + \delta(t).$$

Now applying the Laplace transform to each equation in the system gives us

$$L(x_1'') = L(-3x_1 + 2x_2 + \delta(t))$$
$$L(x_2'') = L(x_1 - 4x_2 + \delta(t))$$

or

$$L(x_1'') = -3L(x_1) + 2L(x_2) + 1$$
$$L(x_2'') = L(x_1) - 4L(x_2) + 1.$$

Using the property $L(x'') = -x'(0) - sx(0) + s^2 L(x)$ and the initial conditions $x_1(0) = 0$, $x_1'(0) = 1$, $x_2(0) = 0$ and $x_2'(0) = 1$, we have

$$-1 + s^2 L(x_1) = -3L(x_1) + 2L(x_2) + 1$$
$$-1 + s2L(x_2) = L(x_1) - 4L(x_2) + 1.$$

or

$$(s^2 + 3)\mathcal{L}(x_1) - 2\mathcal{L}(x_2) = 2$$
$$-\mathcal{L}(x_1) + (s^2 + 4)\mathcal{L}(x_2) = 2.$$

Solving this system for $\mathcal{L}(x_1)$ and $\mathcal{L}(x_2)$, we find

$$\mathcal{L}(x_1) = \frac{8}{3} \cdot \frac{1}{s^2+2} - \frac{2}{3} \cdot \frac{1}{s^2+5}$$
$$\mathcal{L}(x_2) = \frac{4}{3} \cdot \frac{1}{s^2+2} + \frac{2}{3} \cdot \frac{1}{s^2+5}.$$

Applying the inverse Laplace transform leads to

$$x_1 = \frac{4\sqrt{2}}{3} \sin\sqrt{2}t - \frac{2\sqrt{5}}{15} \sin\sqrt{5}t$$
$$x_2 = \frac{2\sqrt{2}}{3} \sin\sqrt{2}t + \frac{2\sqrt{5}}{15} \sin\sqrt{5}t.$$

9. Let q_1 and q_2 be the amount of salt in the first tank and the second tank, respectively. We have

$$\frac{dq_1}{dt} = \text{the rate of salt going into tank 1 - the rate of salt going out of tank 1,}$$

$$\frac{dq_2}{dt} = \text{the rate of salt going into tank 2 - the rate of salt going out of tank 2.}$$

Using the flow rates given, it follows that

$$\frac{dq_1}{dt} = \frac{q_2 \text{ lb}}{100 \text{ gal}} \cdot 3\frac{\text{gal}}{\text{min}} - \frac{q_1 \text{ lb}}{50 \text{ gal}} \cdot 3\frac{\text{gal}}{\text{min}} \qquad \text{and}$$

$$\frac{dq_2}{dt} = \frac{q_1 \text{ lb}}{50 \text{ gal}} \cdot 3\frac{\text{gal}}{\text{min}} - \frac{q_2 \text{ lb}}{100 \text{ gal}} \cdot 3\frac{\text{gal}}{\text{min}}$$

or

$$\frac{dq_1}{dt} = -\frac{3}{50}q_1 + \frac{3}{100}q_2$$

$$\frac{dq_2}{dt} = \frac{3}{50}q_1 - \frac{3}{100}q_2.$$

Now applying the Laplace transform to each equation in the system gives us

$$\mathcal{L}(q_1') = \mathcal{L}\left(-\frac{3}{50}q_1 + \frac{3}{100}q_2\right)$$

$$\mathcal{L}(q_2') = \mathcal{L}\left(\frac{3}{50}q_1 - \frac{3}{100}q_2\right)$$

or

$$\mathcal{L}(q_1') = -\frac{3}{50}\mathcal{L}(q_1) + \frac{3}{100}\mathcal{L}(q_2)$$

$$\mathcal{L}(q_2') = \frac{3}{50}\mathcal{L}(q_1) - \frac{3}{100}\mathcal{L}(q_2).$$

Using the property $\mathcal{L}(q') = -q(0) + s\mathcal{L}(q)$ and the initial conditions $q_1(0) = 4$, $q_2(0) = 0$, we have

$$-4 + s\mathcal{L}(q_1) = -\frac{3}{50}\mathcal{L}(q_1) + \frac{3}{100}\mathcal{L}(q_2)$$

$$s\mathcal{L}(q_2) = \frac{3}{50}\mathcal{L}(q_1) - \frac{3}{100}\mathcal{L}(q_2).$$

or

$$\left(s+\frac{3}{50}\right)\mathcal{L}(q_1)-\frac{3}{100}\mathcal{L}(q_2) = 4$$

$$-\frac{3}{50}\mathcal{L}(q_1)+\left(s+\frac{3}{100}\right)\mathcal{L}(q_2) = 0.$$

Solving this system for $\mathcal{L}(q_1)$ and $\mathcal{L}(q_2)$, we find

$$\mathcal{L}(q_1) = \frac{4\left(s+\frac{3}{100}\right)}{s\left(s+\frac{9}{100}\right)}$$

$$\mathcal{L}(q_2) = \frac{6}{25}\cdot\frac{1}{s\left(s+\frac{9}{100}\right)}.$$

Finding the partial fraction decomposition of each of the expressions above we have

$$\mathcal{L}(q_1) = \frac{4}{3}\cdot\frac{1}{s}+\frac{8}{3}\cdot\frac{1}{s+\frac{9}{100}}$$

$$\mathcal{L}(q_2) = \frac{8}{3}\cdot\frac{1}{s}-\frac{8}{3}\cdot\frac{1}{s+\frac{9}{100}}.$$

Applying the inverse Laplace transform leads to

$$q_1 = \frac{4}{3}+\frac{8}{3}e^{-\frac{9}{100}t}$$

$$q_2 = \frac{8}{3}-\frac{8}{3}e^{-\frac{9}{100}t}.$$

11. Since Figure 6.10 is identical to Figure 6.4 we have the nonhomogeneous system

$$\begin{bmatrix} i_1' \\ i_2' \end{bmatrix} = \begin{bmatrix} -\frac{R}{L} & \frac{R}{L} \\ -\frac{R}{L} & \frac{R}{L}-\frac{1}{RC} \end{bmatrix} = \begin{bmatrix} i_1 \\ i_2 \end{bmatrix} + \begin{bmatrix} \frac{1}{L}E(t) \\ \frac{1}{L}E(t) \end{bmatrix}.$$

Expressing $E(t)$ in terms of a step function we have $E(t) = 100(1-u_1(t))$. With $R=2$, $L=2$ and $C=\frac{1}{4}$ we have the system

$$\begin{bmatrix} i_1' \\ i_2' \end{bmatrix} = \begin{bmatrix} -1 & 1 \\ -1 & -1 \end{bmatrix} = \begin{bmatrix} i_1 \\ i_2 \end{bmatrix} + \begin{bmatrix} 50(1-u_1(t)) \\ 50(1-u_1(t)) \end{bmatrix}.$$

Letting $v_1 = q_1$, $v_2 = q_1' = i_1$, $v_3 = q_2$ and $v_4 = q_2' = i_2$ we have the first order system

$$\begin{aligned} v_1' &= v_2 \\ v_2' &= -v_2+v_4+50(1-u_1(t)) \\ v_3' &= v_4 \\ v_4' &= -v_2-v_4+50(1-u_1(t)) \end{aligned}$$

with the initial conditions $v_1(0) = 0$, $v_2(0) = 0$, $v_3(0) = 0$, $v_4(0) = 0$. Now applying the Laplace transform to each equation in the system gives us

$$\begin{aligned} \mathcal{L}(v_1') &= \mathcal{L}(v_2) \\ \mathcal{L}(v_2') &= \mathcal{L}(-v_2+v_4+50(1-u_1(t))) \\ \mathcal{L}(v_3') &= \mathcal{L}(v_4) \\ \mathcal{L}(v_4') &= \mathcal{L}(-v_2-v_4+50(1-u_1(t))) \end{aligned}$$

or

$$\begin{aligned} \mathcal{L}(v_1') &= \mathcal{L}(v_2) \\ \mathcal{L}(v_2') &= -\mathcal{L}(v_2)+\mathcal{L}(v_4)+\frac{50}{s}-\frac{50}{s}e^{-s} \\ \mathcal{L}(v_3') &= \mathcal{L}(v_4) \\ \mathcal{L}(v_4') &= -\mathcal{L}(v_2)-\mathcal{L}(v_4)+\frac{50}{s}-\frac{50}{s}e^{-s}. \end{aligned}$$

Using the property $L(v') = -v(0) + sL(v)$ and the initial conditions we have

$$
\begin{aligned}
sL(v_1) &= L(v_2) \\
sL(v_2) &= -L(v_2) + L(v_4) + \frac{50}{s} - \frac{50}{s}e^{-s} \\
sL(v_3) &= L(v_4) \\
sL(v_4) &= -L(v_2) - L(v_4) + \frac{50}{s} - \frac{50}{s}e^{-s}.
\end{aligned}
$$

or

$$
\begin{aligned}
sL(v_1) - L(v_2) &= 0 \\
(s+1)L(v_2) - L(v_4) &= \frac{50}{s} - \frac{50}{s}e^{-s} \\
sL(v_3) - L(v_4) &= 0 \\
L(v_2) - (s+1)L(v_4) &= \frac{50}{s} - \frac{50}{s}e^{-s}
\end{aligned}
$$

Solving this system for $L(v_1)$, $L(v_2)$, $L(v_3)$ and $L(v_4)$, we find

$$
\begin{aligned}
L(v_1) &= 50\frac{s+2}{s^2(s^2+2s+2)} - 50e^{-s}\frac{s+2}{s^2(s^2+2s+2)} \\
L(v_2) &= 50\frac{s+2}{s(s^2+2s+2)} - 50e^{-s}\frac{s+2}{s(s^2+2s+2)} \\
L(v_3) &= 50\frac{1}{s(s^2+2s+2)} - 50e^{-s}\frac{1}{s(s^2+2s+2)} \\
L(v_4) &= 50\frac{1}{s^2+2s+2} - 50e^{-s}\frac{1}{s^2+2s+2}
\end{aligned}
$$

Here since only v_1 and v_3 represent the charges q_1 and q_2, respectively, (v_2 and v_4 represent currents) we only need to find the partial fraction decompositions of $L(v_1)$ and $L(v_3)$.

$$
\begin{aligned}
L(v_1) &= -\frac{25}{s} + \frac{50}{s^2} + \frac{25}{s^2+2s+2} - e^{-s}\left(-\frac{25}{s} + \frac{50}{s^2} + \frac{25}{s^2+2s+2}\right) \\
L(v_3) &= \frac{25}{s} - \frac{25(s+2)}{s^2+2s+2} - e^{-s}\left(\frac{25}{s} - \frac{25(s+2)}{s^2+2s+2}\right).
\end{aligned}
$$

Applying the inverse Laplace transform we find the charges

$$
\begin{aligned}
q_1 &= -25 + 50t + 25e^{-t}\cos t - 25e^{-t}\sin t + u_1(t)\left(75 - 50t - 25e^{1-t}\cos(t-1) + 25e^{1-t}\sin(t-1)\right) \\
q_3 &= 25 - 25e^{-t}\cos t - 25e^{-t}\sin t + u_1(t)\left(-25 + 25e^{1-t}\cos(t-1) + 25e^{1-t}\sin(t-1)\right).
\end{aligned}
$$

13. From Section 6.6 we have the system

$$
V' = \begin{bmatrix} 0 & 1 & 0 & 0 \\ -\frac{k_1+k_2}{m_1} & -\frac{f_1}{m_1} & \frac{k_2}{m_1} & 0 \\ 0 & 0 & 0 & 1 \\ \frac{k_2}{m_2} & 0 & -\frac{k_2}{m_2} & -\frac{f_2}{m_2} \end{bmatrix} V + \begin{bmatrix} 0 \\ \frac{1}{m_1}h_1(t) \\ 0 \\ \frac{1}{m_2}h_2(t) \end{bmatrix},
$$

where $v_1(t) = x_1(t)$, $v_2(t) = x_1'(t)$, $v_3(t) = x_2(t)$ and $v_4(t) = x_2'(t)$. Since the two masses are at rest and in the equilibrium position at $t = 0$ we have $v_1(0) = v_2(0) = v_3(0) = v_4(0) = 0$.

We are given that $f_1 = f_2 = 0$, $m_1 = m_2 = 1$, $k_1 = 3$, $k_2 = 2$, $h_1(t) = 0$ and $h_2(t) = \delta(t-2)$. Thus, we have

$$
V' = \begin{bmatrix} 0 & 1 & 0 & 0 \\ -5 & 0 & 2 & 0 \\ 0 & 0 & 0 & 1 \\ 2 & 0 & -2 & 0 \end{bmatrix} V + \begin{bmatrix} 0 \\ 0 \\ 0 \\ \delta(t-2) \end{bmatrix}.
$$

Applying the Laplace transform to each equation in the system gives us

$$\begin{aligned}
\mathcal{L}(v_1') &= \mathcal{L}(v_2) \\
\mathcal{L}(v_2') &= \mathcal{L}(-5v_2 + 2v_3) \\
\mathcal{L}(v_3') &= \mathcal{L}(v_4) \\
\mathcal{L}(v_4') &= \mathcal{L}(2v_1 - 2v_3 + \delta(t-2))
\end{aligned}$$

or

$$\begin{aligned}
\mathcal{L}(v_1') &= \mathcal{L}(v_2) \\
\mathcal{L}(v_2') &= -5\mathcal{L}(v_2) + 2\mathcal{L}(v_3) \\
\mathcal{L}(v_3') &= \mathcal{L}(v_4) \\
\mathcal{L}(v_4') &= 2\mathcal{L}(v_1) - 2\mathcal{L}(v_3) + e^{-2s}.
\end{aligned}$$

Using the property $\mathcal{L}(v') = -v(0) + s\mathcal{L}(v)$ and the initial conditions we have

$$\begin{aligned}
s\mathcal{L}(v_1) &= \mathcal{L}(v_2) \\
s\mathcal{L}(v_2) &= -5\mathcal{L}(v_2) + 2\mathcal{L}(v_3) \\
s\mathcal{L}(v_3) &= \mathcal{L}(v_4) \\
s\mathcal{L}(v_4) &= 2\mathcal{L}(v_1) - 2\mathcal{L}(v_3) + e^{-2s}.
\end{aligned}$$

or

$$\begin{aligned}
s\mathcal{L}(v_1) - \mathcal{L}(v_2) &= 0 \\
5\mathcal{L}(v_1) + s\mathcal{L}(v_2) - 2\mathcal{L}(v_3) &= 0 \\
s\mathcal{L}(v_3) - \mathcal{L}(v_4) &= 0 \\
-2\mathcal{L}(v_1) + 2\mathcal{L}(v_3) + s\mathcal{L}(v_4) &= e^{-2s}
\end{aligned}$$

Solving this system for $\mathcal{L}(v_1)$, $\mathcal{L}(v_2)$, $\mathcal{L}(v_3)$ and $\mathcal{L}(v_4)$, we find

$$\begin{aligned}
\mathcal{L}(v_1) &= e^{-2s}\frac{2}{s^4 + 7s^2 + 6} \\[4pt]
\mathcal{L}(v_2) &= e^{-2s}\frac{2s}{s^4 + 7s^2 + 6} \\[4pt]
\mathcal{L}(v_3) &= e^{-2s}\frac{s^2 + 5}{s^4 + 7s^2 + 6} \\[4pt]
\mathcal{L}(v_4) &= e^{-2s}\frac{s^3 + 5s}{s^4 + 7s^2 + 6}.
\end{aligned}$$

Here since v_1 and v_3 represent the positions x_1 and x_2, respectively, (v_2 and v_4 represent velocities) we only need to find the partial fraction decompositions for v_1 and v_3. Finding these decompositions we have

$$\begin{aligned}
\mathcal{L}(v_1) &= e^{-2s}\left(\frac{2}{5}\cdot\frac{1}{s^2+1} - \frac{2}{5}\cdot\frac{1}{s^2+6}\right) \\[4pt]
\mathcal{L}(v_3) &= e^{-2s}\left(\frac{4}{5}\cdot\frac{1}{s^2+1} + \frac{1}{5}\cdot\frac{1}{s^2+6}\right).
\end{aligned}$$

Applying the inverse Laplace transform leads to

$$\begin{aligned}
x_1 &= u_2(t)\left(\frac{2}{5}\sin(t-2) - \frac{\sqrt{6}}{15}\sin\sqrt{6}(t-2)\right) \\[8pt]
x_2 &= u_2(t)\left(\frac{4}{5}\sin(t-2) + \frac{\sqrt{6}}{30}\sin\sqrt{6}(t-2)\right).
\end{aligned}$$

Chapter 8

Power Series Solutions to Linear Differential Equations

8.1 Introduction to Power Series Solutions

Exercises 8.1, pp. 383-384.

1. Follow Example 1 in the text. Since we are given an initial condition at $x = 0$, we use $x_0 = 0$. Our desired solution will have the form $y = \sum_{n=0}^{\infty} a_n x^n$. Substituting this power series for y and the formula given in Theorem 8.2 for y', we have

$$y' + y = \sum_{n=0}^{\infty} (n+1)a_{n+1}x^n + \sum_{n=0}^{\infty} a_n x^n = 0.$$

This equation holds if and only if $(n+1)a_{n+1} + a_n = 0$ for $n = 0, 1, 2, \cdots$. Solving for a_{n+1} gives us $a_{n+1} = \frac{-a_n}{n+1}$, which is a recurrence relation. We calculate a few terms to see if we can notice a pattern:

$$a_1 = \frac{-a_0}{1},$$

$$a_2 = \frac{-a_1}{2} = \frac{-(-a_0)}{2!},$$

$$a_3 = \frac{-a_2}{3} = \frac{-(-(-a_0))}{3!},$$

and so on. We can write

$$a_n = \frac{(-1)^n a_0}{n!}.$$

Since we know from the initial condition that $a_0 = 2$, we have $y = 2\sum_{n=0}^{\infty}(-1)^n x^n/n! = 2e^{-x}$.

3. Follow Example 2 in the text. Since we are given an initial condition at $x = 1$, we use $x_0 = 1$. Our desired solution will have the form $y = \sum_{n=0}^{\infty} a_n(x-1)^n$. Substituting this power series for y and the formula given in Theorem 8.2 for y', we have

$$y' - (2x-2)y = \sum_{n=0}^{\infty} (n+1)a_{n+1}(x-1)^n - 2\sum_{n=0}^{\infty} a_n(x-1)^{n+1} = 0.$$

The powers of $x-1$ do not agree in this equation, since the exponent in the first power series starts at 0, while in the second power series the exponent starts at 1. We rewrite the first power series to get the exponents to start at 1 in the summation part, which gives us

$$a_1 + \sum_{n=1}^{\infty} (n+1)a_{n+1}(x-1)^n - 2\sum_{n=0}^{\infty} a_n(x-1)^{n+1} = 0.$$

We now reindex the first summation so that the index n starts at 0 rather than 1:

$$a_1 + \sum_{n=0}^{\infty} (n+2)a_{n+2}(x-1)^{n+1} - 2\sum_{n=0}^{\infty} a_n(x-1)^{n+1} = 0.$$

This equation holds if and only if $(n+2)a_{n+2} - a_n = 0$ for $n = 0, 1, 2, \cdots$. Solving for a_{n+2} gives us $a_{n+2} = \frac{a_n}{n+2}$, which is a recurrence relation. We calculate a few terms to see if we can notice a pattern:

$$a_1 = 0,$$

$$a_2 = \frac{a_0}{2},$$

$$a_3 = 0,$$

$$a_4 = \frac{a_2}{4} = \frac{a_0}{2 \cdot 4!},$$

and so on. Write a few more if you want to see the pattern more clearly. It follows that only coefficients of the form a_{2n} for $n = 1, 2, 3, \cdots$ are nonzero and that

$$a_{2n} = \frac{a_0}{2n!}.$$

Since we know from the initial condition that $a_1 = 1$, we have $y = \sum_{n=0}^{\infty} x^n / 2^n n! = 2e^{(x-1)^2/2}$.

5. Follow Example 3 in the text. Since we are given an initial condition at $x = 0$, we use $x_0 = 0$. Our desired solution will have the form $y = \sum_{n=0}^{\infty} a_n x^n$. Substituting this power series for y and applying the formula given in Theorem 8.2 twice for y'', we have

$$y'' - y = \sum_{n=0}^{\infty} (n+2)(n+1)a_{n+2}x^n - \sum_{n=0}^{\infty} a_n x^n = 0.$$

This equation holds if and only if $(n+2)(n+1)a_{n+2} - a_n = 0$ for $n = 0, 1, 2, \cdots$. Solving for a_{n+2} gives us $a_{n+2} = \frac{a_n}{(n+2)(n+1)}$, which is a recurrence relation. We calculate a few terms to see if we can notice a pattern:

$$a_2 = \frac{a_0}{2!},$$

$$a_3 = \frac{a_1}{3!},$$

$$a_4 = \frac{a_2}{4 \cdot 3} = \frac{a_0}{4!},$$

$$a_5 = \frac{a_1}{5!},$$

and so on. We can write

$$a_n = \frac{a_0}{n!}$$

or

$$a_n = \frac{a_1}{n!}$$

depending if n is even or odd. Since we know from the initial condition that $a_0 = 1$ and $a_1 = y'(0) = 0$, our series will have only even-numbered terms. Thus, we have $y = \sum_{n=0}^{\infty} x^{2n} / (2n)!$.

7. Following Example 3 in the text (with y' substituted for y, we have

$$y'' + y' = \sum_{n=0}^{\infty} (n+2)(n+1)a_{n+2}x^n + \sum_{n=0}^{\infty} (n+1)a_{n+1}x^n = 0.$$

Combining the two power series and equating terms, we have

$$(n+2)(n+1)a_{n+2} = -(n+1)a_{n+1},$$

so $a_{n+2} = -a_{n+1}/(n+2)$. After calculating a few terms to be able to notice a pattern, we see that, starting at a_2, $a_n = \frac{(-1)^{n+1}a_1}{n!}$. Since $a_0 = y(0) = 1$ and $a_1 = y'(0) = (-1)$, we obtain $y = \sum_{n=0}^{\infty} (-x^n)/n!$.

9. Follow Example 2 in the text, but in this problem $x_0 = 0$, and we substitute $y = \sum_{n=0}^{\infty} a_n x^n$ into $y' - xy = 0$ to obtain

$$y' - xy = \sum_{n=0}^{\infty}(n+1)a_{n+1}x^n - \sum_{n=0}^{\infty}a_n x^{n+1} = 0.$$

The powers of x do not agree in this equation, since the exponent in the first power series starts at 0, while in the second power series the exponent starts at 1. We rewrite the first power series to get the exponents to start at 1 in the summation part, then reindex the first summation so that the index n starts at 0 rather than 1:

$$a_1 + \sum_{n=0}^{\infty}(n+2)a_{n+2}x^{n+1} - \sum_{n=0}^{\infty}a_n x^{n+1} = 0.$$

Therefore, $a_{n+2} = a_n/(n+2)$. After writing a few terms of this recurrence and substituting the initial conditions given, we have $y = \sum_{n=0}^{\infty}(x^{2n})/(2^n(2n)!)$.

11. Follow Example 3 in the text, replacing y with $4y$ to obtain $y = (1/2)\sum_{n=0}^{\infty}(-1)^n((2x)^{2n+1})/(2n+1)!)$.

13. $a_0 = y(0) = 1; a_1 = y'(0) = 2xy(0) = 2x(1) = 2 \cdot 0 \cdot 1 = 0$, since $x_0 = 0$. To calculate $a_2 = y''(0)/2!$, first find $y'' = 2xy' + 2y$ by the product rule applied to the formula given for y', then evaluate it at $y(0), y'(0)$, and $x = 0$ to obtain $y''(0) = 2$. So $a_2 = 2/2! = 1$. Continue to find a_3, a_4 and so on in the same manner, ending with the MacLaurin series $y = 1 + 0x + x^2 + 0x^3 + x^4/2 + \cdots$.

15. $a_0 = y(0) = 1; a_1 = y'(0) = 0$. Since $y'' = 4(x-2)y$, we calculate $y''(0)$ by substituting 0 for x and $y(0) = 1$ for y, giving us $a_2 = y''(0)/2! = -8/2 = -4$. To calculate $a_3 = y'''(0)/3!$, first find $y''' = 4(x-2)y' + 4y$ by the product rule applied to the formula for y'', then evaluate it at $y(0), y'(0)$, and $x = 0$ to obtain $y'''(0) = -4$. So $a_3 = -4/3! = -2/3$. Continue to find a_3, a_4 and so on in the same manner, ending with the MacLaurin series $y = 1 + 0x - 4x^2 - (2/3)x^3 + \cdots$.

8.2   Series Solutions for Second Order Linear Differential Equations

Exercises 8.2, pp. 392-393

1. We substitute

$$y = \sum_{n=0}^{\infty} a_n x^n$$

and

$$y'' = \sum_{n=0}^{\infty}(n+2)(n+1)a_{n+2}x^n$$

into the differential equation $y'' - xy = 0$:

$$y'' - xy = y'' = \sum_{n=0}^{\infty}(n+2)(n+1)a_{n+2}x^n - \sum_{n=0}^{\infty}xa_n x^n = \sum_{n=0}^{\infty}(n+2)(n+1)a_{n+2}x^n - \sum_{n=0}^{\infty}a_n x^{n+1}.$$

Combining the power series and reindexing (see Example 1 in the text for details on how to do this), we have

$$2a_2 + \sum_{n=0}^{\infty}((n+3)(n+2)a_{n+3} - a_n)x^{n+1} = 0,$$

giving us $2a_2 = 0$ and $(n+3)(n+2)a_{n+3} - a_n = 0$. Consequently,

$$a_3 = a_0/3!,$$

$$a_4 = a_1/12 = 2a_1/4!,$$

$$a_5 = a_2/20 = 0,$$

$$a_6 = a_3/(6 \cdot 5) = 4a_0/6!,$$

$$a_7 = a_4/(7 \cdot 6) = (2 \cdot 5)a_1/7!,$$

$$a_8 = 0,$$

$$a_9 = a_6/(9 \cdot 8) = (4 \cdot 7)a_0/9!,$$

and so on. After figuring out the patterns involved, we see that

$$y = a_0 + (a_0/3!)x^3 + \sum_{n=2}^{\infty} \frac{4 \cdot 7 \cdots (3n-2)}{(3n)!} a_0 x^{3n} + a_1 x + (2a_1/4!)x^4 + \sum_{n=2}^{\infty} \frac{2 \cdot 5 \cdots (3n-1)}{(3n+1)!} a_1 x^{3n+1}.$$

Using $a_0 = y(0) = 1$ and $a_1 = y'(0) = 0$, we have

$$y = a_0 + (a_0/3!)x^3 + \sum_{n=2}^{\infty} \frac{4 \cdot 7 \cdots (3n-2)}{(3n)!} a_0 x^{3n} = 1 + (1/3!)x^3 + \sum_{n=2}^{\infty} \frac{4 \cdot 7 \cdots (3n-2)}{(3n)!} x^{3n}.$$

3. We substitute

$$y = \sum_{n=0}^{\infty} a_n x^n$$

and

$$y'' = \sum_{n=0}^{\infty} (n+2)(n+1)a_{n+2}x^n$$

into the differential equation $y'' - xy' + 2y = 0$:

$$y'' - xy' + 2y = y'' = \sum_{n=0}^{\infty}(n+2)(n+1)a_{n+2}x^n - \sum_{n=0}^{\infty}x(n+1)a_n x^n + 2\sum_{n=0}^{\infty}a_n x^n$$

$$= \sum_{n=0}^{\infty}(n+2)(n+1)a_{n+2}x^n - \sum_{n=0}^{\infty}(n+1)a_n x^{n+1} + 2\sum_{n=0}^{\infty}a_n x^n.$$

Combining the power series and reindexing (see Example 2 in the text for details on how to do this), we have

$$\sum_{n=0}^{\infty}((n=3)(n+2)a_{n+3} - (n+1-2)a_{n-1}) = 0$$

Consequently, $a_{n+3} = (a_{n+1}(n-1))/(n+3)(n+2)$. After figuring out the patterns involved, we see that

$$y = a_0(1-x^2) + a_1(x - x^3/3! - x^5/5! - \sum_{n=1}^{\infty}(1)(3) \cdots (2n+1)x^{2n+5}/(2n+5)!)$$

Applying the initial conditions of $y(0) = a_0 = 0$ and $y'(0) = a_1 = 2$, we have

$$y = 2(x - x^3/3! - x^5/5! - (3/7!)x^7 - \cdots - ((1)(3) \cdots (2n+1)/(2n+5)!)x^{2n+5}.$$

5. If you worked Exercise 3 in this set, you already did most of the work for this problem. We substitute

$$y = \sum_{n=0}^{\infty} a_n x^n$$

and

$$y'' = \sum_{n=0}^{\infty} (n+2)(n+1)a_{n+2}x^n$$

into the differential equation $y'' - 2y' + 2xy = 4 - 2x$:

$$y'' - xy' + 2y = y'' = \sum_{n=0}^{\infty}(n+2)(n+1)a_{n+2}x^n - \sum_{n=0}^{\infty}2(n+1)a_n x^n + 2\sum_{n=0}^{\infty}a_n xx^n$$

$$= \sum_{n=0}^{\infty}(n+2)(n+1)a_{n+2}x^n - \sum_{n=0}^{\infty}(n+1)2a_n x^{n+1} + 2\sum_{n=0}^{\infty}a_n xx^n.$$

Combining the power series and reindexing (see Example 2 in the text for details on how to do this), we have

$$\sum_{n=0}^{\infty}((n=3)(n+2)a_{n+3} - (n+1-2+2)a_{n-1}) = 4 - 2x.$$

Note that this means all coefficients of powers of x, except for x^0 and x^1, must be zero.

7. If you worked Exercise 1 in this set you already did most of the work for this problem. We repeat many of the details here. We substitute

$$y = \sum_{n=0}^{\infty} a_n x^n$$

and

$$y'' = \sum_{n=0}^{\infty} (n+2)(n+1)a_{n+2}x^n$$

into the differential equation $y'' - xy = 0$:

$$y'' - xy = y'' = \sum_{n=0}^{\infty}(n+2)(n+1)a_{n+2}x^n - \sum_{n=0}^{\infty} xa_n x^n = \sum_{n=0}^{\infty}(n+2)(n+1)a_{n+2}x^n - \sum_{n=0}^{\infty} a_n x^{n+1}.$$

Combining the power series and reindexing (see Example 1 in the text for details on how to do this), we have

$$2a_2 + \sum_{n=0}^{\infty} \left((n+3)(n+2)a_{n+3} - a_n\right)x^{n+1} = 0,$$

giving us $2a_2 = 0$ and $(n+3)(n+2)a_{n+3} - a_n = 0$. Consequently,

$$a_3 = a_0/3!,$$

$$a_4 = a_1/12 = 2a_1/4!,$$

$$a_5 = a_2/20 = 0,$$

$$a_6 = a_3/(6 \cdot 5) = 4a_0/6!,$$

$$a_7 = a_4/(7 \cdot 6) = (2 \cdot 5)a_1/7!,$$

$$a_8 = 0,$$

$$a_9 = a_6/(9 \cdot 8) = (4 \cdot 7)a_0/9!,$$

and so on. After figuring out the patterns involved, we see that

$$y = a_0 + (a_0/3!)x^3 + \sum_{n=2}^{\infty} \frac{4 \cdot 7 \cdots (3n-2)}{(3n)!} a_0 x^{3n} + a_1 x + (2a_1/4!)x^4 + \sum_{n=2}^{\infty} \frac{2 \cdot 5 \cdots (3n-1)}{(3n+1)!} a_1 x^{3n+1}.$$

9. If you worked Exercise 1 in this set you already did most of the work for this problem. We repeat many of the details here. We substitute

$$y = \sum_{n=0}^{\infty} a_n x^n$$

and

$$y'' = \sum_{n=0}^{\infty} (n+2)(n+1)a_{n+2}x^n$$

into the differential equation $y'' - xy = 0$:

$$y'' - xy = y'' = \sum_{n=0}^{\infty}(n+2)(n+1)a_{n+2}x^n - \sum_{n=0}^{\infty} xa_n x^n = \sum_{n=0}^{\infty}(n+2)(n+1)a_{n+2}x^n - \sum_{n=0}^{\infty} a_n x^{n+1}.$$

Combining the power series and reindexing (see Example 1 in the text for details on how to do this), we have

$$2a_2 + \sum_{n=0}^{\infty} \left((n+3)(n+2)a_{n+3} - a_n\right)x^{n+1} = 0,$$

giving us $2a_2 = 0$ and $(n+3)(n+2)a_{n+3} - a_n = 0$. Consequently,

$$a_3 = a_0/3!,$$

$$a_4 = a_1/12 = 2a_1/4!,$$
$$a_5 = a_2/20 = 0,$$
$$a_6 = a_3/(6 \cdot 5) = 4a_0/6!,$$
$$a_7 = a_4/(7 \cdot 6) = (2 \cdot 5)a_1/7!,$$
$$a_8 = 0,$$
$$a_9 = a_6/(9 \cdot 8) = (4 \cdot 7)a_0/9!,$$

and so on. After figuring out the patterns involved, we see that

$$y = a_0 + (a_0/3!)x^3 + \sum_{n=2}^{\infty} \frac{4 \cdot 7 \cdots (3n-2)}{(3n)!} a_0 x^{3n} + a_1 x + (2a_1/4!)x^4 + \sum_{n=2}^{\infty} \frac{2 \cdot 5 \cdots (3n-1)}{(3n+1)!} a_1 x^{3n+1}.$$

11. If you worked Exercise 3 in this set, you have already done most of the work for this problem. We substitute

$$y = \sum_{n=0}^{\infty} a_n x^n$$

and

$$y'' = \sum_{n=0}^{\infty} (n+2)(n+1)a_{n+2} x^n$$

into the differential equation $y'' - xy' + 2y = 0$:

$$y'' - xy' + 2y = y'' = \sum_{n=0}^{\infty} (n+2)(n+1)a_{n+2}x^n - \sum_{n=0}^{\infty} x(n+1)a_n x^n + 2\sum_{n=0}^{\infty} a_n x^n$$

$$= \sum_{n=0}^{\infty} (n+2)(n+1)a_{n+2}x^n - \sum_{n=0}^{\infty} (n+1)a_n x^{n+1} + 2\sum_{n=0}^{\infty} a_n x^n.$$

Combining the power series and reindexing (see Example 2 in the text for details on how to do this), we have

$$\sum_{n=0}^{\infty} ((n=3)(n+2)a_{n+3} - (n+1-2)a_{n-1}) = 0$$

Consequently, $a_{n+3} = (a_{n+1}(n-1))/(n+3)(n+2)$. After figuring out the patterns involved, we see that

$$y = a_0(1-x^2) + a_1(x - x^3/3! - x^5/5! - \sum_{n=1}^{\infty} (1)(3)\cdots(2n+1)x^{2n+5}/(2n+5)!).$$

13. Substitute the MacLaurin series for $\cos x$:

$$\cos x = 1 - x^2/2 + x^4/4! - x^6/6! + \cdots,$$

$$y'' = \sum_{n=0}^{\infty} (n+2)(n+1)a_{n+2}x^n,$$

and

$$y = \sum_{n=0}^{\infty} a_n x^n$$

to get

$$\sum_{n=0}^{\infty} (n+2)(n+1)a_{n+2}x^n - (1 - x^2/2 + x^4/4! - x^6/6! + \cdots) \sum_{n=0}^{\infty} a_n x^n = 0.$$

After combining these power series, the idea is to equate coefficients and solve for the coefficients $a_0, a_1, a_2, \cdots$. In order to get the first four terms of the power series solution, we only need to look at terms involving $1, x, x^2$, and x^3. So we can view the equation as

$$\sum_{n=0}^{\infty} (n+2)(n+1)a_{n+2}x^n - \sum_{n=0}^{\infty} a_n x^n + (x^2/2)\sum_{n=0}^{\infty} a_n x^n.$$

Equating coefficients as in Example 4 in the textbook, we see that $a_2 = a_0/2$ and $a_3 = a_1/6$. This gives us

$$y = a_0 + a_1 x + (a_0/2)x^2 + (a_1/6)x^3.$$

15. Substitute the MacLaurin series for e^x:

$$e^x = 1 + x + x^2/2 + x^3/3! + x^4/4! + \cdots,$$

$$y'' = \sum_{n=0}^{\infty} (n+2)(n+1)a_{n+2}x^n,$$

and

$$y = \sum_{n=0}^{\infty} a_n x^n$$

to get

$$(e^x = 1 + x + x^2/2 + x^3/3! + x^4/4! + \cdots)(\sum_{n=0}^{\infty}(n+2)(n+1)a_{n+2}x^n + \sum_{n=0}^{\infty} a_n x x^n = 0.$$

After combining these power series, the idea is to equate coefficients and solve for the coefficients $a_0, a_1, a_2, \cdots$. In order to get the first four terms of the power series solution, we only need to look at terms involving $1, x, x^2$, and x^3. Proceed as in Problem 13 to arrive at $a_0 + a_1 x - (a_0/6)x^3 + ((a_0 - a_1)/12)x^4$.

17. $v(v+1) = 3/4$. Using the formulas in the boxes on page 392 of the textbook,

$$a_2 = ((-3/4)/2)a_0 = (-3/8)a_0,$$

$$a_3 = (2 - (3/4))a_1/6 = (5/24)a_1,$$

$$a_4 = ((0+2)(0+3) - (3/4))a_2/(0+4)(0+3) = (-21/16)a_0,$$

and so on. We can continue in this way, generating all the terms of $y = a_0 + a_1 x - (3/8)a_0 x^2 + \cdots$.

19. $v(v+1) = 2$. Using the formulas in the boxes on page 392 of the textbook,

$$a_2 = ((-2/2)a_0 = -a_0,$$

$$a_3 = (2 - 2))a_1/6 = 0,$$

$$a_4 = ((0+2)(0+3) - (2a_2)/(0+4)(0+3) = (-1/3)a_0,$$

and so on. We can continue in this way, generating all the terms of $y = -a_0 - (1/3)a_0 x^2 + \cdots$.

21. $v(v+1) = 20$. Using the formulas in the boxes on page 392 of the textbook,

$$a_2 = ((-20)/2)a_0 = (-10)a_0,$$

$$a_3 = -3a_1,$$

$$a_4 = (-7/6)a_0,$$

and so on. We can continue in this way, generating all the terms of y.

8.3 Euler Type Equations

Exercises 8.3, p. 397

1. The equation $r(r-1) + \alpha r + \beta = 0$ for this problem is $r(r-1) + (-3)r + 3 = 0$. Solving for r, we have

$$r^2 - r - 3r + 3 = r^2 - 4r + 3 = (r-1)(r-3) = 0.$$

So $r = 1$ or $r = 3$. The general solution is therefore $y = c_1 x^1 + c_2 x^3$.

3. The equation $r(r-1) + \alpha r + \beta = 0$ for this problem is $r(r-1) + 1r + 1 = 0$. Solving for r, we have $r^2 + 1 = (r-i)(r+i) = 0$, whose roots are i and $-i$. The general solution is therefore $y = c_1 \cos(\ln x) + c_2 \sin(\ln x)$.

5. Multiplying through by x^2, we have $x^2y'' + 5xy' + 4y = 0$. The equation $r(r-1) + \alpha r + \beta = 0$ for this problem is

$$r(r-1) + 5r + 4 = 0.$$

   Solving for r, we have $r^2 + 4r + 4 = (r+2)^2 = 0$, which has $r = -2$ as a repeated root. The general solution is therefore $y = \frac{c_1}{x^2} + \frac{c_2 \ln x}{x^2}$.

7. The equation $r(r-1) + \alpha r + \beta = 0$ for this problem is $r(r-1) + 3r + 2 = 0$. Solving for r, we have $r^2 + 2r + 2 = 0$, whose solutions (by the quadratic formula) are $-1 + i$ and $-1 - i$. Therefore the general solution is $y = \frac{c_1 \cos(\ln x))}{x} + \frac{c_2 \sin(\ln x)}{x}$.

9. Multiplying through by x^2, we have $x^2y'' - xy' + 2y = 0$. The equation $r(r-1) + \alpha r + \beta = 0$ for this problem is

$$r(r-1) - r + 2 = 0.$$

   Solving for r, we have $r^2 - 2r + 2 = 0$. Using the quadratic formula, we find that $r = 1 + i$ or $1 - i$. Therefore the general solution is $y = c_1 x \cos(\ln x) + c_2 x \sin(\ln x)$. Calculating y' and substituting the initial conditions for y and y', we find that the solution to this initial value problem is $y = x \cos(\ln x) - x \sin(\ln x)$.

11. Adapt the method of the discussion on pages 394 and 395 of the textbook, or use the solution methods of Section 8.2

13. Substituting u for $x - 1$, we rewrite this a $u^2y'' + 5uy' + 3y = 0$. The equation $r(r-1) + \alpha r + \beta = 0$ for this problem is $r(r-1) + 5r + 3 = r^2 + 4r + 3 = (r+1)(r+3) = 0$, whose solutions are $r = -1$ and $r = -3$. So its general solution is $y = c_1 u^{-1} + c_2 u^{-3}$. Replacing u with $x - 1$, we have $y = c_1(x-1)^{-1} + c_2(x-1)^{-3}$.

8.4   Series Solutions Near a Regular Singular Point

Exercises 8.4, pp. 410-411

1. Writing this differential equation in the form $y'' + p(x)y' + q(x)y = 0$, we have $y'' + \frac{\sin x}{x^2}y' + \frac{1}{x^2}y = 0$. Since x^2 is in the denominator, there is a regular singular point when $x = 0$.

3. Written in the form $y'' + p(x)y' + q(x)y = 0$, this differential equation is $y'' + \frac{3x-1}{x(x-1)}y' + \frac{1}{x(x-1)}y = 0$. Then $p(x) = \frac{3x-1}{x(x-1)}$, $xp(x) = \frac{3x-1}{(x-1)}$, and $p_0 = 1$. Also, $q(x) = \frac{1}{x(x-1)}$, $x^2q(x) = \frac{x}{(x-1)}$, and $q_0 = 0$. Finally, $F(r) = r(r-1) + r = r^2$, so $r^2 = 0$ is the indicial equation.

5. Written in the form $y'' + p(x)y' + q(x)y = 0$, this differential equation is $y'' + \frac{3}{x}y' + \frac{2}{6}y = 0$. Then $xp(x) = 3$, and $p_0 = 3$. Also, $x^2q(x) = x^4$, and $q_0 = 0$. The indicial equation is $r(r-1) + 3r = r^2 + 2r$, so $r_1 = 0$ and $r_2 = -2$, and we know there is a solution of the form $x^{-2}\sum_{n=0}^{\infty} a_n x^n$. Following Example 2 in the text, we obtain $y = x^{-2}(c_1 \cos x^2 + c_2 \sin x^2)$.

7. Written in the form $y'' + p(x)y' + q(x)y = 0$, this differential equation is $y'' + \frac{4}{6x}y' + \frac{1}{4x^2}y = 0$. Then $xp(x) = 2/3$, and $p_0 = 2/3$. Also, $x^2q(x) = (1/3)x^2$, and $q_0 = 0$. The indicial equation is $r(r-1) + (2/3)r = r^2 - (1/3)r$, so $r_1 = 1/3$ and $r_2 = 0$, and we are in the case of Theorem 8.7. We follow Example 1 in the text, if possible using Maple or another automated system to obtain the solution, which has large numbers in it.

9. Proof.

11. A Bessel equation has the form $x^2y'' + xy' + (x^2 - v^2)y = 0$. In this problem, we have $x^2y'' + xy' + (x^2 - (9/4))y = 0$, so $v = 3/2$. Since $2v = 3$ is a positive integer, we are in the case of Theorem 8.9. Substituting the power series form of y and its derivatives into the original equation, equating coefficients, and following Example 4 in the text, we obtain the solution $y = x^{-3/2}(c_1 \cos x + x \sin x) + c_2(\sin x - x \cos x)$.

Chapter 9

Inner Product Spaces

9.1   Inner Product Spaces

Exercises 9.1, pp. 419-421

1. We use the formulas given in the text, substituting $v \cdot u$ for $< v, u >$.

$$< v, u > v \cdot u = 2(-3) + (-1)(1) = -6 - 1 = -7.$$

$$\|v\| = \sqrt{2^2 + (-1)^2} = \sqrt{5}.$$

$$\|u\| = \sqrt{(-3)^2 + 1^2} = \sqrt{10}.$$

$$\theta = \cos^{-1} \frac{< v, u >}{\|v\| \|u\|} = \cos^{-1} \frac{-7}{\sqrt{5}\sqrt{10}} = \frac{-7}{5\sqrt{2}}.$$

3. We use the formulas given in the text, substituting $v \cdot u$ for $< v, u >$.

$$< v, u > = v \cdot u = 3(6) + (-1)(-2) + (1)(1) + 0(3) = 18 + 2 + 1 + 0 = 21.$$

$$\|v\| = \sqrt{3^2 + (-1)^2 + 1^2 + 0^2} = \sqrt{11}.$$

$$\|u\| = \sqrt{(6)^2 + (-2)^2 + 1^2 + 3^2} = \sqrt{50} = 5\sqrt{2}.$$

$$\theta = \cos^{-1} \frac{< v, u >}{\|v\| \|u\|} = \cos^{-1} \frac{21}{\sqrt{11}\sqrt{50}} = \frac{21}{5\sqrt{22}}.$$

5. We use the formulas given in the text, substituting $\int_0^1 v(x)u(x)dx$ for $< v, u >$.

$$< v, u > = \int_0^1 v(x)u(x)dx = \int_0^1 (x^2 - x)(3x + 5) = \int_0^1 (3x^3 + 2x^2 - 5x)dx = \frac{-13}{12}.$$

$$\|v\| = \sqrt{< v, v >} = \sqrt{\int_0^1 v(x)v(x)dx} = \sqrt{\int_0^1 (x^2 - x)(x^2 - x)dx} = \sqrt{\int_0^1 (x^4 - 2x^3 + x^2 dx} = \sqrt{\frac{1}{30}}.$$

$$\|u\| = \sqrt{< u, u >} = \sqrt{\int_0^1 (3x + 5)(3x + 5)dx} = \sqrt{int_0^1 (9x^2 + 30x + 25)dx} = \sqrt{43}.$$

$$\theta = \cos^{-1} \frac{< v, u >}{\|v\| \|u\|} = \cos^{-1} \frac{\frac{-13}{12}}{\sqrt{\frac{1}{30}}\sqrt{43}} = \frac{-13}{12}\sqrt{\frac{30}{43}}.$$

7. Property 1: Since $\int_a^b f(x)g(x)dx = \int_a^b g(x)f(x)dx$ by the commutative property of multiplication of polynomials over the real numbers, we have $< f,g >=< g,f >$.
Property 2: $< f+h,g >$

$$= \int_a^b (f(x)+h(x))g(x)dx = \int_a^b ((f(x)g(x)+h(x)g(x))dx$$

$$= \int_a^b f(x)g(x)dx + \int_a^b h(x)g(x)dx$$

$=< f,g > + < h,g >$.
Property 3: $< cf,g >= \int_a^b cf(x)g(x)dx = c\int_a^b f(x)g(x)dx = c < f,g >$.
Property 4: $< f,f >= \int_a^b (f(x))^2 dx$. Since $(f(x))^2 \geq 0$ for any x, $\int_a^b (f(x))^2 dx$ is also greater than or equal to 0 and is 0 only if $f(x) = 0$. Hence $< f,f >= 0$ if and only if $f = 0$.

9. Verify Properties 1, 2, 3, and 4 of the inner product.

11. Verify Properties 1, 2, 3, and 4 of the inner product.

13. Part a: Show that if the partial sums have a limit, the terms of the sequence must approach 0.
Part b: Use the hint given.
Part c: Show that l_2 is closed under addition and scalar multiplication.
Part d: Verify Properties 1, 2, 3, and 4 of the inner product.

15. Part (a): We use mathematical induction on n. The result holds for $n = 1$ by the definition of inner product. Assume that the result holds for $n = t - 1$; that is,

$$< c_1v_1 + c_2v_2 + \cdots + c_{t-1}v_{t-1}, u >= c_1 < v_1, u > + c_2 < v_2, u > + \cdots + c_{t-1} < v_{t-1}, u > .$$

Then for $n = t$,

$$< c_1v_1 + c_2v_2 + \cdots + c_tv_t, u >$$
$$= < (c_1v_1 + c_2v_2 + \cdots + c_{t-1}v_{t-1}) + c_tv_t, u >$$
$$= < c_1v_1 + c_2v_2 + \cdots + c_{t-1}v_{t-1}, u > + c_t < v_t, u > \text{ (by the definition of inner product)}$$
$$= c_1 < v_1, u > + c_2 < v_2, u > + \cdots + c_{t-1} < v_{t-1}, u > + c_t < v_t, u > \text{ (by the induction hypothesis.)}$$

Part (b): Imitate part (a) or apply the requirement that $< u,v >=< v,u >$ to the result of part (a).

17. $0 \leq < v+u, v+u >$

$$=< v,v+u > + < u,v+u >=< v,v > + < v,u > + < u,v > + < u,u >= ||v||^2 + 2 < v,u > + ||u||^2 = 2 < v,u > + 2.$$

Solving the inequality $0 \leq 2 < v,u > + 2$, we see that $< v,u > \geq -1$.

19. To show that two vectors are orthogonal, we show that their inner product equals 0. Using the definition of inner product for $C[-1,1]$, we have

$$< 2\sqrt{3}x, 3\sqrt{5}x^2 - \sqrt{5} >= \int_{-1}^1 (2\sqrt{3}x)(3\sqrt{5}x^2 - \sqrt{5})dx = \int_{-1}^1 (6\sqrt{15}x^3 - 2\sqrt{15}x)dx = 0.$$

21. Assume v and u are orthogonal; that is, $< v,u >=< u,v >= 0$. Then $||v+u||^2 =< v+u, v > + < v+u, u >=< v, v+u > + < u, v+u >=< v,v > + < v,u > + < u,v > + < u,u >=< v,v > + 0 + 0 + < u,u >= ||v||^2 + ||u||^2$.

23. To show that T is a linear transformation, we show that T preserves vector addition and scalar multiplication.
Vector addition: Let u and v be vectors of V. Then $T(u+v) =< u+v, w >$, which by the definition of inner product is equal to $< w, u+v >$. By Theorem 9.1 part 2, $< w, u+v >=< w,u > + < w,v >=< u,w > + < v,w >= T(u) + T(v)$.
Scalar multiplication: Let v be a vector of V and let c be a scalar. Then $T(cv) =< cv, w >=< w, cv >$ by the definition of inner product. By Theorem 9.1 part 3, $< w, cv >= c < w,v >= c < w,v >= cT(v)$.
Since T preserves addition and scalar multiplication, T is a linear transformation. To find the kernel of T, we find all vectors v such that $T(v) = 0$. Since $T(v) =< v, w >$, this means all vectors v such that $< v,w >= 0$–in other words, all vectors orthogonal to w.

25. Part a: Problem 24 gives the definition of parallel vectors: the angle between them is 0 or π. The angle between $\text{proj}_v(u)$ and v is given by $\theta = \cos^{-1}\frac{<\text{proj}_v(u),v>}{||\text{proj}_v(u)||\,||v||}$. First, let's calculate $||\text{proj}_v(u)||$. Since $\text{proj}_v(u)$ is defined to be $\frac{<u,v>v}{||v||^2}$,

$$||\text{proj}_v(u)|| = ||\frac{<u,v>v}{||v||^2}|| = \sqrt{<\frac{<u,v>v}{||v||^2},\frac{<u,v>v}{||v||^2}>}.$$

Applying Theorem 9.1 part 3 twice, this becomes

$$\sqrt{(\frac{<u,v>}{||v||^2})^2<v,v>} = \frac{|<u,v>|}{||v||^2}\sqrt{<v,v>} = \frac{|<u,v>|}{||v||^2}||v||.$$

Notice the absolute value sign; it will be important later. Substituting this result and the definition of $\text{proj}_v(u)$ into the formula for θ, we have

$$\theta = \cos^{-1}\frac{<\text{proj}_v(u),v>}{||\text{proj}_v(u)||\,||v||} = \cos^{-1}\frac{<\frac{<u,v>v}{||v||^2},v>}{||\frac{<u,v>v}{||v||^2}||\,||v||}.$$

Applying Theorem 9.1 part 3, we can simplify this to

$$\theta = \cos^{-1}\frac{\frac{<u,v>}{||v||^2}<v,v>}{\frac{|<u,v>|}{||v||^2}||v||\,||v||}$$

$$= \cos^{-1}\frac{\frac{<u,v>}{||v||^2}||v||^2}{\frac{|<u,v>|}{||v||^2}||v||^2}$$

$$= \cos^{-1}(\pm1) = 0 \text{ or } \pi.$$

Thus, the angle θ between $\text{proj}_v(u)$ and v is 0 or π, so these two vectors are parallel.
Part b: First note that $\text{proj}_v(u)$ is a scalar multiple of v, so $\text{proj}_v(u)$ is an element of the vector space V. To show that the function $T: V \to V$ is a linear transformation, we must show that T preserves vector addition and scalar multiplication. Vector addition:

$$T(u+w) = \text{proj}_v(u+w) = \frac{<u+w,v>}{||v||^2}v = \frac{(<u,v>+<w,v>)}{||v||^2}v = \frac{(<u,v>v+<w,v>v)}{||v||^2}$$
$$= \frac{<u,v>}{||v||^2}v + \frac{<w,v>}{||v||^2}v = \text{proj}_v(u)+\text{proj}_v(w) = T(u)+T(w).$$

Scalar multiplication: Let u be a vector in V and let c be a scalar. Then $T(cu) = \text{proj}_v(cu) = \frac{<cu,v>}{||v||^2}v = \frac{c<u,v>}{||v||^2}v = cT(u)$.
Part c: The vectors in the kernel of T are those vectors u of V such that $T(u) = 0$. To find them, suppose that

$$0 = T(u) = \text{proj}_v(u) = \frac{<u,v>}{||v||^2}v.$$

Note that $||v||$ must be nonzero–otherwise the definition of $\text{proj}_v(u)$ doesn't make sense. That means v is not the zero vector. In that case, the only way $T(u)$ can be zero is if $<u,v>=0$. This is true when u and v are orthogonal. So, $ker(T)$ is the set of all vectors of V that are orthogonal to v.
Part d: Recall that u is an eigenvector of T if $T(u) = \lambda u$ for some scalar λ.

9.2 Orthonormal Bases

Exercises 9.2, pp. 429-430

1. Part a: To show that the vectors v_1, v_2, and v_3 are orthogonal:

$$<v_1,v_2> = v_1 \cdot v_2 = (1/2)\cdot(1/2)+(1/2)\cdot(-5/6)+(\sqrt2/2)\cdot(\sqrt2/6) = 0.$$

Similarly, $<v_1,v_3>$ and $<v_2,v_3>=0$.
To show that v_1,v_2, and v_3 are unit vectors:

$$\|v_1\| = \sqrt{<v_1,v_1>} = \sqrt{(1/2)(1/2)+(1/2)(1/2)+(\sqrt{2}/2)(\sqrt{2}/2)} = 1.$$

Similarly, $\|v_2\|$ and $\|v_3\|$ are both equal to 1.
To show that the vectors v_1,v_2, and v_3 are linearly independent: Suppose $c_1v_1+c_2v_2+c_3v_3=0$. Then

$$c_1 \begin{bmatrix} 1/2 \\ 1/2 \\ \sqrt{2}/2 \end{bmatrix} + c_2 \begin{bmatrix} 1/2 \\ -5/6 \\ \sqrt{2}/6 \end{bmatrix} + c_3 \begin{bmatrix} \sqrt{2}/2 \\ \sqrt{2}/6 \\ 2/3 \end{bmatrix} = \begin{bmatrix} 0 \\ 0 \\ 0 \end{bmatrix}.$$

The determinant of the matrix formed by concatenating v_1,v_2, and v_3 is nonzero; hence this system has only the trivial solution and the three vectors are linearly independent.
To show that v_1,v_2, and v_3 span $\Re^3$: Since we have three linearly independent vectors in a vector space of dimension 3, we know they span.
Since v_1,v_2, and v_3 form a basis, are mutually orthogonal, and are all unit vectors, they form an orthonormal basis for $\Re^3$.
Part b: By Theorem 9.5, we know that

$$v = <v,v_1>v_1+<v,v_2>v_2+<v,v_3>v_3.$$
$$<v,v_1> = v\dot v_1 = (-3)(1/2)+6(1/2)-5(\sqrt{2}/2) = (3-5\sqrt{2})/2.$$

The other coefficients are calculated similarly, giving $v = \left((3-5\sqrt{2})/2\right)v_1 + \left((39+5\sqrt{2})/6\right)v_2 + \left((20-3\sqrt{2})/6\right)v_3$.

3. Let $u_1 = \begin{bmatrix} 1 \\ 2 \end{bmatrix}$ and let $u_2 = \begin{bmatrix} 1 \\ -1 \end{bmatrix}$.

Step 1: $\|u_1\| = \sqrt{1^2+2^2} = \sqrt{5}$. $v_1 = u_1/\sqrt{5} = \begin{bmatrix} 1/\sqrt{5} \\ 2/\sqrt{5} \end{bmatrix}$.

Step 2: Let $k_1 = <v_1,u_2> = (1/\sqrt{5})(1)+(2/\sqrt{5})(-1) = -1/\sqrt{5}$. Let $w_2 = u_2 - k_1v_1 = \begin{bmatrix} 6/5 \\ -3/5 \end{bmatrix}$. Then

$$v_2 = w_2/\|w_2\| = \begin{bmatrix} 2/\sqrt{5} \\ -1/\sqrt{5} \end{bmatrix}.$$

5. Let $u_1 = \begin{bmatrix} 3 \\ 0 \\ 0 \end{bmatrix}$ and let $u_2 = \begin{bmatrix} 2 \\ -3 \\ 4 \end{bmatrix}$, and $u_3 = \begin{bmatrix} 1 \\ -15 \\ -5 \end{bmatrix}$.

Step 1: $\|u_1\| = \sqrt{3^2+0^2+0^2} = 3$. $v_1 = u_1/\|u_1\| = \begin{bmatrix} 1 \\ 0 \\ 0 \end{bmatrix}$.

Step 2: Let $k_1 = <v_1,u_2> = 2$. Let $w_2 = u_2 - k_1v_1 = \begin{bmatrix} 0 \\ -3 \\ 4 \end{bmatrix}$. Then $v_2 = w_2/\|w_2\| = \begin{bmatrix} 0 \\ -3/5 \\ 4/5 \end{bmatrix}$.

Step 3: Let $k_1 = <v_1,u_3> = 1$. Let $k_2 = <v_2,u_3> = 5$. Let $w_3 = u_3 - k_1v_1 - k_2v_2 = \begin{bmatrix} 0 \\ 12 \\ -9 \end{bmatrix}$. Then

$$v_3 = w_2/\|w_2\| = \begin{bmatrix} 0 \\ 4/5 \\ -3/5 \end{bmatrix}.$$

7. $\|u_1\| = \sqrt{1\cdot 1+(-1)\cdot(-1)+1\cdot 1} = \sqrt{3}$. $v = u/\|u\| = \begin{bmatrix} 1/\sqrt{3} \\ -1/\sqrt{3} \\ 1/\sqrt{3} \end{bmatrix}$. $w = \begin{bmatrix} -1/\sqrt{3} \\ 1/\sqrt{3} \end{bmatrix}$. Then

$$ww^T = \begin{bmatrix} 1/3 & -1/3 \\ -1/3 & 1/3 \end{bmatrix}.$$

Next, calculate $1/(a_1+1) = \sqrt{3}/(\sqrt{3}+1)$. The matrix

$$-I + \frac{1}{a_1+1}ww^T = \begin{bmatrix} (-3-\sqrt{3})/6 & (-3+\sqrt{3})/6 \\ (-3+\sqrt{3})/6 & (-3-\sqrt{3})/6 \end{bmatrix}.$$

Finally, we put all the part together to get the matrix

$$A = \begin{bmatrix} 1/\sqrt{3} & -1/\sqrt{3} & 1/\sqrt{3} \\ -1/\sqrt{3} & (-3-\sqrt{3})/6 & (-3+\sqrt{3})/6 \\ 1/\sqrt{3} & (-3+\sqrt{3})/6 & (-3-\sqrt{3})/6 \end{bmatrix}.$$

The columns of A are the three vectors in the orthonormal basis.

9. We start with the standard basis for P_2 of $1, x, x^2$. Applying the Gram-Schmidt process, we let $u_1 = 1, u_2 = x, u_3 = x^2$. Step 1: Using the given definite integral as the inner product, $\|u_1\| = \sqrt{<u_1, u_1>} =$, where

$$<u_1, u_1> = <1, 1> = \int_{-1}^{1} (1)(1)dx = 1 - (-1) = 2,$$

so $\|u_1\| = 2$, and $v_1 = 1/\sqrt{2}$. Continue to apply the Gram-Schmidt process, always remembering that the inner product is the given definite integral.

11. Let $u_1 = \begin{bmatrix} 1 \\ 0 \end{bmatrix}$ and let $u_2 = \begin{bmatrix} 0 \\ 1 \end{bmatrix}$. This is the standard basis for $\Re^2$.
Step 1: Using the weighted inner product from Exercise 8b, Section 9.1,

$$\|u_1\| = \sqrt{(2)(1)(1) + 0(3)(0)} = \sqrt{2}. \ v_1 = u_1/\sqrt{2} = \begin{bmatrix} 1/\sqrt{2} \\ 0 \end{bmatrix}.$$

Step 2: Let $k_1 = <v_1, u_2> = (1/\sqrt{2})(2)(0) + 0(3)(0) = 0$. Let $w_2 = u_2 - k_1 v_1 = \begin{bmatrix} 0 \\ 1 \end{bmatrix}$. Then $v_2 = w_2/\|w_2\| = \begin{bmatrix} 0 \\ 1/\sqrt{3} \end{bmatrix}$.
(Be sure to use the weighted inner product to calculate $\|w_2\|$.)

13. To show that $\begin{bmatrix} \cos\alpha & \pm\sin\alpha \\ \sin\alpha & \mp\cos\alpha \end{bmatrix}$ is an orthogonal matrix, we must show that its columns form an orthonormal basis for $\Re^2$.
To show the columns are orthogonal:

$$<c_1, c_2> = c_1 \cdot c_2 = (\cos\alpha)(\sin\alpha) + (\sin\alpha)(-\cos\alpha) = 0,$$

or

$$(\cos\alpha)(-\sin\alpha) + (\sin\alpha)(\cos\alpha) = 0.$$

To show the columns are unit vectors: $(\pm\sin\alpha)^2 + (\mp\cos\alpha)^2 = \sin^2\alpha + \cos^2\alpha = 1$. To show the columns form a basis for $\Re^2$: Since the determinant of the matrix above is nonzero (it is either 1 or -1), the two columns are linearly independent and hence form a basis for $\Re^2$.

15. Since A is orthogonal, A^T is the inverse of A, so $A^T A = AA^T = I$. Similarly, if B is orthogonal, then $B^T B = BB^T = I$. Recall that $(AB)^T = B^T A^T$. Then

$$(AB)(AB)^T = (AB)(B^T A^T) = A(BB^T)A^T = AIA^T = AA^T = I.$$

Since $(AB)(AB)^T = I$, AB is orthogonal.

17. Part a: Let $u_1 = \begin{bmatrix} 1 \\ -1 \\ 1 \end{bmatrix}$ and let $u_2 = \begin{bmatrix} 2 \\ 0 \\ 1 \end{bmatrix}$.

Step 1: $\|u_1\| = \sqrt{1^2 + 1^2 + 1^2} = \sqrt{3}$. $v_1 = u_1/\|u_1\| = \begin{bmatrix} 1/\sqrt{3} \\ -1/\sqrt{3} \\ 1/\sqrt{3} \end{bmatrix}$.

Step 2: Let $k_1 = <v_1, u_2> = 2$. Let $w_2 = u_2 - k_1 v_1 = \begin{bmatrix} 1 \\ 1 \\ 0 \end{bmatrix}$. Then $v_2 = w_2/||w_2|| = \begin{bmatrix} 1/\sqrt{2} \\ 1/\sqrt{2} \\ 0 \end{bmatrix}$.

Part b:

$$v_1 \times v_2 = \begin{vmatrix} i & j & k \\ 1/\sqrt{3} & -1/\sqrt{3} & 1/\sqrt{3} \\ 1/\sqrt{2} & 1/\sqrt{2} & 0 \end{vmatrix}$$

$$= i \begin{vmatrix} -1/\sqrt{3} & 1/\sqrt{3} \\ 1/\sqrt{2} & 0 \end{vmatrix} - j \begin{vmatrix} 1/\sqrt{3} & 1/\sqrt{3} \\ 1/\sqrt{2} & 0 \end{vmatrix} + k \begin{vmatrix} 1/\sqrt{3} & -1/\sqrt{3} \\ 1/\sqrt{2} & 1/\sqrt{2} \end{vmatrix}$$

$$= i(1/\sqrt{6}) - j(1/\sqrt{6}) + k(2/\sqrt{6}) = \begin{bmatrix} 1/\sqrt{6} \\ -1/\sqrt{6} \\ 2/\sqrt{6} \end{bmatrix}.$$

By the properties of the cross product, this vector is perpendicular (orthogonal) to the plane of v_1 and v_2.

9.3   Schur's Theorem and Symmetric Matrices

Exercises 9.3, p. 439

1. The only eigenvalue of A is $\lambda = 2$. The vector $u_1 = \begin{bmatrix} 1 \\ 1 \end{bmatrix}$ forms a basis for E_2. As you did in Exercise 7 of Section 9.2, find an orthonormal basis for $\Re^2$ starting with u_1. Use the vectors v_1, v_2 from the Gram-Schmidt process as the columns of the matrix P. The triangular matrix $B = P^T A P$.

3. The only eigenvalue of A is $\lambda = -1$. The vector $u_1 = \begin{bmatrix} 1 \\ 1 \\ 2 \end{bmatrix}$ forms a basis for E_{-1}. As you did in Exercise 7 of Section 9.2, find an orthonormal basis for $\Re^2$ starting with u_1. Use the vectors v_1, v_2, v_3 from the Gram-Schmidt process as the columns of the matrix P_1. The matrix $A_1 = P_1^T A P_1$. A_1 has all zeros in the first column, except for entry $(1,1)$. Use the same process on the $(1,1)$ minor B_1 of A_1, obtaining a 2×2 matrix Q to triangularize B_1. Then matrix P_2 is the 3×3 matrix whose $(1,1)$ entry is 1, with all zeros elsewhere in column 1 and row 1, and whose $(1,1)$ minor is Q. Finally, $P = P_1 P_2$ and $B = P^T A P$.

5. Solve $Y' = BY$ where B is the triangular matrix we found in Exercise 3. The solution matrix is denoted Z. Then $Y = PZ$, where P is the orthogonal matrix we found in Exercise 3.

7. This matrix has eigenvalues -1 and 3, with basis for E_{-1} given by $\begin{bmatrix} 1 \\ -1 \end{bmatrix}$ and basis for E_3 given by $\begin{bmatrix} 1 \\ 1 \end{bmatrix}$. Apply the Gram-Schmidt process to the first vector to construct an orthonormal basis v_1 for E_{-1} and to the second vector to construct an orthonormal basis v_2 for E_3. Arrange v_1 and v_2 as the columns of a matrix P, and $P^T A P$ will be the diagonal matrix $diag(-1, 3)$.

9. This matrix has eigenvalues 2 and 4, with basis for E_2 given by $\begin{bmatrix} 0 \\ 0 \\ 1 \end{bmatrix}, \begin{bmatrix} 1 \\ 1 \\ 0 \end{bmatrix}$ and basis for E_4 given by $\begin{bmatrix} -1 \\ 1 \\ 0 \end{bmatrix}$. Apply the Gram-Schmidt process to the first two vectors to construct an orthonormal basis v_1, v_2 for E_2 and to the third vector to construct an orthonormal basis v_3 for E_4. Arrange v_1, v_2, and v_3 as the columns of a matrix P, and $P^T A P$ will be the diagonal matrix $diag(2, 2, 4)$.

11. This matrix has eigenvalues 0 and $4\sqrt{2}$, with basis for E_0 given by $\begin{bmatrix} -\sqrt{2} \\ 0 \\ 1 \end{bmatrix}, \begin{bmatrix} -1 \\ 1 \\ 0 \end{bmatrix}$ and basis for $E_{4\sqrt{2}}$ given by $\begin{bmatrix} 1 \\ 1 \\ \sqrt{2} \end{bmatrix}$. Apply the Gram-Schmidt process to the first two vectors to construct an orthonormal basis v_1, v_2 for E_0 and to the third vector to construct an orthonormal basis v_3 for $E_{4\sqrt{2}}$. Arrange v_1, v_2, and v_3 as the columns of a matrix P, and $P^T A P$ will be the diagonal matrix $diag(0, 0, 4\sqrt{2})$.

13. Solve $Y' = BY$ where B is the diagonal matrix we found in Exercise 9. The solution matrix is denoted Z. Then $Y = PZ$, where P is the orthogonal matrix we found in Exercise 9.